Lecture Notes in Earth Sciences 76

Springer-Verlag Berlin Heidelberg GmbH

Jörg Trappe

Phanerozoic Phosphorite Depositional Systems

A Dynamic Model for a Sedimentary Resource System

Springer

Author

PD Dr. Jörg Trappe
Geological Institute, University of Bonn
Nussallee 8, D-53115 Bonn, Germany
E-mail: trappe@geo.uni-bonn.de

"For all Lecture Notes in Earth Sciences published till now please see final pages of the book"

Cataloging-in-Publication data applied for

Die Deutsche Bibliothek - CIP-Einheitsaufnahme

Trappe, Jörg:
Phanerozoic phosphorite depositional systems : a dynamic model for a sedimentary resource system / Jörg Trappe.

(Lecture notes in earth sciences ; 76)
ISBN 978-3-540-63581-9 ISBN 978-3-540-69604-9 (eBook)
DOI 10.1007/978-3-540-69604-9

ISSN 0930-0317
ISBN 978-3-540-63581-9

Originally published by Springer-Verlag Berlin Heidelberg New York in 1998

Typesetting: Camera ready by author
SPIN: 10570455 32/3142-543210 - Printed on acid-free paper

Preface

The knowledge on sedimentary phosphate resources has drastically increased during the last years. New methods, interdisciplinary research, and comparative field studies were applied to the phenomenon "phosphorites". Perspectives have changed with our understanding of a dynamic Earth. This enormous database, new ideas, and the opportunity to work in various phosphate deposits gave way for the development of a general genetic model for this deposystem. This dynamic approach addresses not only one type of a deposit, but for the first time the entirety of the deposystem and its manifold appearances and geological settings. The formation of phosphorites and phosphate rocks is now understood by a box model - a modular system of pathways linking genetic stages of resource formation in a permanently changing world - contrasting the traditional view of a single process genesis.

This textbook is designed for geologists from academia and industry. It presents the state-of-the-art knowledge on phosphorites with additional explaining case studies in the context of the hosting deposystem evolution and its temporal and spatial variations. The interdisciplinary approach in this book tries to understand and explain the geologic context of a sedimentary phenomenon. But it is not the intention of this work to present again numerous generic data or their compilations. This was done in the past in numerous regional and specialist studies. The amount of data anyway would be far beyond the frame of a single book. The latter applies also to the interpretation of Precambrian phosphorites, which were formed under very uncertain environmental conditions and differ significantly from the Phanerozoic occurrences.

The significance of phosphorus in the biosphere as an essential element for the bioproduction emphasizes the importance of mineral fertilizers for the world food production. An increasing world population and the limitation of agriculture land makes the availability of P-fertilizers, and consequently of phosphate rock resources, absolutely necessary. Phosphate rock is not only an important resource for agriculture and industry, these rocks also serve as important tools for the interpretation of sedimentary processes in all marine environments and for the understanding of the formation of organogenic sediments. The interpretation of the depositional processes in phosphorite deposystems helps to determine the dynamics of fine-grained sediments. The environmental changes recorded in these rocks document the secular variations in the global phosphorus cycle, the burial and regeneration of P, and the interrelationship of global ecological mechanisms in the system Earth.

The depositional history of phosphorites, commonly embedded in multi-lithologic deposystems, is probably one of the most complex of all sediments. Just recently, we achieved a better understanding of these sediments as the result of the enormous efforts made within international working groups under the roof of the International Geological Correlation Program (IGCP) of the UNESCO, respectively the IGCP programs 156 "Phosphorites" and 325 "Paleogeographic reconstructions with phosphorites and other authigenic minerals". Regardless to the large number of monographs and larger volumes of regional and thematical compilations (BENTOR 1980, COOK & SHERGOLD 1986, NOTHOLT et al. 1989, BURNETT & RIGGS 1990, NOTHOLT & JARVIS 1990, JARVIS et al. 1994, GLENN et al. 1994, KRAJEWSKI et al. 1994), an interdisciplinary synthesis remained open. The intensive geochemical research fundamentally increased our knowledge on phosphogenesis, i.e. the mineralization of sedimentary phosphate. But the complex sedimentary processes (phosphorite genesis) and the enormous spectrum of depositional environments of the ancient deposits remained only broadly touched. The latter include the important sedimentary processes, which finally lead to economic phosphorite deposits. The importance of these processes was apparently recognized with the consequent application of microfacies analysis methods.

After an early phase of descriptive research, process oriented research began in the late 60ies with the fundamental work of the McKELVEY group in the Phosphoria formation (McKELVEY et al. 1956, 1959). So far, single processes have been understood or mechanisms were investigated with a single methodical approach. Integrated studies focused on special deposits. The research focused on the finding of the single process of phosphogenesis. This point of view has just recently changed. A milepost for this change of understanding was the box model of LUCAS & PRÉVôT (1993).

The approach in this book intents to go forward the next step and investigates pathways of phosphogenesis in an interdisciplinary approach of sedimentology, geochemistry, sequence stratigraphy, and paleoecology to understand the various phosphorite depositional systems. Especially, the sedimentological and sequence stratigraphic approach made a number of reinterpretations necessary and opened new perspectives on the evolution of phosphorite deposystems. In a research project over the last couple of years, the controlling factors and their complex interrelationships were determined in economic deposits in North America, Australia, Germany, Morocco, Estonia, from observations during field trips, and from a large sample collection. The studied deposystems were interpreted in terms of paleogeography and sequence stratigraphy. The sedimentological studies were completed by geochemical investigations (trace elements, REE), where reworking, sediment mixing, and weathering conditions allow a useful application for paleoenvironmental interpretation. Besides the own field investigations and analyses, stratigraphic, sedimentological and geochemical data were interpreted from an own literature database of roughly 2000 contributions, from which about 700 are found referenced in this study.

A proper database to paleogeographic setting, climate, and sea-level changes plays a significant role in this strategy of investigation. The described examples try to cover all major types of economic and non-economic phosphate deposits. These investigations, partly addressed in the "Case Studies" are based on the detailed sedimentological (microfacies), stratigraphic (inculding sequence stratigraphic methods), partly also geochemical investigations of about 200 outcrop sections and cores from the Permian Phosphoria Rock Complex in the Western United States, from the Mississippian Delle Phosphatic interval in the Great Basin/USA, from the Neogene phosphate system in Florida, from the Cambrian phosphate-bearing interval in the Georgina Basin/Australia, from the Early Carboniferous phosphogenic interval in Germany, from the Ordovician phosphorites in Estonia, and from the Cretaceous/Paleogene system in Morocco. All these sections and cores were sampled in great detail and were investigated by microfacies techniques in thin sections. The latter were developed from carbonate petrology techniques (FLÜGEL 1982). Additional thin sections were provided temporarily by the Australian Geological Survey, the U.S. Geological Survey, and the Florida Geological Survey.

The research was generously supported by serveral grants of the German Research Foundation (DFG) to the author (Tr273/1-1,2,3 and Tr273/4-1,2). Many colleagues from abroad extensively contributed to the completion of this investigation with their assistance and support during several, partly long visits in North America, Australia, North Africa, especially C.F. Vondra (Iowa State University), Bill Burnett (Florida State University), E. K. Maughan †, N. J. Silberling, K. M. Nichols (U.S. Geological Survey), J. H. Shergold and P. Southgate (Australian Geological Survey), R. Raudsep (Estonian Geol. Survey). During the course of this work, I received further encouragement by L. Prévôt-Lucas, and J. Lucas (University Strasbourg), W. Zimmerle (Celle), B. Erdtmann and K. Germann (Berlin), S. Riggs (East Carolina University), and many other colleagues.

I have to give very special thanks to my colleagues at the Geological Institute of the University in Bonn, who supported the work with patience and extensive scientific and technical assistance, in particular J. Stets, J. Thein and P. Wurster †. The geochemical investigations in Bonn would not have been possible without the help of R. Klingel, and the staff of the Geological Institute. The extensive thin section preparations were done in Bonn by W. Merten and P. Reiser, and benefited specially from the assistance of W. Gebhardt, further at Iowa State University by B. Tanner. I. Gebhardt completed most of the drawings with great effort. Finally, I have to thank my family, especially my wife Martina and my two children. Without their patience and enthusiasm during the long periods of field investigations, this work would not have been completed.

Jörg Trappe

Table of contents

PART I
PHOSPHORUS IN NATURE: WHAT'S THIS ALL ABOUT

1.1
The role of phosphorus in the living world

The element phosphorus is essential to all forms of life on earth. The element plays a fundamental role in many metabolic processes and as a major constituent of skeletal material. Phosphorus is a structural element in the DNA, RNA, and a number of enzymes (e. g. ATP, ADP, AMP). Phospholipids participate in the construction of cell membranes and bone marrow, phosphagenes are involved in the processes of muscular contraction, and phosphoglyceric acid contributes to photosynthesis. Bone material and especially teeth of vertebrates contain phosphorus minerals. Some modern invertebrates, but a much larger number of ancient groups, build their skeletons of phosphate minerals (see listings in LOWENSTAM 1972 and LOWENSTAM & WEINER 1989). The average human body demands roughly 650 g P. Some crops, bacteria and viruses need relative more phosphorus for their living activity.

With an average of 0.01% phosphorus in the earth crust, the element ranks number 11 by its abundance. The occurrence in the crust is mainly bound to dispersed organic matter, to the mineral apatite in magmatic rocks, and to the detrital grain spectrum in sediments. The phosphorus contents of most sedimentary rocks are higher than the earth average.

The fundamental role of phosphorus in all biological processes and the limited availability on the earth's surface qualifies the element for a role as a regulating element in the biosphere. Phosphorus controls as a limiting nutrient the amount of bioproduction in all environments. This casual relationship between bioproduction and phosphorus availability defines the importance of P-fertilizer and the significance of phosphate mineral resources.

1.2
The global phosphorus cycle and P-fluxes

Important for the understanding of the development of phosphorite deposystems is the identification of the major phosphorus reservoirs and the main phosphorus fluxes in the biosphere. GARRELS et al. (1975), SANDSTORM (1982), and later MACKENZIE et al. (1993) presented quantitative models for the global P-cycle in the modern world. These still very general approaches give rough estimates of the main P-fluxes in the modern world. The models have to ignore

Table 1.1. Phosphate contents of some selected biota, soils, and sediments.

Selected examples of P_2O_5 contents in weight %	[%P x 2.2914 = %P_2O_5]
human body (~1500g)	1 (5)
bone (fish)	up to 35 (7)
shell of inarticulate brachiopods	34-42 (1)
bacteria and viruses	6.8-11.5 (6)
cyanophyta (dry weight)	0.5-1.2 (1)
milk	2.0 (4)
crops	
wheat	1.03 (3)
maize	0.91 (3)
rice	0.80 (3)
soybeans	1.60 (3)
mushrooms	2.63 (3)
tomatoes	1.48 (3)
soil	0.02-0.09 (4)
guano	4.0 (5)
sandstone	0.08-0.27 (2)
shale	0.11-0.33 (2)
carbonate	0.03-0.18 (2)
phosphate rock	
sedimentary deposits	20-30
igneous apatite deposits	18 (5)

(1) VINOGRADOV (1953), (2) McKELVEY (1973), (3) GERVEY (1973)
(4) HENIN (1982), (5) SLANSKY (1986), (6) BOWEN (1966),
(7) KIZEWETTER (1973)

the variations in different environments as well as seasonal and climatic effects. These fluxes may have shifted drastically during the global environmental changes during earth's history (BATURIN et al. 1995, FÖLLMI 1995, 1996), but models for ancient configurations are not available by the present stage of knowledge. First approaches to ancient flux rates are achieved by the investigations of FILIPPELLI & DELANEY (1992, 1994a,b), FILIPPELLI et al. (1994), FILIPPELLI (1997), and FÖLLMI (1995, 1996), who calculated the net phosphate burial in ancient phosphate-rich marine sediments. But calculations from the phosphate sink on variations of global flux rates are problematic. These estimates are based on the assumption that the P-sink in general is not significantly affected by the environmental factor of P-preservation rates in marine sediments and diagenetic overprint. Furthermore, the behavior of the P-reservoir masses during global changes is uncertain.

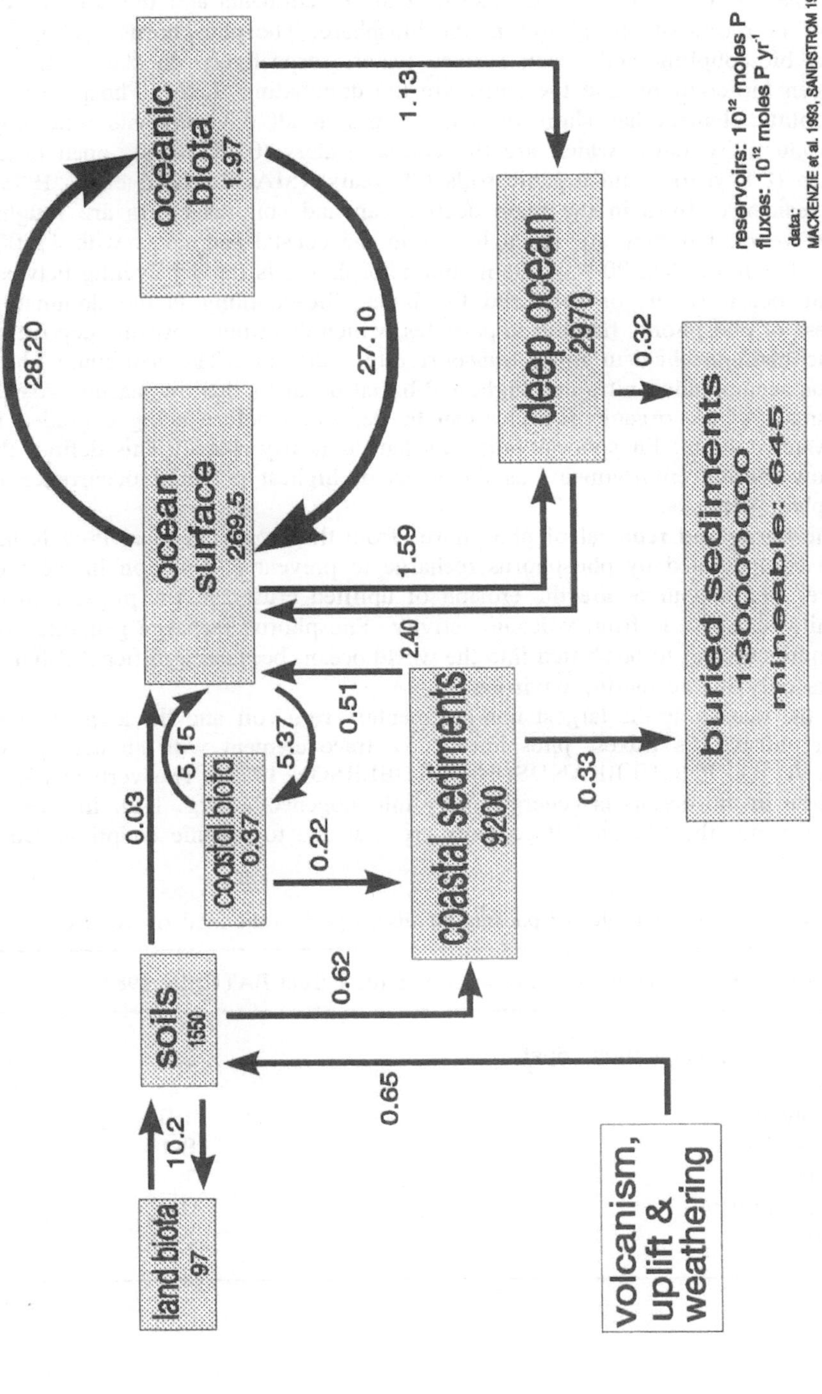

Fig. 1.1. The global phosphorus cycle for the modern biosphere indicating the major phosphorus reservoirs (boxes) and the fluxes between reservoirs.

The state-of-the-art model of the global phosphate cycle (Fig. 1.1) clearly identifies the deep oceans, the shallow marine sediments and the soils as the major reservoirs of phosphorus in the biosphere. The phosphorus cycling between bio-coupling and ocean surface water, respectively on the continents between bio-coupling and the soils, are the dominating fluxes. Phosphorus as a biophile element has short residence times in all environments with high bioproductivity rates, which are the coastal waters (0.75 years), open ocean surface (8.8 years), and organic soils (39 years) (MACKENZIE et al. 1993). The residence times in the main deep ocean and soil reservoirs are roughly 1 500 years, but they are much longer in the coastal sediments with 11 000 years. Far more than 90% of the mobile phosphorus is rapidly cycling between marine ocean water, or soil, and the biota. Bio-coupling is the dominating process of phosphorus fixation in particles, which determines organic deposition as the main mechanism for P-transport into sediments. The maximum phosphorus accumulation rates and highest P-burial occur in shallow marine, coastal sea areas, where organic particles can be deposited before being degraded in the water column. Phosphorus burial on land is nearly absent. This defines the shallow marine environments as the areas of highest potential occurrence of phosphate deposits.

The burial and removal of phosphorus from the biogeochemical P-cycle has to be compensated by phosphorus recharge to prevent P-depletion in the biosphere. Major sources are the erosion of uplifted crust and the production of juvenile phosphorus from volcanic activity. Phosphorus recharge generated on the continents has to be shifted into the world ocean, because significant P-burial occurs only in the marine environment.

In the oceans as the largest non-sedimentary reservoir and the area of maximum phosphorus fluxes, phosphorus is a trace element with an average of about 70 ppb P (GULBRANDSON & ROBERSON 1973). The vertical P-distribution in the oceans is generally very inhomogeneous (Fig. 1.2). In oceanic surface water the liberated P contents are low due to organic coupling. Deep

Table 1.2. Sources and rates of phosphorus discharge into the modern world ocean.

Discharge of P into the world ocean in 10^6t/yr (data from BATURIN 1982)	
Atmospheric precipitation, dust	3-4
Rivers	
dissolved	1.5
suspended	9-14
Glaciers	1
Coastal abrasion	0.2
Volcanism	2-3

oceanic water is rich in P because of the sink of dead organic material. In the latter, degradation exceeds bioproductivity and liberated P is concentrated with time. Thermoclinal and haloclinal stratification in modern oceans prevent mixing with surface water and the storage of a significant amount of liberated P.

Major exchange of deeper oceanic water with surface water occurs in upwelling zones along continental margins, mainly along westcoasts in low latitudes. Ancient continent configurations may also have allowed dynamic upwelling due to intense current activity and upwelling along elevated areas (SHELDON 1980). Extreme nutrient supply in these zones induced maximum bioproductivity and subsequently high organic matter sink. A well defined stratification in the water column gives way for organic matter preservation by oxygen deficiency due to reduced vertical mixing. These areas are also characterized by increased P sink into the sediment. With the regional variation of bioproduction, the phosphorus content of the oceanic surface water also varies by the factor 10 to 20. In general, the inorganic and organic dissolved phos-

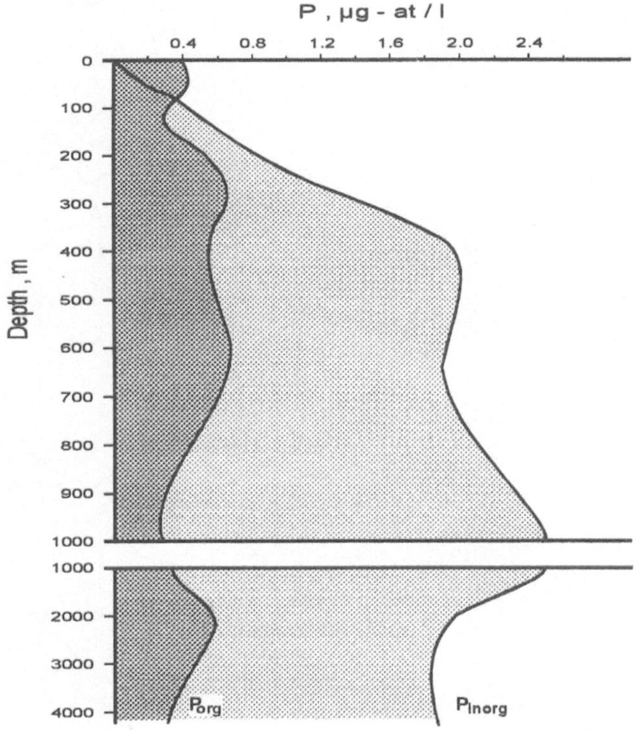

Fig. 1.2. Vertical profile of dissolved inorganic and organic phosphorus in equatorial Pacific ocean water. (Reprinted from BATURIN, Phosphorites on the sea floor, 1982, p. 22, with kind permission from Elsevier Science - NL Sara Burgerhartstraat 25, 1055 KV Amsterdam, The Netherlands).

phorus contents of oceanic surface water increase in polar and coastal areas (BATURIN 1982).

The phosphate distribution in soils as the second important reservoir varies even more. Soil profiles in central Europe show usually decreasing P_2O_5 contents with depth. Maximum weight percentage is about 20%, but in most soils the values reach only a few percents (MÜCKENHAUSEN 1977, FÖLLMI 1996).

PART II
PHOSPHATE MINERALS AND SEDIMENTS

2 Mineralogy of sedimentary apatites

Phosphorus occurs in over 200 mineral species. The most common forms are minerals of the apatite family, a calcium phosphate with various substitutions. In igneous rocks and metamorphic rocks, most apatites are approaching a fluorapatite composition, less abundant are apatites with mixtures of fluorine with hydroxyl and chlorine (McCONNELL 1973, McCLELLAN & VAN KAUWENBERGH 1990). In sedimentary rocks, the apatite variety francolite, a carbonate fluorapatite (CFA), is the dominating mineral phase. Most francolite fabrics in sediments have a cryptocrystalline to microcrystalline texture, which was long thought to be an amorphous fluorine-bearing calcium phosphate material known as collophane. Modern techniques could identify the cryptocrystalline or microcrystalline nature of the microscopic pseudo-isotropic mineral phase. Both minerals fluorapatite and francolite have similar, but distinctively different, x-ray diffraction pattern (McCONNELL 1938). Francolite was later defined to be a carbonate fluorapatite with more than 1% fluorine and appreciable amounts of CO_2 (McCONNELL 1973). But this definition has led to some confusions because sedimentary francolites contain more fluorine (3.77 %) than does fluorapatite. These high fluorine contents are called "excess fluorine" in francolite (McCLELLAN & VAN KAUWENBERGH 1990).

2.1 Structure and substitutions in natural francolites

The structure of sedimentary apatites allows a wide range of substitutions, which potentially may reflect in part the geochemical environment of precipitation or subsequent alteration during diagensis or weathering.

A simplified structural formula for francolite can be written in first approximation as follows:

$$Ca_{10}[(PO_4)_{6-x}(CO_3)_x]F_{2+x}$$

The value x is variable and ranges in most natural francolites from 0.39-1.36 with an average of 0.75, but can exceed 1.90 in some occurrences (LEHR et al. 1967, SLANSKY 1986).

Besides substitutions of PO_4^{3-} by CO_3^{2-} and F, usually Ca^{2+} is also substituted by Na^+ and Mg^{2+}, and is strictly related to the CO_3^{2+} substitution (GULBRANDSEN 1960, 1966, LEHR et al. 1967). Additionally, the replacement of F by OH is fairly frequent and is thought to be responsible for a poor correlation of the CO_3/PO_4 molar ratio (LEHR et al. 1967). Including all of

the following structural formula:

$$(Ca,Na,Mg)_{10}(PO_4)_{6-x}(CO_3)_xF_y(F,OH)_2$$

in which x varies between 0 and 1.5 and y between 0.33 and $0.5x$. The moles of Na are approximately x-y, and are greater than the moles of Mg, of which the number is mostly greater than 2.

The structure allows other substitutions which usually are smaller in extend. Less than 0.2% of the calcium sites are frequently occupied by potassium, strontium, uranium, thorium and REE. SO_4^{2-} can substitute PO_4^{3-} instead of CO32- (SLANSKY 1986, McCLELLAN & VAN KAUWENBERGH 1990). A number of substituents have not the same valency as the original ion and coupled substitutions are necessary to maintain charge balance (JARVIS et al. 1994).

The structure of apatite is extensively discussed in SLANSKY (1986). Ca ions and O atoms form columns which are interconnected P atoms and attached O atoms. These Ca-O columns are parallel to the c axis of the crystal and define hexagonal prisms with a central channel in which fluorine atoms are situated as well as other Ca ions on marginal channel position (Fig. 2.1). The unit cell of the fluorapatite lattice is a regular prism with a rhomboid base with a = 9.37 Å and c = 6.88 Å.

The altered dimensions of the ideal unit cell directly correspond to some substitutions. The parameter a decreases with the increase of the CO_3/PO_4 molar ratio and c tends to increase slightly. GULBRANDSEN (1970) proposed a method to determine the weight percent of CO_2 from the parameter a-value expressed in the x-ray diffractogram. The reduction in a may be concealed by the substitution of F by OH. The occurrence of Cl and OH for F in the channels switches the crystallographic symmetry from hexagonal to monoclinic or triclinic

Table 2.1. Possible substitutions in the francolite structure. Compositional important substitutions are indicated in bold (from JARVIS et al. 1994).

Constituent ion	Substituing ion
Ca^{2+}	**Na^+**, K^+, Ag^+
	Mg^{2+}, Sr^{2+}, Ba^{2+}, Cd^{2+}, Mn^{2+}, Zn^{2+}
	Bi^{3+}, Sc^{3+}, Y^{3+}, REE^{3+}
	U^{4+}
PO_4^{3-}	**CO_3^{2-}, SO_4^{2-}**, CrO_4^{2-}
	CO_3, F^{3-}, CO_3, OH^{3-}, AsO_4^{3-}, VO_4^{3-}
	SiO_4^{4-}
F^-	OH^-, Cl^-, Br^-
	O^{2-}

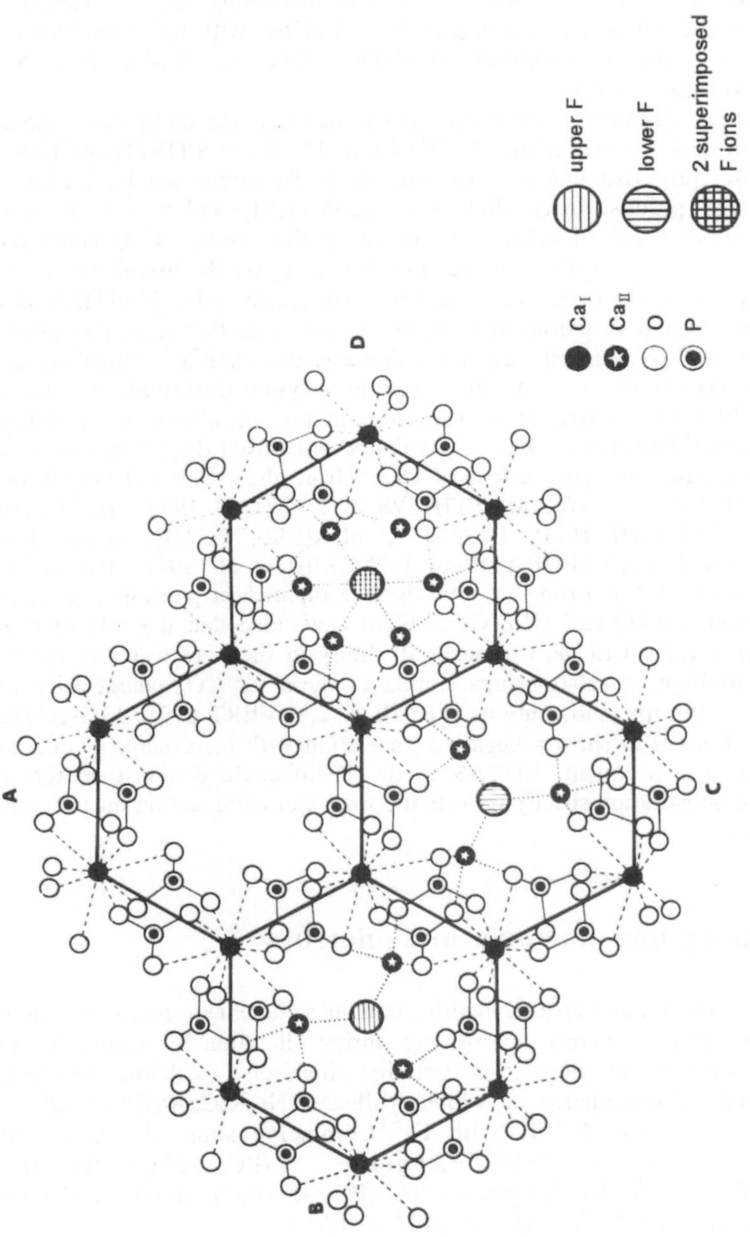

Fig. 2.1. Projection of the apatite structure onto plane 001 (fom ALTSCHULER et al. 1958 and SLANSKY 1986). In position A, the F and Ca_{II} In position A, the F and Ca_{II} ions have been ignored, in positions B and C only one of the F ions are shown. All ions, including the two superimposed F ions and the six Ca_{II} ions, are shown in positionD.

(McCONNELL 1973, YOUNG 1975). The increasing CO_3^{2-}, respectively reduction of the parameter *a*, is negatively correlated with the crystalline size, but positively with the solubility (SLANSKY 1986, McCLELLAN & VAN KAUWENBERGH 1990).

The type and degree of substitutions should be partly driven by the geochemical reservior during precipitation. RUSSELL & TRUEMAN (1971) and LUCAS et al. (1987) proposed that the Na contents in francolite can be used as an indicator for paleosalinity, but final geological evidence is pending. GULBRANDSEN (1970) noted a variation in the average CO_2 contents of francolite from the Phosphoria Formation depending on the lithofacies association and consequently to the depositional environment. Also NATHAN et al. (1990) found similar variations in the Upper Createaceous-Paleocene phosphorites in Israel. In well oxygenated carbonate lithofacies associations, francolites show higher substitutions of CO_2 than in the oxygen-deficient lithofacies. McARTHUR (1985) relates these variations in the Phosphoria by a different grade of burial diagenesis. There is no doubt that burial diagenesis, metamorphism and weathering cause a loss of CO_3^{2-} from the lattice (BLISKOVSKIY 1976, GUSEV et al. 1976, MATTHEWS & NATHAN 1977, McARTHUR 1980, McCLELLAN 1980, LUCAS et al. 1980, ZANIN et al. 1985, McCLELLAN & SAAVEDRA 1986, PANCZER et al. 1989, Da ROCHA ARAUJO et al. 1992). From the investigation of modern phosphate sediments GLENN et al. (1988) and GLENN (1990b) concluded that the rate of CO_3^{2-} substitution is related to the carbonate alkalinity in the pore water of the host sediment. Ambiguous relationships similar to these for CO_2 characterize also the degree of fluorine substitutions (PRICE & CALVERT 1978, McARTHUR 1990), which seem to show a negative correlation with increasing organic contents of the host sediment. LUCAS et al. (1990) could demonstrate that the degree of Sr substitution strictly reflects the sea-water composition during apatite precipitation.

2.2 Other phosphate minerals in sediments

Francolite is the typical apatite modification in sedimentary rocks. A number of other phosphate minerals are forming during alteration by weathering and precipitation during late diagenesis. Lateritic alteration of calcium fluorapatite leads several Al-phosphates of which Millisite $(Na,K)CaAl_6(PO_4)_4(OH)_9$ x $3H_2O$, the Fe-variety Pallite with Fe^{3+} replacing some Al, Crandallite $Ca_2Al_6(PO_4)_4(OH)_{10}$ x $2H_2O$, Augelite $Al_2(PO_4)_2(OH)_3$, Wavellite $Al_3(PO_4)_2(OH)_3$ x $5H_2O$, Turquoise $CuAl_6(PO_4)_4(OH)_8$ x $4H_2O$ are the most common species (FLICOTEAUX & LUCAS 1984).

The Fe-phosphate Vivianite $Fe_3(PO_4)_2$ x $8H_2O$ occurs in weathering profiles, but is more common in reducing lacustrine environments (NRIAGU & DELL 1974) in association with other exotic Fe-phosphates. In insular phosphate deposits, guano derived phosphate deposits ammonium-phosphate minerals are common (CULLEN 1988). Dahllite, a hydroxyl-rich fluorine-poor apatite variety forms during late diagenesis (TRAPPE & VONDRA 1995). Other rare sedimentary phosphate minerals are summarized in NRIAGU & MOORE (1984), McCONNELL (1973), SLANSKY (1986).

3 Geochemistry of sedimentary phosphate particles: tools for understanding phosphate mineral precipitation

The geochemistry of sedimentary apatite is affected by kinetic mechanisms, thermodynamic factors, as well as secular variations of seawater, and the pathway of formation. Subsequent effects of weathering or metamorphism are discussed separately. The geochemistry of unaltered francolite in sedimentary phosphate particles gives clues to the chemical environment of formation. Other techniques help to date the formation of the grains. The amount of literature on the geochemistry of sedimentary apatite is immense. Recently, JARVIS et al. (1994) summarized and reviewed the subject in a state-of-the-art report.

3.1 Major elements

The major element chemistry of unaltered francolite is mainly determined by the principal substitutions and displays only little variations of the composition of 32% P_2O_5, 52% CaO, and 4% F (McARTHUR 1978, 1985, JARVIS et al. 1994). Minor constituents are 1.2 ± 0.2% Na, 0.25 ± 0.02% Sr, 0.36 ± 0.03% Mg, 6.3 ± 0.3% CO_2, 2.7 ± 0.3% SO_4. Consequently, the major element compositions cannot be used to determine the environmental conditions during phosphogenesis. Ambiguous implications offer the grade of CO_3, F and Na substitution. These are also reflected in the major element composition due to major element exchange to maintain charge balance. These are already discussed in course of the apatite substitutions (see chapt. 2.1).

3.2 Trace elements

The various possibilities of substitution in apatite enables numerous trace element incorporations into the lattice, e.g. Ag, Cd, Mo, U, Zn, Y and REE (the latter discussed in chapt. 3.4, see below). Some trace elements are furthermore enriched in sedimentary phosphorite rock to considerable degree, others show normal abundance or are depleted (Tabl. 3.2). Their abundance varies largely in different deposits (GULBRANDSEN 1966, TOOMS et al. 1969, NATHAN et al. 1979, ALTSCHULER 1980, PRÉVôT 1990, PIPER & MEDRANO 1995, JARVIS et al. 1994). The elements Br, Cu, Cr, Ni, Pb, and V show a strong

Table 3.1. Major element composition of phosphate rock reference material derived from phosphorite ore concentrates. Compositions from POTTS et al. 1992; BCR: Community Bureau of Reference, Belgium; NIST: National Institute of Standards and Technology, USA. Total Fe expressed as Fe_2O_3; n.d. = not determined.

Element	NIST 120c Western Phosphate Field %	BCR 32 Morocco %	NIST 694 Florida %
SiO_2	5.5	2.09	11.2
TiO_2	0.103	0.0285	0.11
Al_2O_3	1.30	0.55	1.8
Fe_2O_3	1.08	0.231	0.79
MnO	0.027	0.0024	0.0116
MgO	0.32	0.403	0.33
CaO	48.02	51.76	43.6
Na_2O	0.52	n.d.	0.86
K_2O	0.147	n.d.	0.51
P_2O_5	33.34	32.98	30.2
CO_2	3.27	5.10	n.d.
F	3.82	4.04	3.20
SO_3	0.92	1.84	n.d.

affiliation to the phosphate fraction in sediments (PRÉVôT & LUCAS 1979, 1980, 1985, PACEY 1985, PIPER 1991, JARVIS 1992). Many trace elements replace Ca^{2+} in the apatite structure (GILINSKAYA 1991, 1993, GILINSKAYA & ZANIN 1983, GILINSKAYA et al. 1993). But the trace elements are not necessarily located within the lattice, they may also be absorbed to the crystal surface or be related to organic matter or sulfides in phosphate particles (JARVIS et al. 1994).

Maximum enrichment ratio reaches cadmium, which is of special environmental concern because of its toxicity (ALTSCHULER 1980, SLANSKY 1986, BATURIN & ORESHKIN 1984). Cadmium is closely associated with zinc. Both elements do not correlate with either francolite or organic matter (PRÉVôT 1990, NATHAN et al. 1991, PIPER 1991, PIPER & MERANO 1995). The occurrence is partially bound to sphalerite (ALTSCHULER 1980, MINSTER & FLEXER 1991). Whereas some authors favored the substitution of Ca^{2+} by these elements, which is problematic because of the much smaller ionic radius of Zn^{2+}, JARVIS et al. (1994) and also PIPER & MERANO (1995) suggest a marine plankton source of these elements.

Trace elements have been used to determine trends to specific geochemical environments (KREJCI-GRAF 1966, LUCAS & PRÉVôT 1985, GERMANN

Table 3.2. Concentration of trace elements ($\mu g\ g^{-1}$) of the "Average Marine Phosphorite" of ALTSCHULER (1980) in comparison to the element distribution of the "Average Shale" of TUREKIAN & WEDEPOHL (1961).

Element	Average phosphorite	Average shale	Normal abundance	Depletion/ enrichment factor
Ag	2	0.07		30
As	23	13	N	
B	16	100		6
Ba	350	580	N	
Be	2.6	3	N	
Cd	18	0.3		60
Co	7	19		3
Cr	125	90	N	
Cu	75	45	N	
Ga	4	19		5
Hg	0.055	0.4		7
La	147	40		4
Li	5	66		13
Mo	9	2.6		4
Ni	53	68	N	
Pb	50	20		2
Sc	11	13	N	
Se	4.6	0.6		8
Sn	3	6		2
Sr	750	300		2
U	120	3.7		30
V	100	130	N	
Y	260	26		10
Yb	14	2.6		5
Zn	195	95		2
Zr	70	160		2

et al. 1984), but investigations in non-reworked phosphorites show that the incorporation of trace elements is distinctively controlled by fractionation processes during phosphate fabric formation. The trace element composition of the phosphate components and the host sediment differ significantly (see Case Study 3). All elements which are not preferentially substituted in the apatite lattice (such as Sr) are commonly depleted in the phosphate fabrics compared to the host shale. Some of these depletions are already indicated by the data compilation of ALTSCHULER (1980) who used granular phosphorites and compared them with the "average shale" of TUREKIAN & WEDEPOL (1961). PIPER & MEDRANO (1995) demonstrated the complexity of the source and location of trace elements in phosphorite rocks.

3.3 Uranium

Uranium is usually showing a dictinct enrichment in phosphorites and was intensively investigated for military and civilian nuclear programs (McKELVEY 1956, ALTSCHULER et al. 1958). The positive correlation between U and P, which is evident in microscopic scale (BURNETT & VEEH 1977, KRESS & VEEH 1980, AVITAL et al. 1983), results from two mechanisms: U^{4+} and also Th^{4+} substitutes Ca^{2+} in the apatite lattice (GILINSKAYA 1991, 1993, GILINSKAYA et al. 1993). Some uranium seems to be complexed by organic matter (LUCAS & ABBAS 1989). Phosphorites contain varies proportions of U(IV) and U(VI) (ALTSCHULER et al. 1958, KOLODNEY & KAPLAN 1970, VEEH et al. 1973, O'BRIEN et al. 1987). The uranium incorporated into the lattice seems to be about entirely U(IV) and is only subsequently converted into U(VI) by oxidation (ALTSCHULER et al. 1958, O'BRIEN et al. 1987). The uranium up-take is thought to occur mainly in the initial phase of phosphate particle formation (JARVIS et al. 1994). The total uranium contents of phosphorites from different deposits vary significantly and range from very few $\mu g\ g^{-1}$ and up to 4000 $\mu g\ g^{-1}$ (e.g. SHELDON 1959, KOZLOV 1975, CATHART 1978, 1992, BATURIN et al. 1982, MOTT & DREVER 1983, ROE & BURNETT 1985, SLANSKY 1986, JARVIS et al. 1994). Extreme high values are restricted to deposits which occur to be drastically altered by postdepositional diagenetic, hydrothermal, metamorphic or most important weathering effects. The U contents of the "average phosphorite" of ALTSCHULER (1980) reach 120 $\mu g\ g^{-1}$ in contrast to 3.7 $\mu g\ g^{-1}$ of the "average shale" of TUREKIAN & WEDEPOHL (1961).

The uranium contents in apatite were tested to be an indicator of diagenesis (STARINSKY et al. 1982) and a paleo-redox marker (BURNETT & GOMBERG 1977), but JARVIS et al. (1994) pointed out, that the U(IV) concentrations do not necessarily indicate anoxic conditions during precipitation.

3.4 Rare earth elements and yttrium

The enrichment of REEs in phosphorites was noticed very early (COSSA 1878) and extensively described by McKELVEY (1950), ALTSCHULER et al. (1967), ALTSCHULER (1980), ILYIN & RATNIKOVA (1976), McARTHUR & WALCH (1984), JARVIS (1984,1992), HEIN et al. (1993), PIPER & MEDRANO (1995) and many other authors. The REE pattern initially was thought to reflect the geochemistry of the environment during formation and the negative Ce-anomaly (ELDERFIELD & GREAVES 1982) was commonly used as a tool for the characterization of the depositional environment.

The bulk of the REE^{3+} and Y^{3+} substitute Ca^{2+} in the francolite structure (JARVIS et al. 1994). ALTSCHULER's average phosphorite records a REE enrichment roughly of factor 2 and a yttrium enrichment of factor 9 in respect to the average shale. In phosphate concretions from the Early Carboniferous Alaunschiefer Formation in Central Germany, the enrichment factor for the REE is about 10 in respect to the enclosing shale (see Case Study 3).

The contents and element distribution of REE and yttrium vary widely in phosphate fabrics from different deposits. The shale-normalized element distribution of REEs and yttrium was thought to record the sea-water pattern from their place of precipitation (GOLDBERG et al. 1963, ALTSCHULER et al. 1967). The compilation of data from various ancient and modern deposits (Fig. 3.1) demonstrates that only a few of the phosphate occurrences actually display one of the natural seawater distributions described by ELDERFIELD & GREAVES (1982) and BERTRAM & ELDERFIELD (1993) with a Ce- or Eu-anomaly and a heavy REE enrichment. This is not only true for granular phosphorites which may have been altered during potential winnowing and transportation, but also for concretionary phosphorites.

Besides a direct seawater source of REE, the elements may also be derived from ferromanganese oxide-organic matter particles, biogenic silica, clay minerals, precursor carbonates. PIPER & MEDRANO (1995) pointed out that the bulk REE composition in phosphorites has detrital, biogenic and authigenic source. The various REE patterns are also interpreted to be a result of fractionation processes during CFA precipitation or are representing porewater compositions (see discussion in JARVIS et al. 1994). McARTHUR & WALSH (1984) found evidence for a process in which phosphorites initially precipitate with low REE and subsequently scavenge REE during their depositional history.

Peculiar are a number of concave REE patterns (Fig. 3.1), which are found in biogenic phosphate fabrics, in some phosphate ooids, and in most phosphate concretions. These MREE-enrichments are considered to be typical for conodonts and fish teeth (WRIGHT et al. 1987, GRANDJEAN & ALBARRÉDE 1989, GRANDJEAN-LÉCUYER et al. 1993). Similar patterns are also found in primary phosphatic shell material in Estonia. But this distribution occurs also in non-skeletal phosphate fabrics. Phosphatic ooids from Middle Ordovician strata in Estonia show MREE enriched patterns (STRUESSON 1995), but not ooidic phosphorites from the Permian Phosphoria Formation. Very abundant is a MREE-enrichment in phosphate concretions which have grown in a shale matrix. These are reported by McARTHUR & WALSH (1984) and KIDDER & EDDY-DILEK (1994) from occurrences in North America and are characteristic for the concretionary phosphorites in Germany (see Fig. 3.1 and Case Study 3). A combination of MREE-enrichment with a negative Ce-anomaly is evident in some phosphorites from the Cambrian in China (SIEGMUND 1995). The concretionary phosphorites all have in common to be the result of phosphate cementations around aggregates of organic matter. From the German occurrences is known that the directly enclosing shales show normal shale-like pattern.

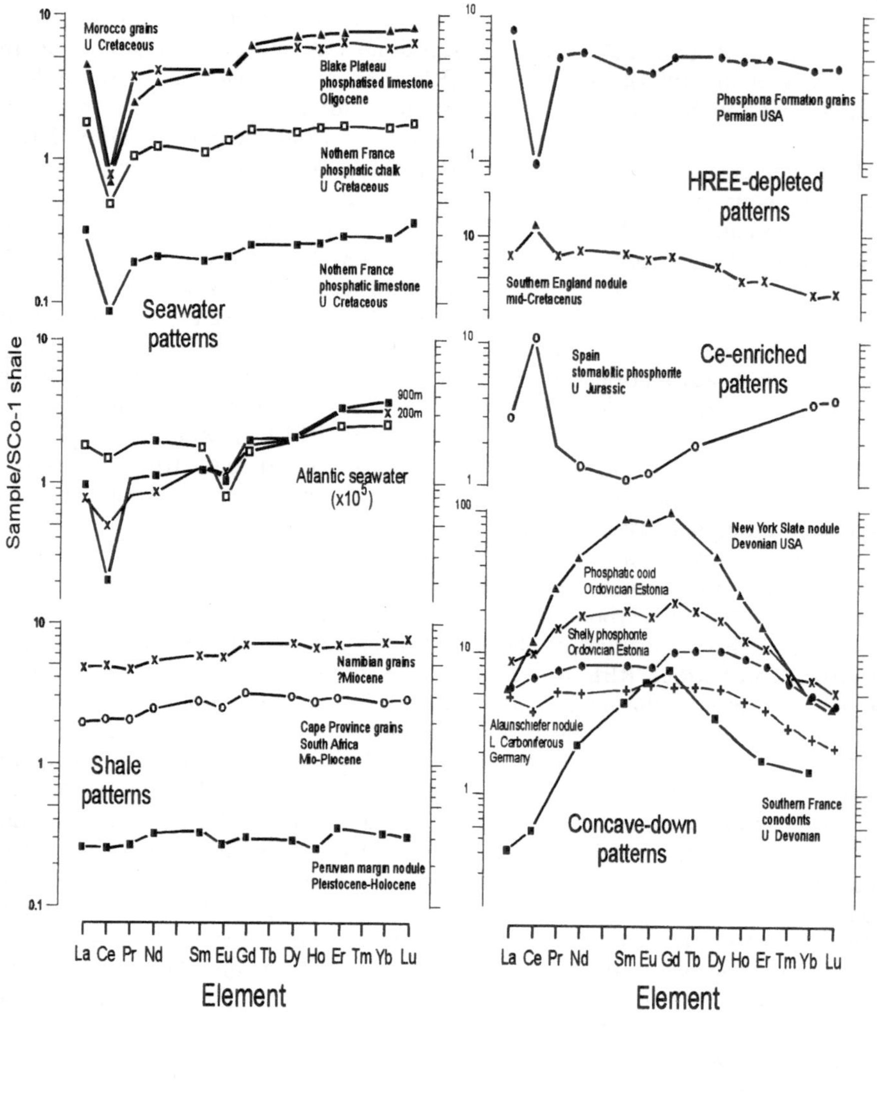

Fig. 3.1. The spectrum of various shale-normalized REE pattern of phosphorites. Compilation of major pattern by JARVIS et al. (1994). In many granular deposits, a distinct Ce-anomaly can be found, which corresponds to the REE distribution of deeper oceanic seawater. Several types of different phosphorites show a MREE-enrichment and HREE-depletion, which indicate a common genetic relationship: triangle: phosphatic ooids from the Ordovician in Estonia (STURESSON 1995), dotts: shelly phosphorites in Estonia, squares: concretionary phosphorites from Germany. See also text for discussion.

KIDDER & EDDY-DILEK (1994) relate the distribution to fecal-pellets, but a general biologic and microbial mediation have to be considered as well.

The poorly understood, but commonly applied REE composition for the reconstruction of the paleoenvironment during phosphogenesis has become very ambiguous with the discovery of more and more untypical patterns. Biological and diffussion processes as well as porewater effects seem to cause selective fractionation which is distinctively documented in the enigmatic positive Ce-anomaly in phosphatic stromatolites (Fig. 3.1). Further overprint has to be expected from the changes of REE composition in the Phanerozoic seawater (WRIGHT et al. 1987). McARTHUR & WALSH (1984) pointed out that with incorporation of shale material, the REE-distribution approaches the shale pattern, which underlines the polygenetic composition of the bulk REE content in phosphate fabrics.

3.5 Organic components

The general occurrence of phosphate rocks is commonly bound to lithofacies associations which are rich in organic matter. But in smaller scale, the phosphate mineral contents in the rock are usually associated with a decrease in the organic matter compound (SANDSTROM 1986, SLANSKY 1986). The organic matter composition of ancient phosphorites was investigated in a few studies (e.g. POWELL et al. 1975, BELAYOUNI & TRICHET 1981,1984, SANDSTROM 1986, 1990, FIKRI et al. 1989, KRAJEWSKI 1989, NATHAN 1990, TRICHET et al. 1990). These investigations indicate that the organic matter is of a marine source and reflects intensive microbial degradation and oxidation. The latter explains the negative correlation of C_{org} and phosphate mineral contents. The more stable humic compounds support the results from the organic matter composition (NATHAN 1990). Biomarkers reveal mostly an algal source of the organic matter (BEIN & AMIT 1982, SANDSTROM 1986).

The organic geochemistry of phosphate particles is still in its infancy and further investigations may help to develop a useful tool to determine various types of organic phosphorus sources, and to follow the degradation and oxidation processes during phosphogenesis.

3.6 Isotopes

The broad spectrum of substitutions in the apatite lattice allows the incorporation of various isotopes which offer a number of instruments for the interpretation of paleoenvironment and diagenesis. The relative stability of francolite compared to other sedimentary minerals has led to an enormous number of studies and applications. The various isotopes occupy the Ca and the PO_4 sites of the apatite lattice (Fig. 3.2).

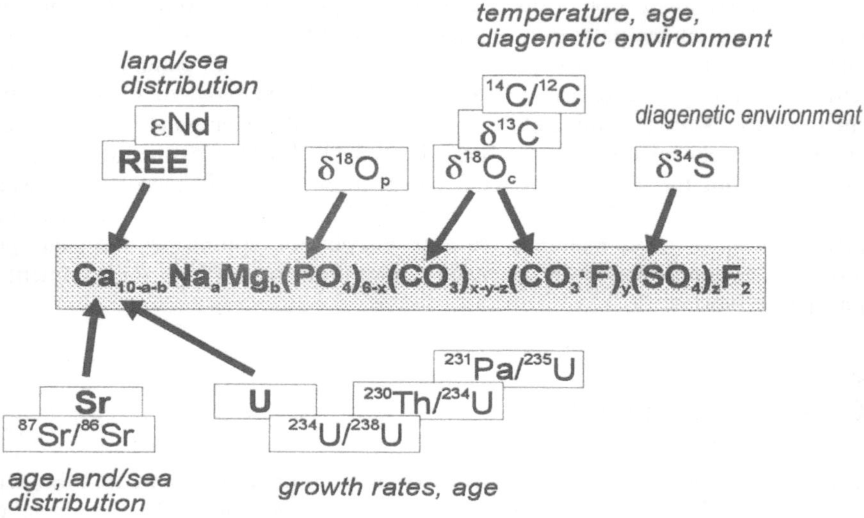

Fig. 3.2. The incorporation of various isotopes in francolite and their application for various paleoenvironmental and dating methods (modified from KOLODNY & LUZ 1992 and JARVIS et al. 1994).

Carbon, oxygen and sulphur isotopes are used to reconstruct the oxygenation stages of the sediments during organic matter degradation and apatite precipitation (McARTHUR et al. 1986)(Fig. 3.3). The application of this method gives good results for modern and Neogene deposits. In older occurrences, the signature of the carbon and oxygen composition is commonly overprinted by diagenetic and burial diagenesis (Fig. 3.4).

The diagenetically very stable Sr isotopic composition is used for dating as well as uranium isotopes with their decay products (BURNETT & VEEH 1977). For modern phosphorites, BURNETT et al. (1988) applied a combination of uranium and ^{14}C dating to determine growth rates of phosphate particles in modern phosphorites. The neodymium isotope composition is used to distinguish different major water masses in the ocean or to draw correlation to main current systems (STILLE et al. 1992).

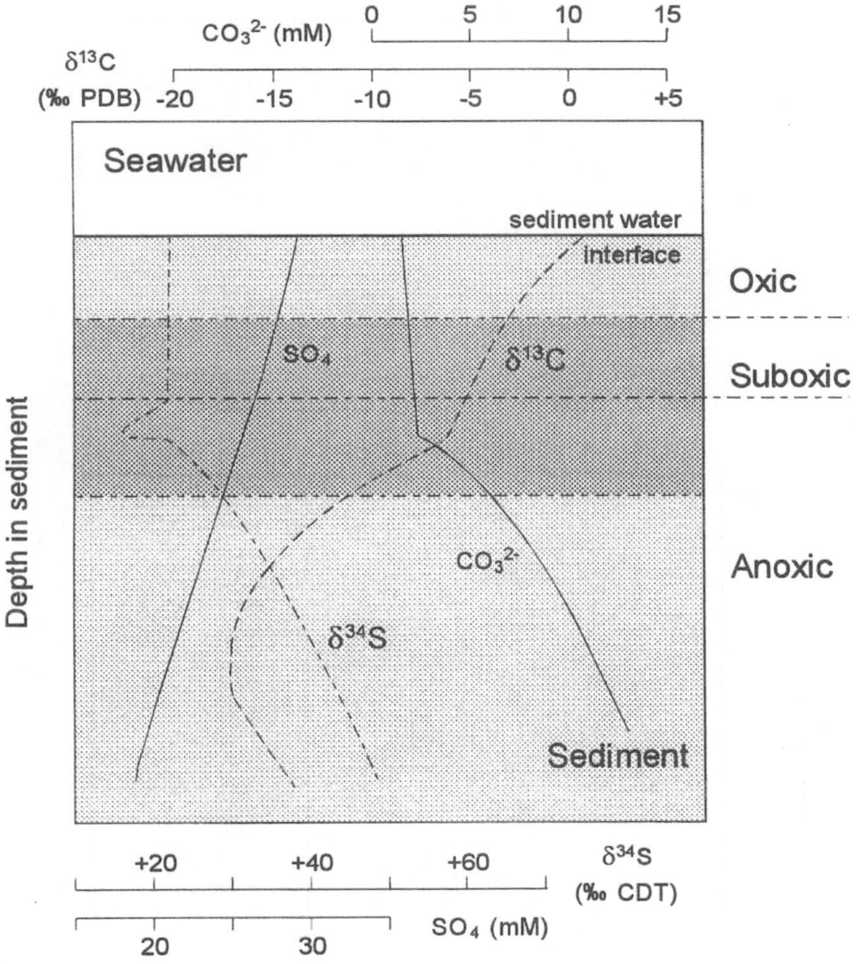

Fig. 3.3. The relationship of $\delta^{34}S$ and $\delta^{13}C$ profiles to the $SO4^{2-}$ and $CO3^{2-}$ contents of a idealized modern soft sediment column. The schematic diagram assumes constant pH conditions and linear increasing inorganic carbon with depth (from JARVIS et al. 1994 based on McARTHUR et al. 1986).

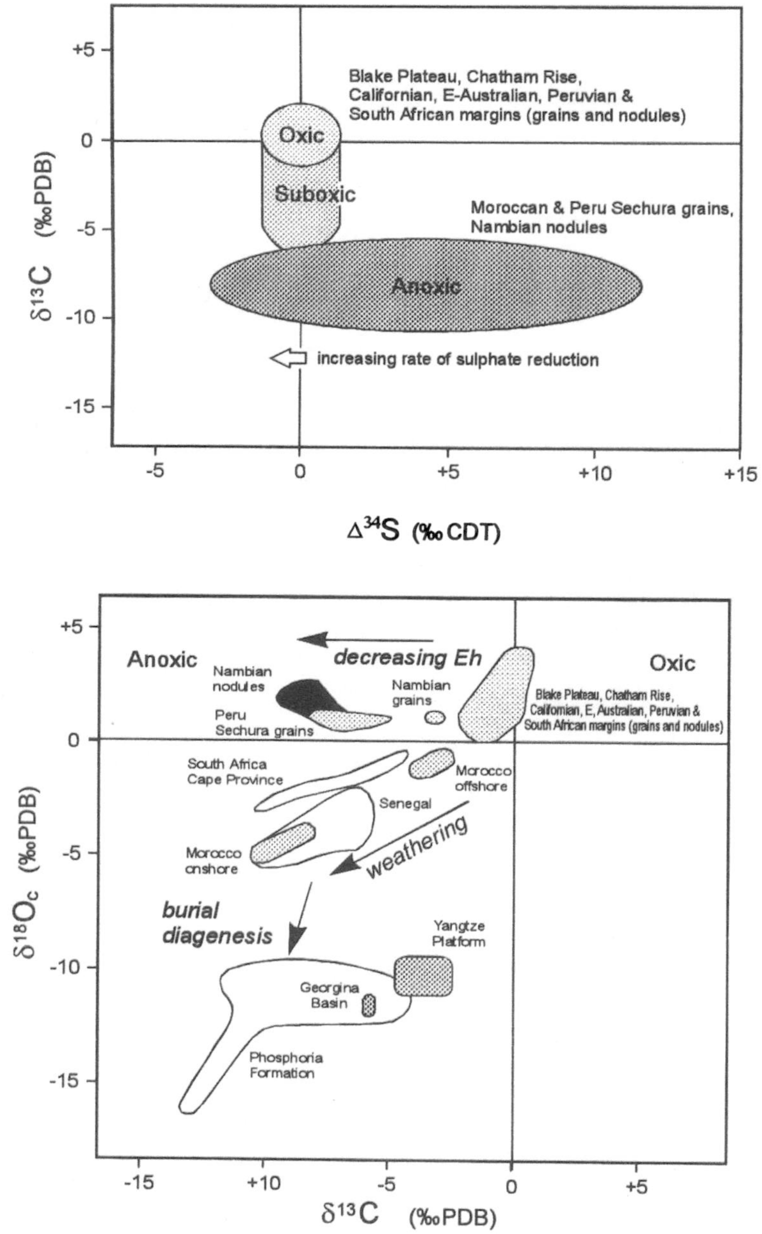

Fig. 3.4. Carbon, oxygen and sulphur isotope composition and stages of organic matter degradation and burial diagenesis (based on McARTHUR et al. 1986, modified from JARVIS et al. 1994, additional data from SIEGMUND 1995).

3.7
Relation of phosphorite and host-sediment geochemistry

The interpretation of geochemical signals from phosphate sediments is complicated by various effects which control the element flux into the bulk sediment, respectively into the different sediment components and fabrics (Fig. 3.5). These complex relationships partly overprint the data for geochemical paleoenvironment analysis or affect ambivalent signals from the various sediment constituents. These effects become rising importance for the interpretation of granular phosphorites which owe their origin from a complex sedimentary history including reworking, transport, and redeposition (see chapt 4).

The incorporation of trace elements and rare earth elements in unaltered francolites is driven by two important parameters. These are supply, but also kinetic effects, which both control the substitution mechanisms (see also chapt. 2). Consequently, the bulk trace element composition of apatite does not purely display the porewater or seawater conditions. The data are consequently not a reliable indicator of the paleoenvironmental conditions in the host sediment during mineral precipitation.

The geochemical signals from the element spectrum of bulk phosphate fabrics are modified by the proportions of the biologic, detrital, and inorganic-authigenic sediment components, which each contributed differently to the total element flux in the sediment (Fig. 3.5). Important are the studies of PIPER (1991), PIPER & MEDRANO (1995) and PIPER & ISAACS (1995) who demonstrated the polygenetic origin of the trace elements and REE compositions as well as their various incorporation during different stages of organic matter degradation. The trace element ratios of the hydrogeneous component, which is indicative for the geochemical environment during precipitation, is overprinted by increasing presence of organic matter and detritus. This is evident for both, a host shale and related phosphate fabrics (see also PRÉVôT 1990).

Important to note are furthermore the discrepancies between phosphate fabric and host sediment composition. Trace element contents and distributions in non-reworked phosphate concretions and in directly enclosing shales show significant variations (see also Case Study 3). These variations result partly from kinetic effects of element incorporation in the apatite lattice during growth of the concretion. The phosphate fabrics commonly have significantly lower C_{org} contents and seem to reflect sites of enhanced organic matter decomposition. Their degradation liberalizes locally particulate element contents (e.g. Fe, Mn, Cu, Zn) which are associated with organic sediment component (see Fig. 3.5). This alters the signals from host sediment and phosphate fabric additionally.

Most available geochemical data are derived from bulk phosphate sediments or from phosphate particle concentrates (e.g. ALTSCHULER 1980). The majority of the studies investigated granular phosphorites in which synsedimentary alteration and exchange of matrix during sediment reorganization generally cause a multi-genetic sediment composition. These analyses principally record

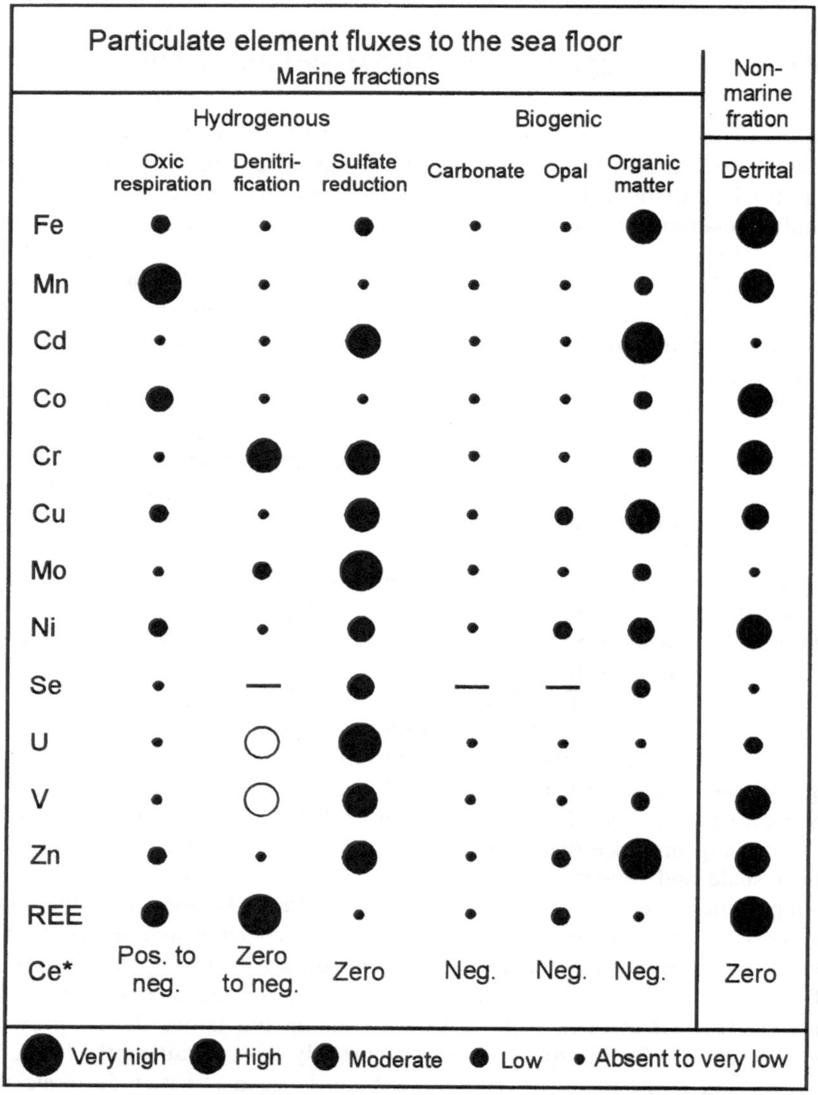

Fig. 3.5. Particular element fluxes into the sediment in the photic zone of high bioproductivity areas with low detrital supply (from PIPER & ISAACS 1995).These areas have a high potential for phosphogenesis. Ce* = Cerium anomaly, defined as log[3Ce/(2La+Nd)], elements normalized to North American Shale Composite (NASC).

a spectrum of various geochemical signals, emphasizing different indicators in individual deposits. The most important effect is the amount of detritus incorporation. Conspicuous data are only achieved in very extensive sequential analyses following the PIPER (1991), PIPER & MEDRANO (1995) and PIPER & ISAACS (1995).

However, in some case, trace element compositions of phosphate sediments have been successfully used to determine individual environments. GERMANN et al. (1987) could distinguish upwelling and fluvial influenced phosphorites by a number of trace element ratios. Also Sr and Ba contents seem to reflect seawater compositions (PRÉVôT & LUCAS 1985). These methods reveal in individual deposits sufficient results, but none of these techniques can be generally used of the paleoenvironmental investigations. This is also true for the paleoredox indicator V/V+Ni ratio (see WIGNALL 1994) and the basic geochemical facies analysis of KRECJI-GRAF (1966). The REE composition, which indeed shows better results, reveals similar problems with the increasing number of data. Some conspicuous REE patterns, which do not correspond to natural sea water distributions, are still enigmatic.

Better indicators for the investigation of the paleoenvironmental conditions during the formation of ancient phosphorites are stable isotopes, but in older deposits diagenetic and weathering effects overprint the primary signature. All geochemical studies have to be combined with extensive sedimentological investigations to exclude significant overprint of geochemical signals by sediment mixing or synsedimentary alteration.

3.8 The effect of weathering and metamorphism on francolite composition

The alteration of the element and isotopic composition is well-documented. The major element and trace-element geochemistry show a decrease of substitutions with increasing age, burial diagenesis, and weathering (LUCAS et al. 1980, McARTHUR 1980, 1985, DA'RIN & ZANIN 1985, NATHAN et al. 1990). This is displayed significantly in the CO_2 contents. Weathering leads also to leaching and redistribution of the REE composition (BONNOT-COUTOIS & FLICOTEAUX 1989, FLICOTEAUX 1990). The carbon and isotopic composition is altered during weathering and burial diagenesis/metamorphism (McARTHUR et al. 1986).

4 Phosphate sediments

4.1 Sedimentology of phosphate rocks

Phosphate rocks consist of one or several phosphate components within a siliciclastic, siliceous, organic or carbonate host sediment. Pure phosphatic sediments are existent, but are very rare. Phosphate rocks comprise stratiform sediment bodies, unconformity-bound pavements, multi-event crusts, and condensed beds. These sediments are composed of granular accumulates, microbial structures, or presumeable inorganic layers and concretions. The phosphate mineral phase consists of a broad spectrum of various phosphate particles and fabrics which differ significantly in the origin, internal texture, and sediment-particle relationship (pristine, in-situ, autochthonous versus detrital, clastic, allochthonous). The fabrics originate from carbonate replacement, presumable inorganic concretionary growth, microbial mediation or possible inorganic precipitation. Sediment structures of the finally preserved phosphorite sediments mostly indicate a clastic accumulation of the grains (granular phosphorites, allochemical phosphorites). Analogous to carbonates, compositional and textural properties of the phosphate bearing sediments allow the reconstruction of the depositional environment and history with microfacies analysis techniques. The depositional history, especially the mechanical concentration process, is significant for the economic value of the deposits.

Detailed petrographic-sedimentological analysis of the apatite - glauconite - pyrite - dolomite paragenesis in the only modern deposits on the shelf off Peru and Chile were presented by VEEH et al. (1973), GLENN & ARTHUR (1988), BAKER & BURNETT (1988), GLENN (1990b) and GARRISON & KASTNER (1990). Furhter sedimentologic studies exist from a number of other very young deposits which were initially thought to be modern, but have been later identified being Quarternary in age (off S' Africa: BATURIN 1969, BIRCH 1977, 1979a,b, 1980, PRICE & CALVERT 1975, BREMMER 1980; off NW' Africa: PIETZNER & RICHTER 1986, SUMMERHAYES et al. 1972; off E'Australia: RIGGS et al. 1989). Phosphate mineralization in these sediments occurs as nodules, crusts, coatings, and spherical grains. The latters, commonly described as pellets, are an inhomogeneous group of spherical grains summarizing ooids, coated grains, structureless grains, intraclasts, lumps, and biogenic grains. GARRISON & KASTNER (1990) and GLENN et al. (1994) recognized the significance of pelletal grains in these occurrences and closed the discrepancy between the grain dominated ancient deposits and the modern sediments off

Peru and Chile which were thought to be dominated by crusts and nodules. Their research could show that structureless phosphate grains, which are very common in ancient deposits, can be formed either by reworking of crusts or by particle growth within the sediment.

Compared to the large number of deposits, modern sedimentologic-petrographic studies of the ancient deposits are still exceptional. SCHRÖTER (1986) and GLENN (1990a,b) presented detailed lithofacies analysis of the Egyptian Late Cretaceous phosphorites. GERMANN et al. (1984, 1985, 1987) tried to correlate the geochemical and petrographic results of that deposit. More attention received the petrography of phosphatic hardgrounds. Important are the detailed investigations of JARVIS (1980) and SOUTHGATE (1983) who described the microfacies of calcareous phosphate suites of phosphatic hardgrounds, phoscretes and coated grains from the Cambrian sequence in the Georgina Basin/Australia and from the Cretaceous of NW' France. FÖLLMI (1989, 1990) presented a detailed study of an equivalent siliciclastic phosphate facies from the mid-Cretaceous of the Helvetian Alps. PRÉVôT (1990) documented the petrography of various phosphate grains from the Moroccan Ganntour deposits. TRAPPE (1989, 1992a) used consequently the carbonate microfacies analysis for both carbonates and phosphorites to interpret the marginal portion of the Morroccan phosphate basin. GARRISON et al. (1990), GRIMM (1994) and SCHWENNICKE (1992, 1994) described the sedimentology of the Miocene phosphorites in Baja California.

Most of the ancient deposits are dominated by a peloidal phosphate facies. Condensed intervals with phosphatic crusts and nodules are much more exceptional in the geological record than the number of research contributions, which are focussing on the more stimulating exotic deposits, may imply. The modern phosphate deposits are located on a narrow and deep shelf along an active margin. The sedimentology and lithofacies of most ancient deposits assign phosphogenic deposystems to dominantly shallow water on broad continental shelves or gulf-like epicontinental sea areas (see chapt. 6). Recently, GLENN et al. (1994) tried to emphasize some of the common characteristics of ancient and modern deposits.

4.1.1
Stratiform phosphate fabrics

Non-granular phosphate fabrics in ancient deposits cover a wide range of different forms. They split into three major groups.

Phosphatization of erosional surfaces along hardgrounds and in condensed sediments (Fig. 4.1):These fabrics are known from many marine sediments and attracted early numerous researchers (BRAID 1978, JARVIS 1980a,b, 1992, PEDLEY & BENNETT 1985, SOUTHGATE 1986a,b, FÖLLMI 1989a,b, 1990, MONCIARDINI 1989, SOUDRY & LEWY 1990, POMONI-PAPAIOANNOU & SOLAKIUS 1991, CARSON & CROWLEY 1993, REHFELD & JANSSEN 1995). Different phosphate fabrics developed either in carbonates or siliciclastic sediments. Erosional surfaces on carbonate rock show a penetration zone parallel to the surface in which carbonate is replaced by phosphate. The partial carbonate replacement and the ultrafabrics of apatite crystallization are described by LAMBOY (1990a,b) from phosphatized chalks in France. In siliciclastic sediments, a mostly complete phosphate cementation within the present pore space occurs irregularly along the discontinuity surface (s. a. FÖLLMI 1989a, 1990). In both depositional systems, these erosional surfaces are usually capped by microbial crusts or linings, furthermore by a pavement of all kinds of granular phosphate fabrics. Non-erosional condensed beds rarely develop distinct stratiform fabrics. These usually comprise various phosphate particles which are partial or complete phosphatizations of bioclasts itself or of their internal sediment, phosphate cementations within porous sediments, or more rarely crusts and stromatolites. The association with glauconite or related minerals is most common (KENNEDY & GARRISON 1975, LUCAS et al. 1978, BURNETT 1980a, ODIN & LETOLLE 1980, LOUTIT et al. 1988, FÖLLMI 1990, O'BRIEN et al. 1990, CARSON & CROWLEY 1993, TRAPPE & ELLENBERG 1994).

Pristine stratiform layers, concretions and lenses (Fig. 4.1): These are structureless, dense fabrics of various shape and size. Layers or thin beds occur only occasionally and are described from Australia (SOUDRY & SOUTHGATE 1989) and from Israel (SOUDRY 1992). More common are phosphate concretions of largely varying dimensions. The latter split in nucleated concretions which grow around organic remains, and non-nucleated impregnations of defined portions of sediment. These non-nucleated fabrics can preserve particles and sediment structures from the original sediment (RUNNELS et al. 1953, TIMMERMANN 1974, 1976, CONKIN & CONKIN 1975, HECKEL 1977, PAPROTH & ZIMMERLE 1980, STRUCKMEYER 1982, KIDDER 1985, SLANSKY 1986, 1989b, BALSON 1990, ZIMMERLE 1992, SOUDRY 1993, TRAPPE et al. 1995, SIEGMUND 1995). A special form of pristine phosphate mineralization is the phosphatization of burrows (SUHR 1991, ZIMMERLE 1994).

Fig. 4.1a. (next page) Photo plate of stratiform phosphate fabrics. (A) Laminated black shale with structureless phosphate concretions which become obvious by their whitish weathering color. Phosphoria Rock Complex, USA, Permian. (B) Laminated phoscrust. Georgina Basin, Australia, Cambrian. (C) Fragment of a bored hardground surface which is phosphatized along its surface. Florida, USA, Neogene.

Fig. 4-1b. (2 pages further) Photo plate of stratiform phosphate fabrics. (A) Microbial crust on an erosional surface. Subatlas Group, Morocco, Paleogene. (B) An erosional surface with borings is thinly coated with a phosphatic blanket and filled with a clastic phosphorite concentrate. Subatlas Group, Morocco, Paleogene. (C) Microsphorite layer of several millimeter thickness with clusters of organic matter. Alaunschiefer Formation, Germany, Early Carboniferous.
See also color copies of these plates in the appendix.

Fig. 4.1a.

A

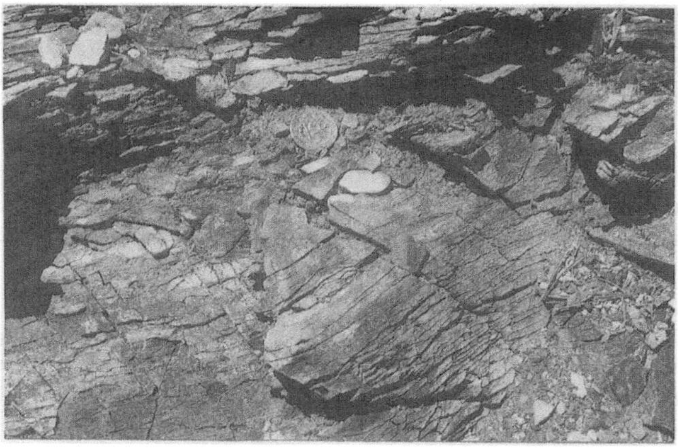

B

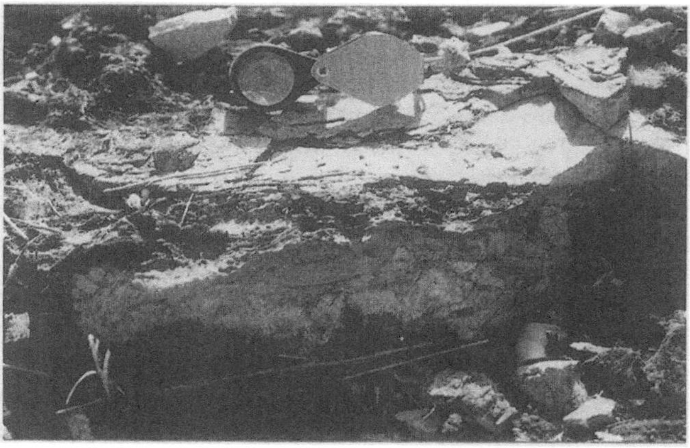

C

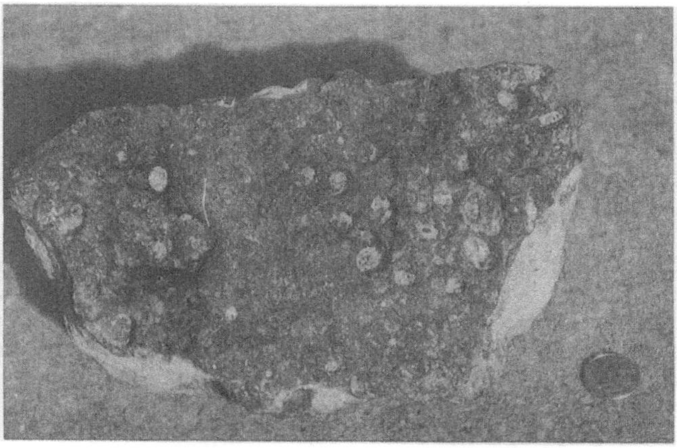

Fig. 4.1b.

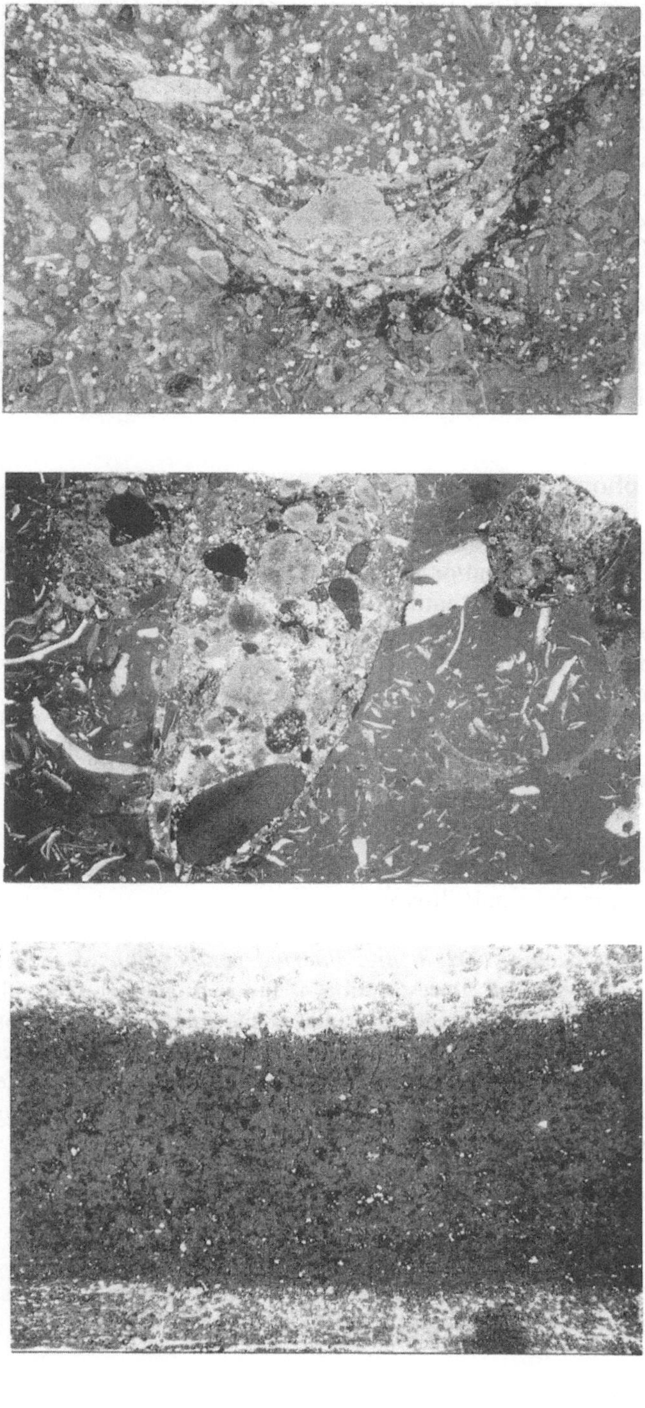

A B C

Phosphatic stromatolites and phoscretes (Fig. 4.1): Microbial mineralization known from carbonate sediments also occur as their phosphate equivalents. Growth forms and lamination types are similar to these of microbial carbonates (BANERJEE 1971, CHAUDAN 1979, SOUTHGATE 1980, 1986a,b, KRAJEWSKI 1981, TEWARI 1981, 1984, 1989, BASHYAL 1984, CHAUHAN & SISODIA 1984, KRASIL'NIKOVA & PAUL, 1984, VERMA 1984, KUMAR & MULLER 1988, FÖLLMI 1989a, 1990, SISODIA & CHAUDAN 1990, POMONI-PAPAIOANNOU & SOLAKIUS 1991, KRAJEWSKI et al. 1994, MARTÍN-ALGARRA & VERA 1994, MARTÍN-ALGARRA & SANCHEZ-NAVAS 1995). Extensive Precambrian phosphatic stromatolite growth contrasts the rare, mostly hardground-related occurrences in the Phanerozoic. Phoscretes are laminated or irregular incrustations which are commonly associated with microbial cementations. Phoscretes develop sporadically on land surfaces (COOK 1972, SOUTHGATE 1986, 1988).

4.1.2
Granular phosphorite rocks

Similar to clastic sediments, most phosphate rocks consist of three components: matrix and/or cement and/or particles. The constituents of the particle fraction are composed in part or entirely of rounded or angular phosphate grains of various types. Non-phosphatic components comprise all other grains known from siliciclastics or carbonate sediments. The matrix consists rarely of phosphate material, more commonly of carbonates, of siliciclastics, of silica, or of clay minerals. All common cements occur also in phosphate rocks including phosphate cements. The macroscopic approach already assigns many phosphate rocks to the group of clastic sediments. The phosphate particles differ significantly in grain sizes and internal texture.

The grain size varies analogous to siliciclastic rocks ranging from coarse gravel-size to fine grained sand/silt-size.

The internal structure, origin and internal composition of the grains differs largely. A grain-size based description of the rock analogous to siliciclastics is not applicable due to a polygenic origin of the grains. The composition of the phosphate fraction can only be investigated by microfacies techniques (see chapt. 4.3). The grain spectrum includes rounded phosphatized rock particles of various origin, primary or secondary phosphatic bioclasts, structureless peloids, various coated grains, intraclasts, reworked concretions.

4.1.3
Textural properties of phosphorite rocks

For the sedimentological characterization of granular phosphorites, which are common in most ancient deposits, granulometric methods derived from clastic

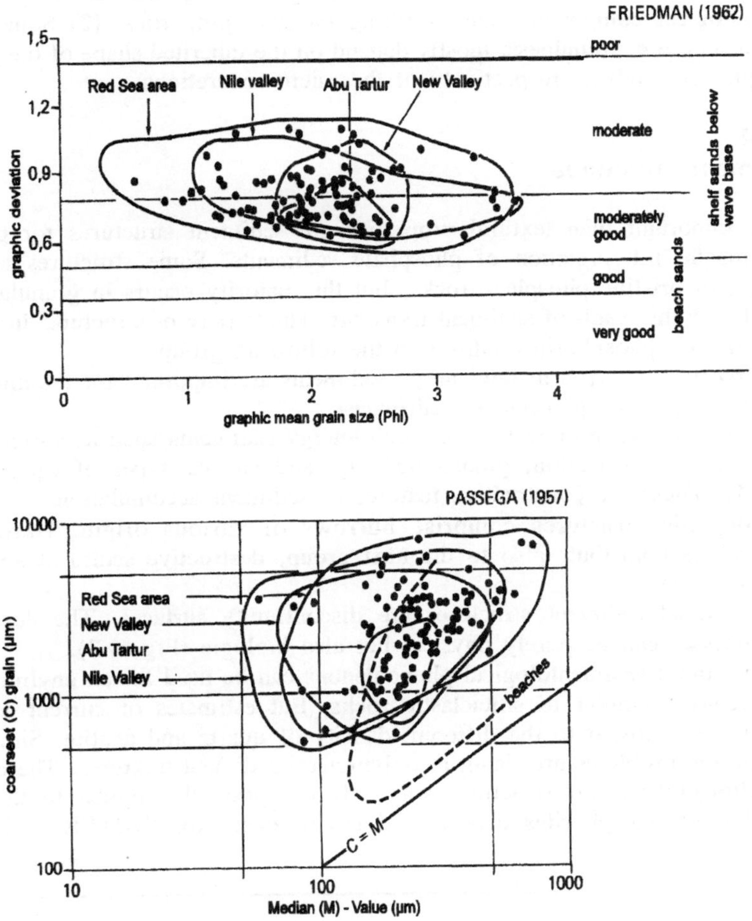

Fig. 4.2. Granulometric parameters from Egyptian phosphorites and their paleoenvironmental interpretation (from SCHRÖTER 1986).

rock analysis have been applied. Granulometric data can be used to determine transport processes (COOK 1967). SCHRÖTER (1986, 1989) used granulometeric techniques for the interpretation of the Egyptian deposits and could determine different depositional environments by their textural sediment composition (Fig. 4.2). All other granulometric studies are fairly rudimentary (SHELDON 1963, CRESSMAN & SWANSON 1964, BOUJO 1976).

Most researchers have not been successful to use these techniques to characterize depositional environments as precise as in siliciclastic rocks. Sorting gives some rough data to the water energy. All other properties can not be used by applying the methods from siliciclastic because of two differences: (1)

Phosphate grains are an extreme polygenetic group of grains with very different mineralogical composition and resulting physical properties. (2) Some of the parameters, e.g. roundness, mostly depend on the inherited shape of the primary phosphogenic fabric, respectively of the micro-concretions.

4.1.4 Sediment structures

More important than textural properties are sediment structures for paleoenvironmental interpretation of phosphate sediments. Some structures are only evident in pristine phosphate rocks, but the majority occurs in granular phosphorites as the result of sediment transport. The variety of structures in pristine and granular phosphorites splits into the following groups:

(1) Stratification types in low energy sediments are important for the interpretation of pristine phosphate occurrences.

(2) Sediment transport features of high energy sediments such as various types of cross-stratification, graded bedding, and various types of ripples (Fig. 4.3). These are constructive features of sediment accumulation.

(3) Biogenic structures comprise burrows of various origin. Hardground dwellers contribute also to the next group, destructive sediment structures (Fig. 4.3).

(4) Erosional sediment structures are discontinuity surfaces. The destructive processes can be purely physical but also biologic (Fig. 4.3).

Most of these sedimentological phenomenons can be used for an environmental interpretation similar to siliciclastic rocks. But estimates of current intensity need to be adjusted to the different density of quartz and apatite. Similar interpretation problems are rising in polymineral sediment mixtures. The analysis of sediment transport structures was only exceptionally applied to the interpretation of phosphorites depositional environment. SCHRÖTER (1986) and

---

Fig. 4.3a. (next page) Sediment structures in phosphate rocks. (A) Large scale cross-bedding in granular phosphorites. Georgina Basin, Australia, Middle Cambrian. (B) Cross-bedding in a fine clastic polymictic phosphorite. Phosphoria Rock Complex, USA, Permian. Base of photograph = 30 mm. (C) Granular phosphorite is concentrated in cross-stratified bedsets. Phosphoria Rock Complex, USA, Permian. Base of photograph = 40 mm.

Fig. 4.3b. (2 pages further) Sediment structures in phosphate rocks. (A) Graded bedding in a granular phosphorite bed. Phosphoria Rock Complex, USA, Permian. (B) A burrow in dolomitic shale is filled with structureless phosclasts. Base of photograph = 20 mm. Phosphoria Rock Complex, USA, Permian. (C) Burrowed bedding surface of a granular phosphorite bed. Phosphoria Rock Complex, USA, Permian. (D) Intensively burrowed sandy phosphorites. N'Florida, Neogene. (E) Erosional surface on carbonate mudstone is phosphatized along its surface. The overlying concentrate includes reworking products of the erosional surface. Phosphoria Rock Complex, USA, Permian. Base of photograph = 50 mm.

Fig. 4.3a.

A

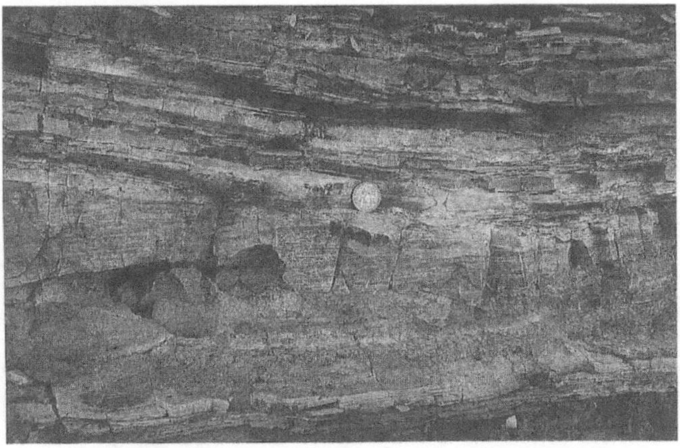

B

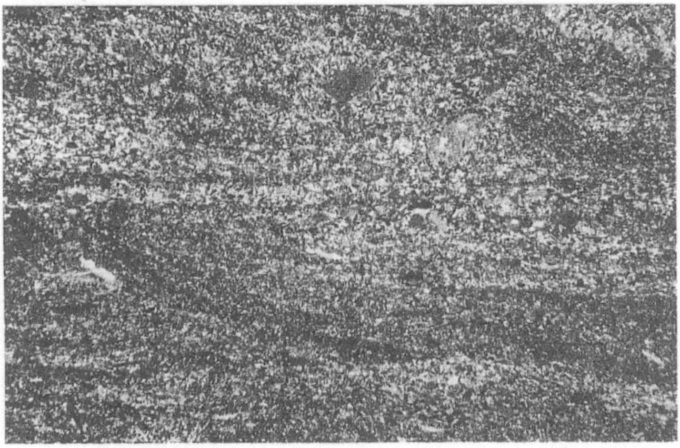

C

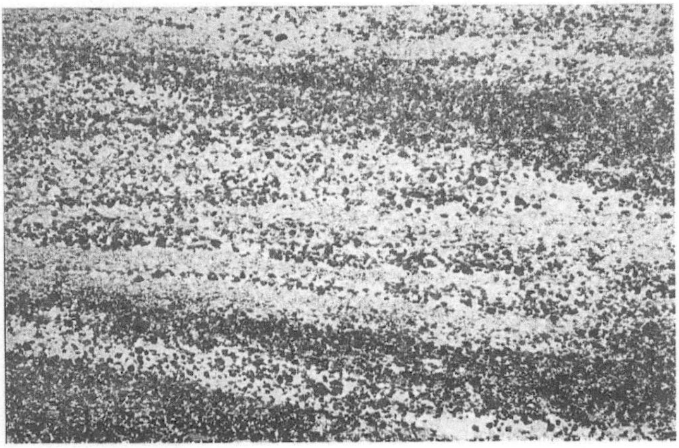

Fig. 4.3b.

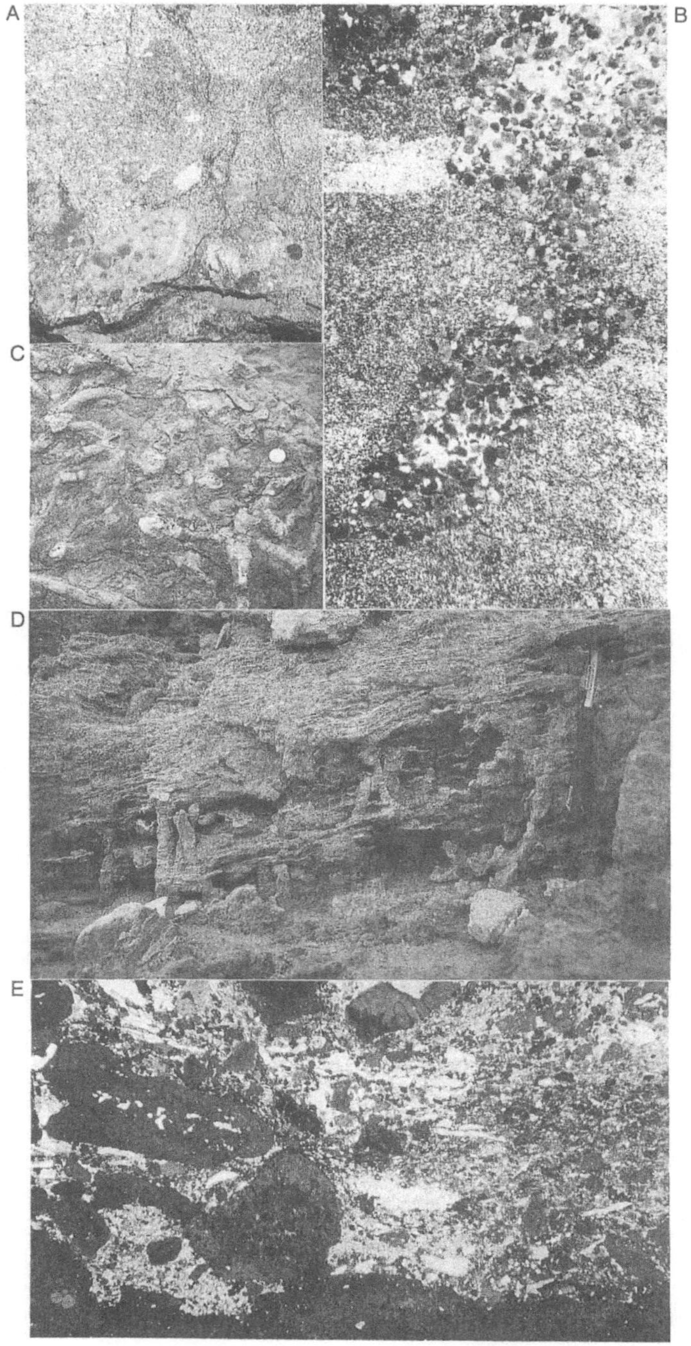

later GLENN (1990a) used these for the reconstruction of the depositional environment of the Cretaceous phosphorites in Egypt, and SCHWENNICKE (1992, 1994) for the Oligocene phosphorite-bearing sequence in Baja California. Stratification patterns are discussed in FÖLLMI et al. (1991).

Particularly, erosional features, mostly hardground suites, attracted special attention of many researchers (JARVIS 1980a,b, 1992, SOUTHGATE 1986a,b, FÖLLMI 1990, CARSON & CROWLEY 1993). These fabrics play a specific role for sequence stratigraphic analyses (SOUTHGATE & SHERGOLD 1991, GLENN et al. 1994, MALLINSON et al. 1994). Intervals of clastic sedimentation starvation, which are also recorded by discontinuity surfaces and corrosional fabrics, represent times of undiluted high organic deposition and can initiate phosphogenic processes.

4.2 Nomenclature for phosphate sediments

4.2.1 Phosphate - phosphorite - phosphatite: a nomenclatoric mess

Phosphate bearing rocks are named differently in respect to the concentration and texture of the phosphate component. The applied terms are also used in different schemes synonymous for each other and contribute to the confusion. The historic development of this chaos is extensively outlined and discussed in SLANSKY (1980a, 1986). The use of the terms "phosphate rock" and "phosphorite" was accepted by most authors and seems to be a helpful tool to differentiate phosphate-bearing rocks from more economically valuable phosphate-rich sediments. The most popular definition for the two terms has turned out to be the 18% P_2O_5 boundary, respectively 50% phosphate mineral component, with "phosphate rocks" defining compostions <18% P_2O_5 and "phosphorite" these of >18% P_2O_5. This definition will be followed in this work. The kind of phosphate mineral is not specified. The name phosphatite in sensu SLANSKY (1980a, 1986) was not commonly accepted.

4.2.2 Earlier classification schemes for phosphatic rocks and phosphorites

Only SLANSKY (1980a, 1986) presented a short overall documentation of various phosphorites and phosphate rock types. Most of the ancient occurrences are characterized by a small number of particle types. Consequently, classifications based on regional descriptions are lacking completeness. Relative comprehensive regional examples are presented by MABIE & HESS (1964), RIGGS (1979a), SCHRÖTER (1986), TRAPPE (1989, 1992a), GLENN & ARTHUR (1990), PRÉVôT (1990), JARVIS (1992) and SIEGMUND (1995). ZANIN (1987) published an atlas of ultra-microstructures in phosphorites.

The wide range of phosphogenic processes and the complex depositional history of phosphorites and common syn- and post-depositional alterations of

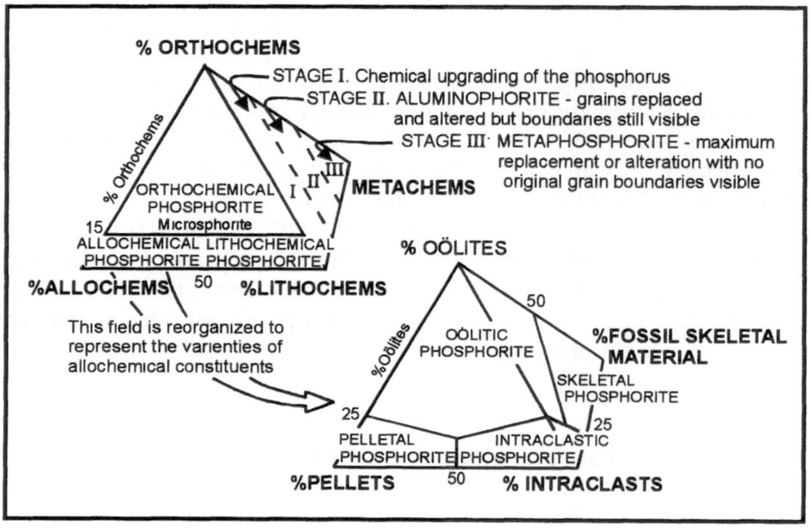

Fig. 4.4. Macroscopic classification scheme of RIGGS (1979a) for pristine (orthochems) and granular (allochems and lithochems) phosphate sediments.

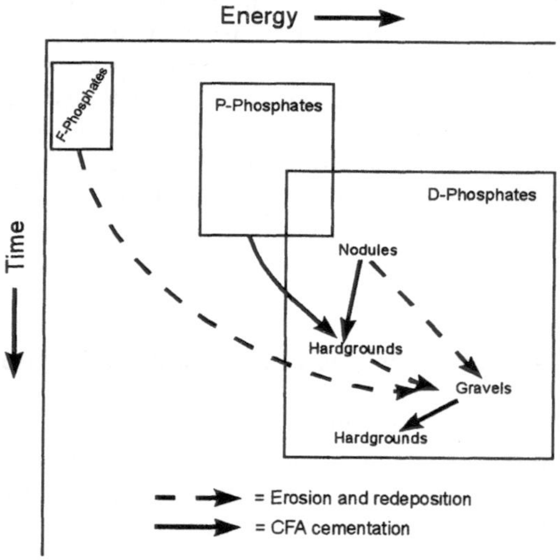

Fig. 4.5. Classification scheme for modern phosphorites from GARRISON & KASTNER (1990). This concept reveals the paleoenvironmental and time relationships of the various phosphate sediments.

the sediments entitles the development of an universal classification system for phosphorites to an unsolvable problem. The various definitions of the terms phosphatite - phosphorites - phosphatic are broadly discussed in RIGGS (1979a,b) and SLANSKY (1986). Very different approaches for a nomenclature scheme were used in the last two decades using macroscopic fabrics, microscopic petrographic tools or granulometric methods. A first macroscopic nomenclature for phosphorites was published by Russian authors (GIMMELFARB et al. 1959), but did not find larger attention in the western world. The earliest petrographic classification system was introduced by MABIE & HESS (1964) for the Phosphoria phosphorites. RIGGS (1979a,b) applied a modified carbonate classification of FOLK (1962) to phosphorites from the Neogene Florida system (Fig. 4.4). He divided phosphorite fabrics into orthochemical and allochemical constituents. Reworked and redeposited components are named lithochems. For chemically altered or weathered components, he introduced the term metachems. The quantitativly most important group of allochems, he split into pellets, intraclasts, and ooids. These, however, reflect only a small spectrum of the potential phosphate particles. COOK & SHERGOLD (1986) made an effort to establish a general phosphorite rock nomenclature which is fundamentally based on the DUNHAM (1962) carbonate classification and emphasized more the texture of phosphorite rock. COOK & ELGUETA (1986) and SOUTHGATE (1986) demonstrated the application of this concept. SLANSKY (1980a, 1986) favored a classification system based on the grain size of the main phosphate fraction and the second fraction, the endogangue (phosphatic) or exogangue (non-phosphatic), the latter being the main non-phosphatic component. He distinguished microsphatite, phosphalutite, phosarenite, -rudite. Recently, GARRISON & KASTNER (1990) created a genetic classification system designed for the specific requirements of the macroscopic description of modern phosphorites on the Peru margin (Fig. 4.5). The subdivision in F-phosphates (friable nodules, peloids or laminae of light colored CFA), D-phosphates (lithified often dark nodules, gravels and hardgrounds) and P-phosphates (phosphatic sands dominated by structureless and coated phosphate grains) is strictly designed for the requirements in modern phosphogenetic environments. Nearly all ancient phosphorites fall into the category of P-phosphates and consequently disqualifies the scheme for a general classification.

None of the nomenclatures was commonly accepted, but since the basic work of RIGGS (1979a,b) most petrographic studies of phosphorites tend to apply modified carbonate classifications to phosphorites. The phosphate fabrics of the modern occurrences, which are partly comparable, but not typical for ancient deposits, are better characterized by the KASTNER & GARRISON (1990) nomenclature.

Similar to carbonates, a reliable description of the composition and texture of ancient phosphate rocks is only possible with microfacies techniques (see chapt. 4.3). For a genetic, textural and compositional nomenclature, the well established carbonate classification of FOLK (1962) or DUNHAM (1962) is completely sufficient (see chapt. 4.2.4, and Fig. 4.6). The characterization of the components can be adopted from the RIGGS scheme from 1979 with slight modifications addressing the specific demand of the investigated specific phosphorite facies.

4.2.3
The "pelletal phosphorite" problem

Most ancient phosphate deposits consist of granular phosphorite rocks. The commonly round and spherical phosphate grains are traditionally described by the term "pellet" or "peloid". The problematic use of this purely descriptive term was first addressed by PRÉVôT (1982). This group of particles consists of a wide range of various phosphatic grains, which differ significantly in internal structure, composition and origin. The term pellet or peloid was derived from the carbonate petrography. In the original use, it was introduced as a purely descriptive term for spherical carbonate particles regardless to their internal texture (see discussion for carbonates in FLÜGEL 1982). In the literature, the term was abundantly mistaken for fecal pellets, which implies a specific genetically defined group of grains. Beside of that, most ancient phosphorites fall into this group, although these grains represent a wide range of polygenic particles, which only have in common to be phosphatic and nearly spherical. The useless terms which inhibit a genetic interpretation of the phosphate facies should be abandoned.

4.2.4
The microfacies classification: a practical approach

The textural composition of phosphorites and phosphate rocks corresponds roughly to carbonate rock. Also the genetic interpretation of a number of constituents conforms mainly with their carbonate equivalents. This allows the application of the well established carbonate microfacies nomenclature for an efficient description of phosphate rocks. In contrast to carbonates, phosphorites or phosphate rocks are usually a multi-mineral depositional system of carbonate, siliciclastics including clay minerals, apatite, and organic matter. Nevertheless, all features can be described with nomenclature applied from carbonate terminology.

The wide range of phosphatic sediments or particles split in two fundamental groups: orthochemical/microbial and allochemical phosphates (in sense of FOLK 1962 for carbonates).

Orthochemical/microbial phosphorites summarize all phosphatic fabrics which reveal in-situ, "pristine" (FÖLLMI & GARRISON 1991) phosphatic precipitates and are consequently not mechanically altered. A strict differebtiation between active or passive microbial mediation and inorganic precipitation has to remain open in most cases. The resulting fabrics are structureless phosphatic mudstones or microsphorite (BATURIN 1971, RIGGS 1979a,b, ILYIN & RATNIKOVA 1981, GLENN & ARTHUR 1988, SCHUFFERT 1988, SOUDRY & SOUTHGATE 1989, GARRISON & KASTNER 1990, GLENN et al. 1994), phosphatic stromatolites, microbial crusts on hardgrounds, and in condensed sections (e. g. BANERJEE 1971, SOUTHGATE 1980, 1986a,b, FÖLLMI 1989a,b, 1990, POMONI-PAPAIOANNOU & SOLAKIUS 1991, KRAJEWSKI et al. 1994, MARTÍN-ALGARRA & SANCHEZ-NAVAS 1995), impregnations of lithified carbonate (e. g. JARVIS 1980a,b, 1992, SOUTHGATE 1986a,b), phoscretes (SOUTHGATE 1986a,b), pristine concretions, lenses, and layers (e.g. HECKEL 1977, PAPROTH & ZIMMERLE 1980, KIDDER 1985, SLANSKY 1986, 1989a, SOUDRY 1992, 1993), microbial layers or coatings (e. g. SOUDRY & CHAMPETIER 1983, LAMBOY 1990a, ABED & FAKHOURI 1990, LEWY 1990), and precipitates in bioclasts (e. g. PIETZNER & RICHTER 1986, FÖLLMI 1989a, LAMBOY 1987b).

Allochemical phosphate grains are all phosphatic fabrics, which are clastic in character. The primary textural relationship to the host sediment is destroyed by reworking and transport. These grains are derived from three different processes:

(1) Bioclasts can be of primary biogenic origin like bioclasts of inarticulate brachiopods (LOWENSTAM 1972, ILYIN & HEINSALU 1990) (Fig.4.7).

(2) Reworking of the above listed orthochemical/microbial phosphorites produces structureless phosphate grains which are commonly described as pellets, peloids, ovoids or ovular grains. These round grains are more precisely named as intraclasts or lithoclasts if they are reworked from older rocks and/or transported into other depositional environments. The spectrum of grains includes also phosphatic ooids, oncoids, pisoids and fecal pellets (Fig.4.7).

(3) Epigenetic replacement of carbonate grains (JARVIS 1980a,b, PRÉVôT & LUCAS 1986, LAMBOY 1993) covers the entire range of the carbonate allochem spectrum. The grains, mostly bioclasts and intraclasts are characterized by marginal or complete phosphate impregnation/replacement under preservation of the primary internal fabrics (Fig. 4.7).

The textural composition of phosphate sediments is also sufficiently addressed by the classifications of DUNHAM (1962) with the extensions of EMBRY & KLOVAN (1971) for coarse grained sediments and boundstones, or the one of FOLK (1962). The use of one of the two schemes depends on the focus of the study and the composition of the deposit. The sediment texture as a result of the energy regime is best described by the DUNHAM classification (Fig. 4.6), the compositional variations are emphasized by FOLK. A further advantage of the DUNHAM classification is also the simple application to many non-carbonate rock compositions.

The name of the phosphate sediment is composed of the rock name for the textural composition, either from DUNHAM or FOLK, and one or two abbreviation(s) of the main grain type(s). Two particle types as the maximum are set in front of the textural term. Phosphatic grain types are defined by the additional short form "phos". For example, a biophosoopackstone would be a packstone in which the two major grain types are bioclasts and phosphatic ooids. If a primary carbonate rock e.g. in hardground horizons is partly or

Phosphate components are allochems and not fixated by microbial mats						Phosphate fabrics are microbial mat derived
Generally smaller grains (<2mm)				More than 10 % larger grains (>2mm)		Microbial mats act as sediment binder or as substate for apatite precipitation
Contains mud matrix (carbonate, phosphate or siliciclastic mud)		Lacks mud matrix (various cements or pores)		Contains mud (carbonate, phosphate, siliciclastic)	Lacks mud (various cements or pores)	
Less than 10% grains	More than 10% grains	Grain-supported				
Mud-supported						
Non-phosphatic mudstone or microsphorite	Phosclast-wackestone	Phosclast-packstone	Phosclast grainstone	Phosclast-floatstone	Phosclast-rudstone	(phosphatic) Bindstone

Fig. 4.6. Textural classification scheme for granular and microbial phosphorites and phosphate-bearing multi-mineral rock composition (based on the carbonate classification of DUNHAM 1962 with extentions by EMBRY & KLOVAN 1971). The phosphate allochem constituents can be precisely named following the outline in the text.

completely replaced, respectively impregnated, by apatite, the term phosphatized should be used to address this rock property. All non-phosphatic grains are treated following the classification of DUNHAM (1962) and FOLK (1962) or according the discussion in FLÜGEL (1982). For many applications, this terminology has to be extended by specific definitions to address uncommon rock compositions and textures. These definitions should be explained before usage.

4.3 Microfacies techniques - an analytical approach to the understanding of phosphorites

Macroscopic investigation methods are generally limited in terms of detailed characterization of the sediment properties for a compositional, paleo-environmental and genetic interpretation. Due to the polygenic origin of phosphate particles and the inherited textural parameter of some of the particles, also granulometric and textural investigation methods fail to address the wide range of phosphate facies and paleo-environmental settings. The necessary data are only provided by microfacies analysis methods applied from carbonate petrography (FLÜGEL 1982). Microfacies descriptions of phosphorite depositional systems are still exceptional (SOUTHGATE 1983, CAROZZI 1989, TRAPPE 1989, 1992a, GLENN 1990b, SIEGMUND 1995). The few studies have shown that this method is the only technique which properly identifies the compositional spectrum, facies architecture, depositional setting and evolution of phosphate deposits.

As mentioned above, phosphate rocks consist of phosphatic and non-phosphatic particles, and/or phosphatic and non-phosphatic matrix and/or cements. The description and interpretation of non-phosphatic constituents, especially the carbonate fraction is broadly discussed in FLÜGEL (1982). The application of these methods to phosphatic components requires a definition of the phosphate constituents which is done in the following chapters.

4.3.1 Phosphatic allochems: the phosphate grains

The group of allochemical constituents covers the wide range of clastic phosphatic grains:

Phoslithoclasts (Fig. 4.7): structureless, primary spherical, rounded or angular phosphate particles which are by textural evidence definitely not in-situ and originate from another depositional environment. Their clastic character results from a process of winnowing or reworking of the original host sediment and furthermore of transportation into the environment where these grains are finally deposited. Similar to the lithoclast definition of carbonate grains, phosphatic clasts which are reworked from significantly older rocks in the same environment of deposition are also assigned to this group. Grains which are marginally phosphatized carbonate clasts, can optionally be named phosphatized lithoclasts.

Phosintraclasts (Fig. 4.7): structureless grains of various shape similar to phoslithoclasts. In contrast to phoslithoclasts, these grains must have been formed in the same environment where they are redeposited. The grains are derived from reworking of phosphatic sediments or, more common, originate from winnowing of portions of sediment which still hosts the phosphate allochems. The process commonly results in a concentration of the phosphate fabrics and is known as "BATURIN cycling" (BATURIN 1971a,b). The optional definition of phosphatized intraclasts is equivalent to lithoclasts.

Phosclasts: Optional definition for structureless grains of an uncertain transport component to replace lithoclasts and intraclasts.

Phosooids (Fig. 4.7): multicoated, uniformly laminated spherical grains of apatitic or phosphatic-polymineral mineralogy. Very regular spherical ooids are rare (MABIE & HESS 1964, HORTON et al. 1980, SWETT & CROWDER 1982, PRÉVôT 1990). Examples from the Phosphoria Formation where partly preserved matrix is also phosphatic, but not the cement, may result from phosphatization of carbonate ooids in lag concentrates. The phosphatization of ooidic

---

Fig. 4.7a. Allochemical phosphate fabrics. (A) Phosphatic oncoid (10 mm) partly amalgamated with a microbial layer. Phosphatic interval at the Pennsylvanian/Permian boundary, Idaho, USA. (B) Phosphate oncoids with large nuclei and only very few coatings. Phosphoria Rock Complex, USA, Permian. (C) Normal ooids in phosphatic mineralogy. Phosphoria Rock Complex, USA, Permian. (D) Ooids with chamositic and phosphatic coatings. Thuringia, Germany, Ordovician. (E) Irregularly coated ooids. Phosphoria Rock Complex, USA, Permian. (F) Foraminifera nuclei are coated by a structureless phosphate envelope. Phosphoria Rock Complex, USA, Permian. All grains ~1 mm in diameter.
Fig. 4.7b. Allochemical phosphate fabrics. (A) Phosphatic lithoclasts of different internal composition and roundness. Phosphoria Rock Complex, USA, Permian. (B) Marginally phosphatized carbonate intraclast. Phosphoria Rock Complex, USA, Permian. (C) Different types of brown-colored phosphatic intraclasts, dominantly slightly phosphatized carbonate mudstone fragments and phosphatized calcareous limestones. Florida, USA, Neogene. Grain sizes in (A) to (C) range from 1 mm to 5 mm.
Fig. 4.7c. Allochemical phosphate fabrics. (A) Phosphatic bone fragment (center of photograph) and phosphate filled foraminifera (right margin of photograph). Florida, USA, Neogene. (B) Phosphatic bone fragments. Subatlas Group, Morocco, Paleogene. Base of photograph = 20 mm. (C) Phosphatized bryozoan fragment. Phosphoria Rock Complex, USA, Permian. Particle size = 10 mm. (D) Phosphatized crinoid fragment. Crossed nicols emphasize the phosphatized areas. Phosphoria Rock Complex, USA, Permian. Particle size = 10 mm. (E) Intensively abraded primary phosphatic bioclasts. Estonia, Ordovician. Particle size = 10 - 15 mm. (F) Primary phosphatic shell fragments occur with phosphatic ooids and phosphatic lithoclasts. Phosphoria Rock Complex, USA, Permian. Base of photograph = 5 mm.
See also color copies of these plates in the appendix.

Fig. 4.7a.

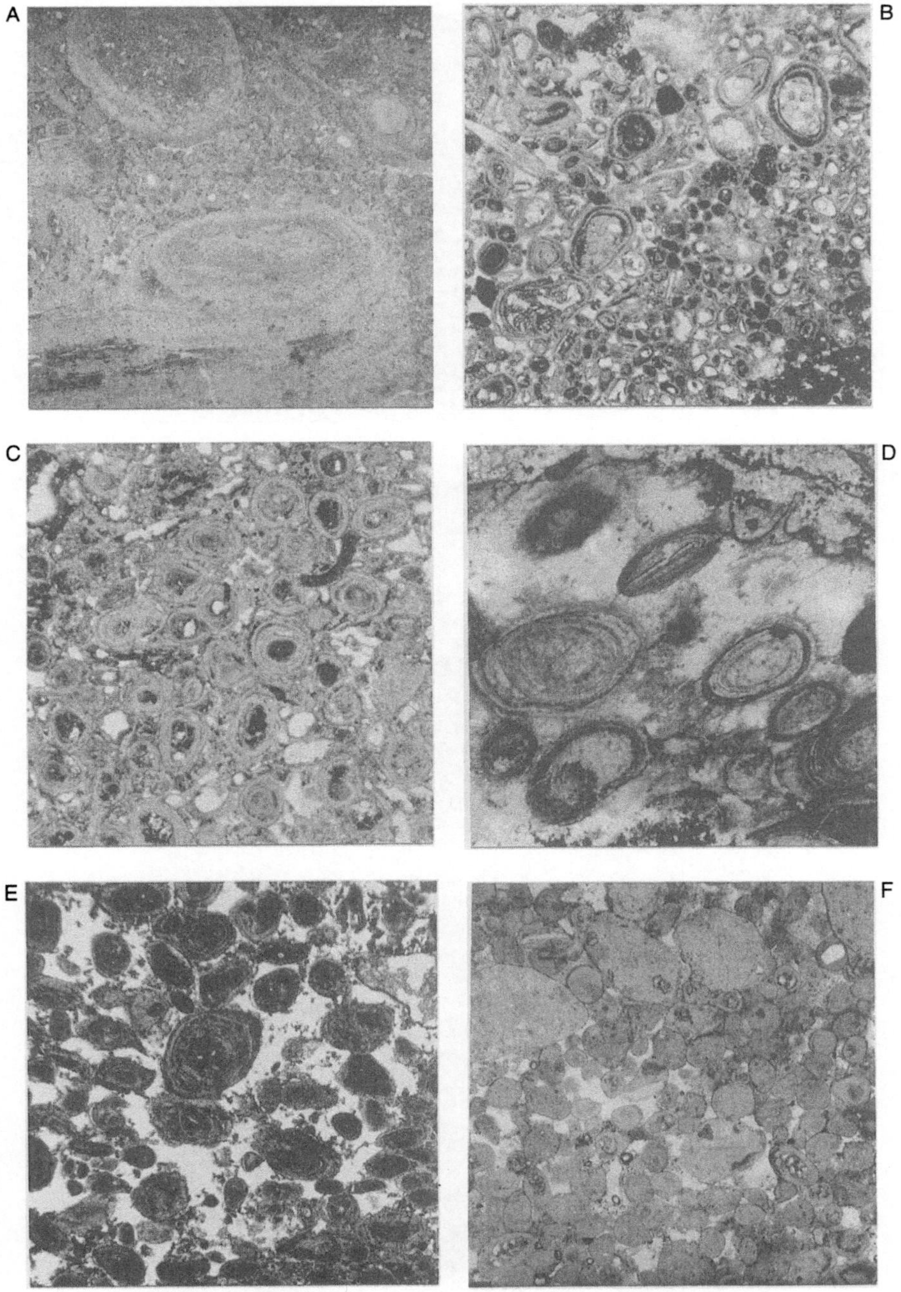

Fig. 4.7b.

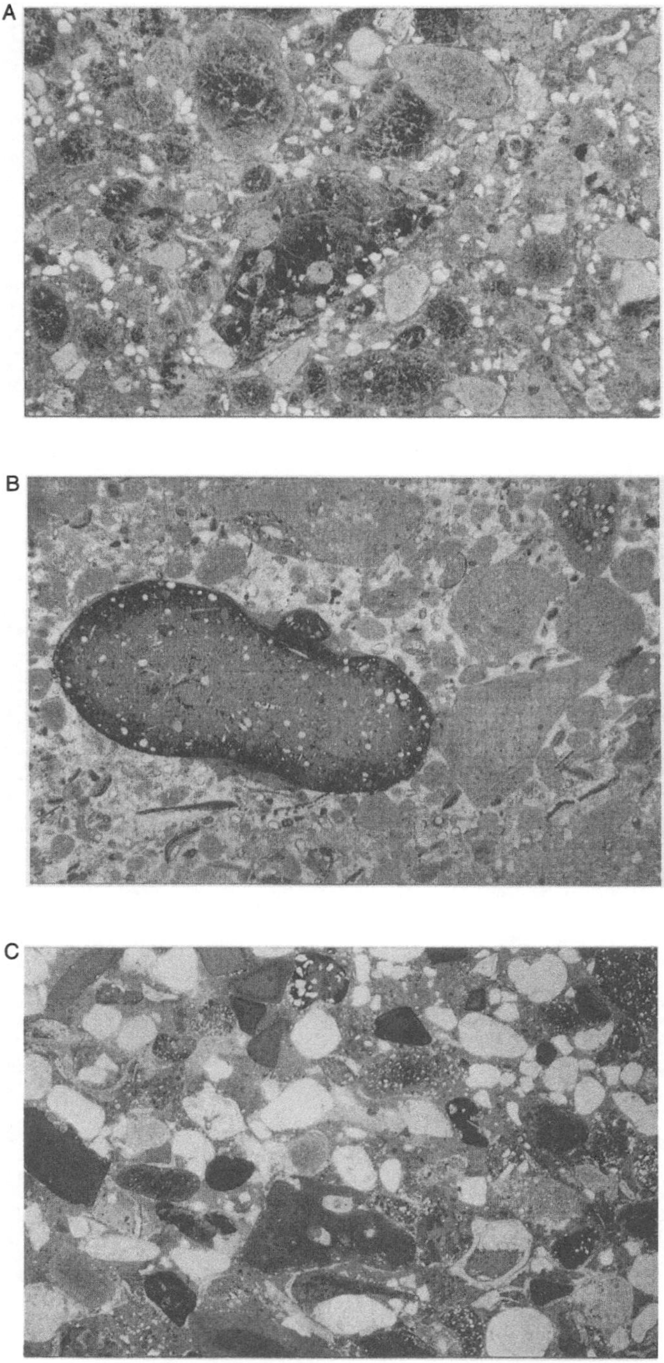

Fig. 4.7c.

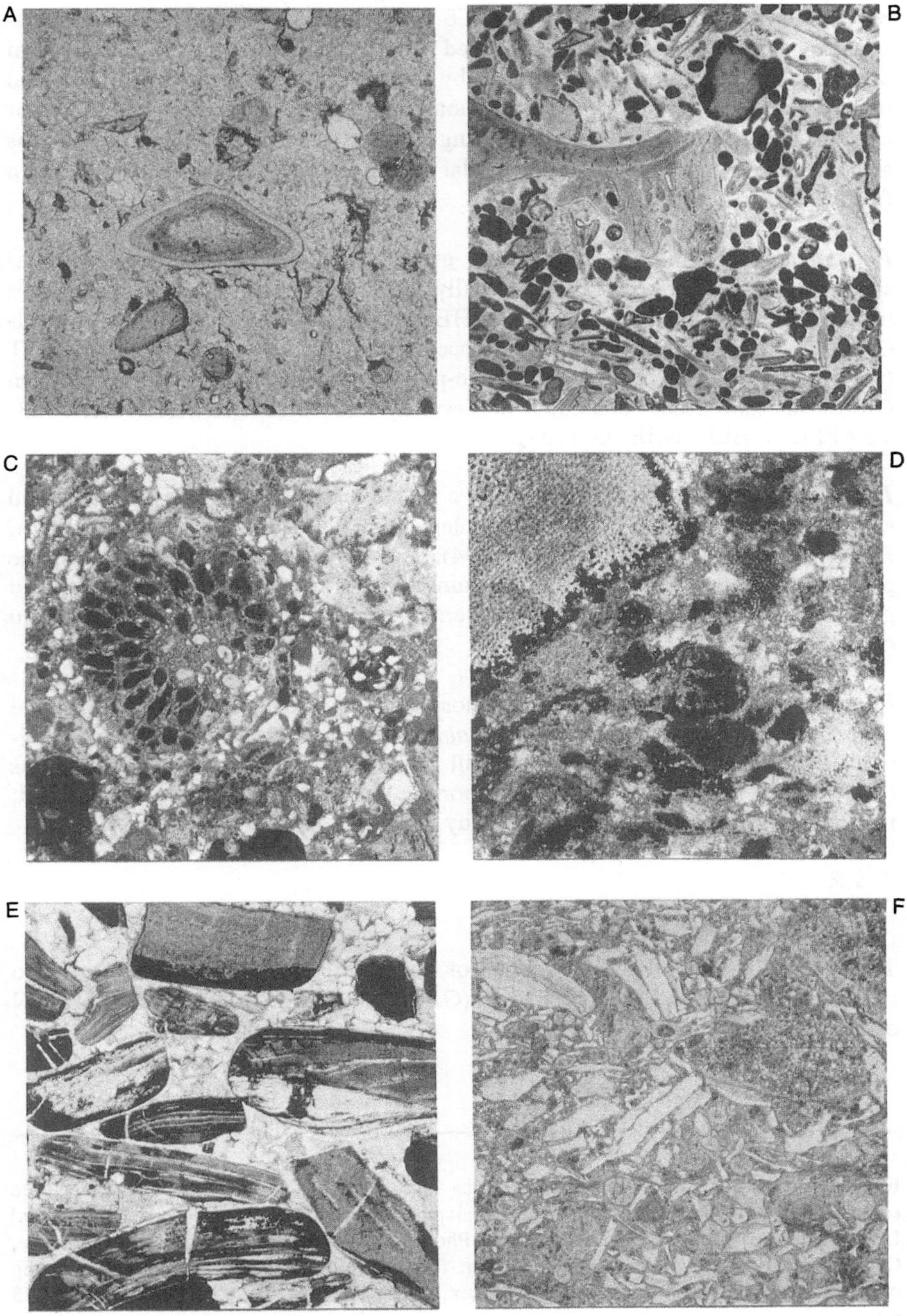

sediments by aggressive P-rich solutions (e.g. guano) is known from insular phosphate occurrences (BRAITHWAITE 1980). To indicate this origin of phosphate mineralogy, the term phosphatized can optionally be used. In depositional environments which are characterized by low net sedimentation rates, microbial phosphate precipitation is very commonly present and produces nucleated or coated grains. This can be a long lasting process (GRIMM 1992). These grains are commonly characterized by irregular laminated envelopes which are known from carbonate oncoids.

Phosoncoids (Fig. 4.7): multicoated grains with irregular coatings and unspecified shape. These grains are usually larger than ooids and can reach sizes up to 20 mm in diameter (MABIE & HESS 1964, HOFMANN 1975, SOUTHGATE 1986a,b). Young to modern occurrences are described by BURNETT (1977). Exceptionally large phosphatic-polymineral oncoids are reported from Ordovician and Cretaceous strata in Germany and Poland (KRAJEWSKI 1983, TRAPPE & ELLENBERG 1994).

Phosphatic coated grains (Fig. 4.7): a nuclei of any mineralogy, shape, and structure is enveloped by a structureless apatite coating (LAMBOY 1987b, PRÉVôT 1990, SCHWENNICKE 1994). In contrast to ooids and oncoids, the phosphate envelope is lacking any lamination. A number of phosphatic coated grains probably originate from an alteration or recrystallization of ooids and oncoids.

Phosbioclasts (Fig. 4.7): primary phosphatic fragments to complete skeletal elements e.g. from *Schmidtites* or *Ungala* (HEINSALU 1990), *Lingula* (SWANSON & CRESSMAN 1967), as well as bone fragments from vertebrates (TRAPPE 1992a). Also primary carbonate bioclasts which are phosphatized, respectively impregnated or replaced by apatite (LAMBOY 1987a).

4.3.2 Orthochemical phosphate fabrics

This group defines all in-situ fabrics of phosphatic mineralogy which are also summarized under the term "pristine" (GARRISON & FÖLLMI 1991, GLENN et al. 1994).

Fig. 4.8. Orthochemical phosphate fabrics and cements. (A) A spiculitic glauconite sandstone is cemented with structureless phosphate forming concretion-like fabrics. Basal Upper Cretaceous, Central Germany. (B) Phosphatic stromatolite. Thuringia, Germany, Ordovician. Base of photograph = 2 mm. (C) Phosphatic microbial mats are binding phosphatic ooids. Phosphoria Rock Complex, USA, Permian. Base of photograph = 15 mm.
See also color copy of this plate in the appendix.

Fig. 4.8.

A

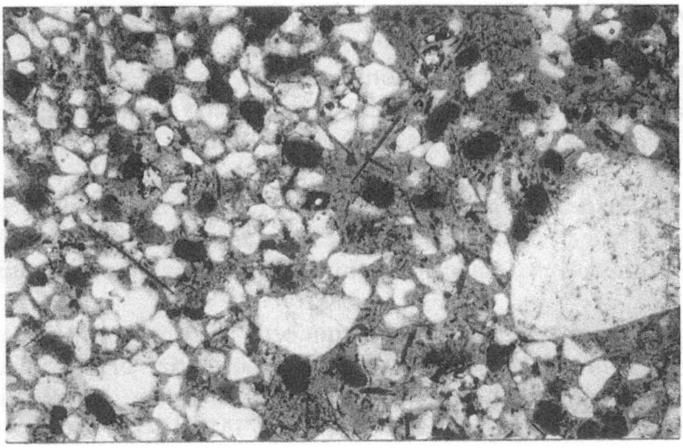

B

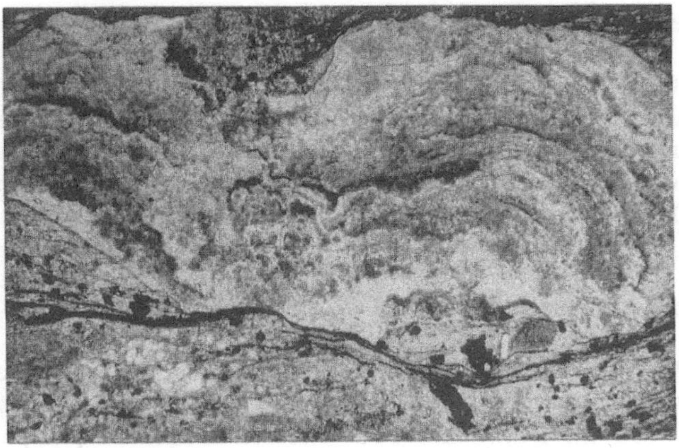

C

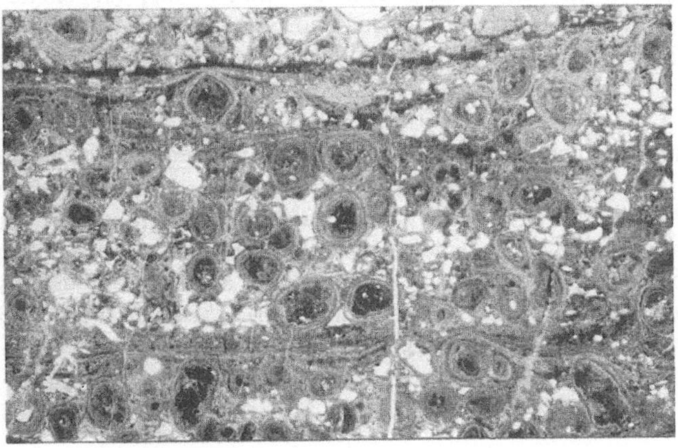

Microsphorites (phosphatic mudstones): cryptocrystalline, structureless layers or lenses of phosphatic mineralogy which are definitely in-situ. Particle are absent or less than 10 vol. %. Bioturbation is possible, but usually lacking. These fabrics are known from ancient deposits and from modern occurrences (DeKEYSER & COOK 1967, GLENN et al. 1994).

Phosphate concretions (Fig. 4.1, 4.8): in-situ nodules in phosphatic mineralogy. Size and shape vary widely. The concretions are structureless, include fabrics of the host sediment or are nucleated (most commonly around organic remains).

Phosphatic bindstones (Fig. 4.1, 4.8): thin beds or crusts, which show a distinct lamination or, if better preserved, fabrics of bacterial or cyanobacterial filaments, including stromatolites and phosphatized biogenic structures, linings in burrows (e.g. ZIMMERLE & EMEIS 1989, ZIMMERLE 1994).

4.3.3 Phosphate mud matrix

A phosphate mud matrix is extremely rare and only present in microsphorite beds. Primary phosphate matrix (microsphorite) in combination with particles is unknown. Sporadically, neomorphic phosphatized micrite is present, e. g. in the Phosphoria Rock Complex or along lithified carbonate hardgrounds (LAMBOY 1993).

4.3.4 Phosphate cements

There is still not much known about these fabrics. Phosphatic cements are common in phosphate concretions where also different generations have been identified (SIEGMUND 1995, TRAPPE et al. 1995). The phosphatic cement fabrics generally correspond to these of carbonate rocks. An early diagenetic cement A generation can be differentiated from a later diagenetic cement B, which fills remaining open pore space. Phosphate cements have been addressed by HEWITT (1980), KRAJEWSKI (1984), FÖLLMI (1989a) and SIEGMUND (1995).

Cement A or rim cement (KRAJEWSKI 1984) is characterized by rapid crystal growth on particles or other internal surfaces. The existence of various types of cements and their environment of formation are still widely unknown. In phosphate concretions early cements form well defined palisades of apatite crystals on all internal surfaces (see Case Study 3). Crystals are usually clear and seem to prefer organic matter as a substrate (SIEGMUND 1995). KRAJEWSKI (1984) described various inorganic and organic rim cements from a condensed sequence.

Cement B is a cryptocrystalline apatite precipitation in remaining open pore space. Abundantly internal voids of fossil remains (intraparticle space) are filled with clear apatite cement of white or honey-brown color (KRAJEWSKI 1983). Another type is the formation of apatite filling completely the interparticle space in siliciclastic sediments (Fig. 4.8)(FÖLLMI 1989a).

4.3.5
Phosphate ultrastructures

With the exception of apatite cements and primary phosphatic skeletal material, which form apatite crystals of 0,001 to 1 mm size, all other sedimentary phosphate fabrics have a pseudo-isotropic appearance resulting from the average crystal size of $< 10\ \mu m$. This cryptocrystalline nature of sedimentary phosphate particles was described by the early researchers by the term collophane (for discussion see McCONNELL 1973). The crystals in most rocks are preferentially columnar-shaped or fibrous and commonly non-oriented. The replacement of inherited sediment fabrics can affect the formation of large pseudomorphic phosphate cement structures (TRAPPE & VONDRA 1995).

MIRTOV et al. (1976), LAMBOY (1982a,b, 1986, 1987a,b, 1990a,b, 1993) and ZANIN (1987) documented various phosphate ultrastructures including microbial fabrics. They could demonstrate the abundance of crystal shapes which correspond to microbial structures. The reasons for the wide range of different, presumably inorganic crystal growth is uncertain (see chapt. 5.5 for detailed discussion).

4.3.6
Composition and texture of phosphate rocks and phosphorites

The textural composition of phosphate sediments corresponds widely to that of carbonate rocks. The components are comparable in shape and origin, matrix and cements are present. The major difference is the multi-mineral composition of all basic constituents: particle, matrix, and cements. In most phosphate

---

Fig. 4-9a. Typical phosphate rock textures and compositions. (A) Phosclastpackstones are the most common granular phosphorite rock type. Phosphoria Rock Complex, USA, Permian. (B) Poorly sorted neomorphic (silicified) phosclastrudstone with common occurrence of phosbioclasts. Subatlas Group, Morocco, Paleogene. (C) Poorly sorted, polymictic phoslithoclastic packstone. Phosphoria Rock Complex, USA, Permian.
Fig. 4-9b. Typical phosphate rock textures and compositions. (A) Phosclastpackstone with large poorly fragmented oyster shell of much larger size. Subatlas Group, Morocco, Paleogene. (B) Phosclastwackestone with high contents of quartz detritus. Florida, USA, Neogene. (C) Phosclastic mudstone to wackestone with pelitic matrix. Particles of various roundness and size are found widely scattered in the matrix. Florida, USA, Neogene. ***See also color copies of these plates in the appendix.***

Fig. 4.9a.

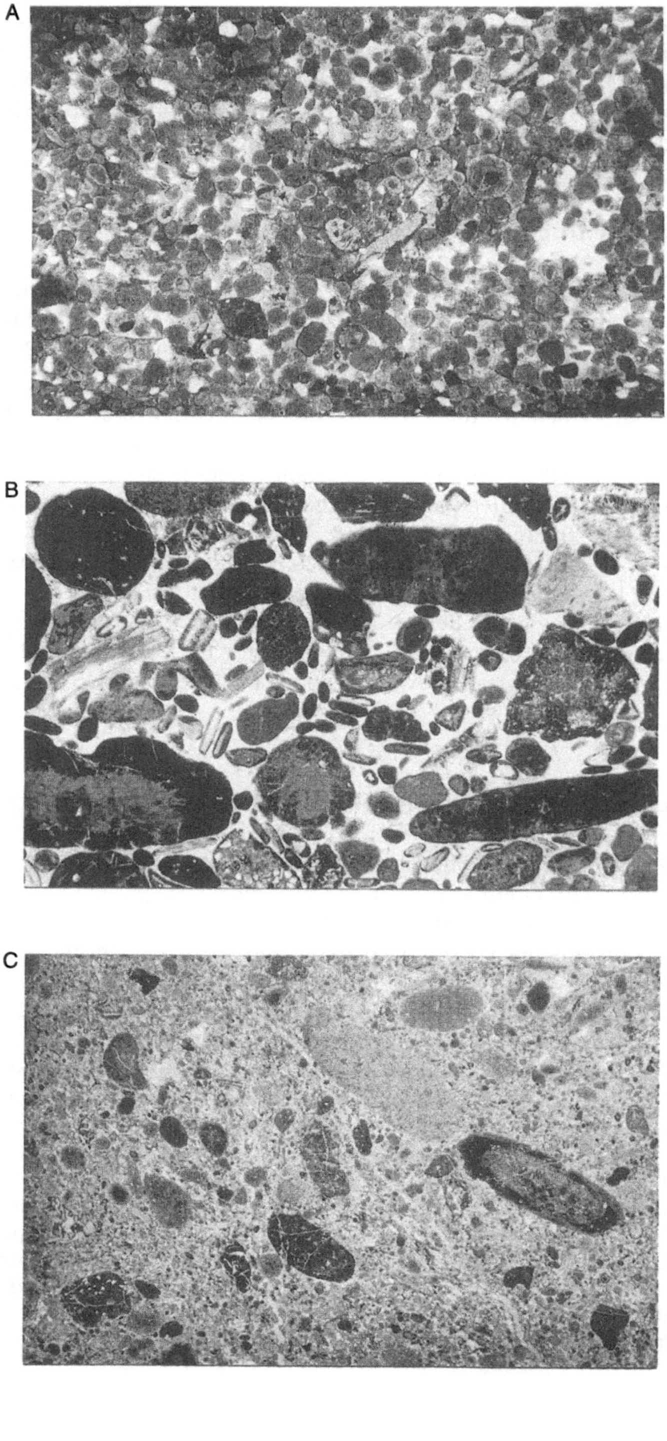

Fig. 4.9b.

A B C

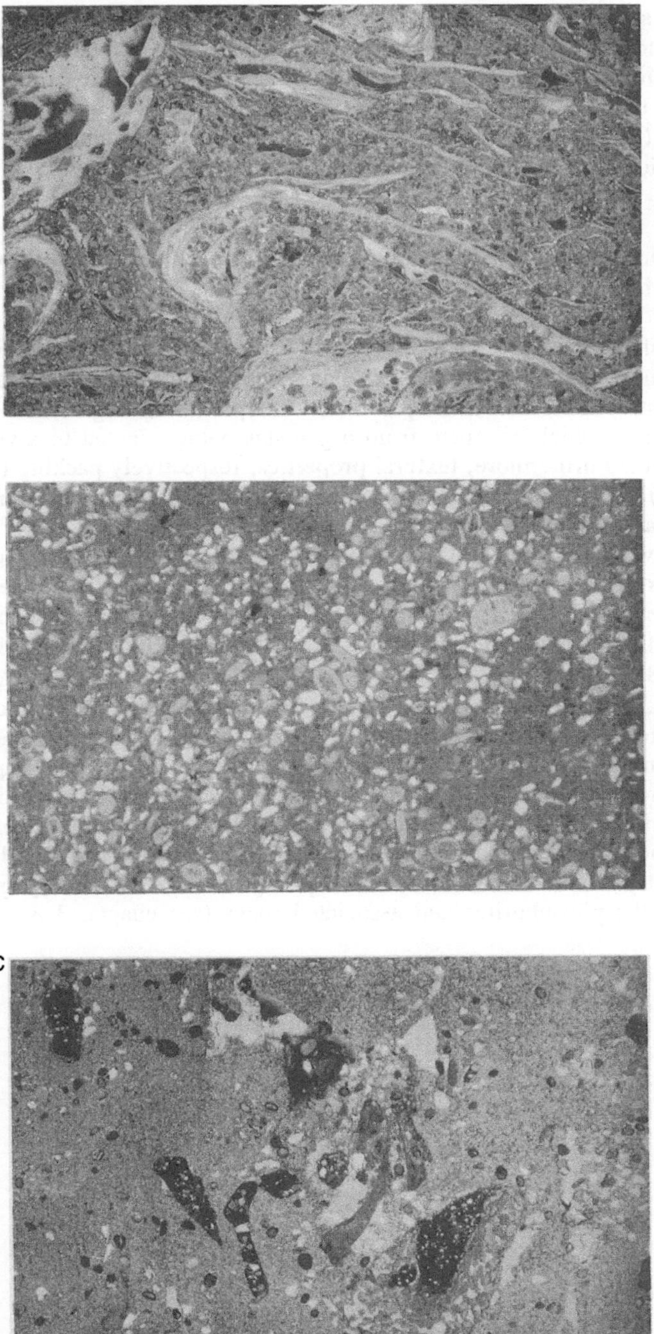

deposits, the phosphate phase is only located in the particle component, orthochemical fabrics are exceptional, cements are relatively rare, phosphatic matrix is formed in most cases by epigenetic replacement. The texture of phosphorites and phosphate rocks covers the complete spectrum from particle-free mudstone to pack/grainstone packing, bindstones, and rarely also floatstones/rudstones (Fig. 4.9). All other physical or biogenic microfabrics, e.g crossbedding or bioturbations, are present (see chapt 4.1).

4.3.7
Depositional environment analysis of phosphate rocks and phosphorites by microfacies techniques

The microfacies analysis of phosphate rocks and phosphorites combines indicative parameters of particle type, composition, texture, cements, matrix, and sediment structures to a multi-parameter interpretation. Significant importance have bioclasts which originate from organisms being adapted to a very specific environment. Furthermore, textural properties, respectively packing (matrix/particle ratio), sorting and roundness of the particles and sediment structures including trace fossils, are of major importance.

Indicative parameters are not only derived from the phosphate component, but also from all non-phosphatic constituents. The method in general follows the outlines of FLÜGEL (1982) for carbonate rocks. There have been only a very few facies models for phosphate deposystems being developed (SOUTHGATE 1983, 1988, CAROZZI 1989, FÖLLMI 1989a,b, 1990, TRAPPE 1989, 1992a, GLENN 1990a, and case studies presented in this book), most data are isolated observations. The lack of a larger number of facies models or even standard microfacies types for phosphate sediments gives great importance to the interpretation of the associated carbonates. Their facies analysis with the well established methods helps to interpret genesis and depositional environment of the phosphate facies. Additional data may be also provided by the granulometry of the siliciclastic or the phosphorites itself, or by geochemical studies of the phosphorites and associated rocks (see chapts. 3 & 4.1).

PART III
THE PROCESSES: PHOSPHOGENESIS AND PHOSPHORITE GENESIS

5 Phosphogenesis

5.1 Fundamentals of phosphogenesis in natural environments: the geochemical approach

The term phosphogenesis summarizes all processes of apatite precipitation/mineralization which lead to the development of pristine (orthochemical, in-situ) phosphate fabrics. All subsequent sedimentary processes such as winnowing, reworking, and sediment transport, which result the formation of allochemical phosphorites, are described under the term phosphorite genesis (chapt. 7).

Phosphogenesis characterizes the development of an environment in which the geochemical system reaches stages of supersaturation for apatite precipitation. Under biological mediation, mineralization may be initiated earlier. The physico-chemical parameters for apatite precipitation define the "apatite precipitation window" (APW) of a geochemical system. These parameters are a steady-state condition, at least for Phanerozoic deposystems, but the individual pathway leading to the formation of such an environment depend on the evolution of the deposystem itself and is altered in adaption to the permanently changing global and local environmental conditions. Very uncertain are the environmental conditions in the Precambrian when large phosphate deposits were formed, but the chemistry of the oceans and atmosphere may have been fundamentally different (SCHIDLOWSKI 1985, SCHIDLOWSKI & WIGGERING 1988, DONNELLY et al. 1990, TUCKER 1992, KEMPE & DEGENS 1985, KEMPE & KAZMIERZAK 1994).

The physico-chemical parameters of sedimentary apatite precipitation are determined through three different approaches: direct measurements in modern deposystems, laboratory experiments, and geochemical investigations of ancient phosphate deposits.

5.2 Historic overview on phosphogenic concepts

The modern research assigns phosphogenesis in sedimentary systems to a broad spectrum of physico-chemical to purely biological processes. The strictly physico-chemical approach of apatite precipitation was fundamentally influenced

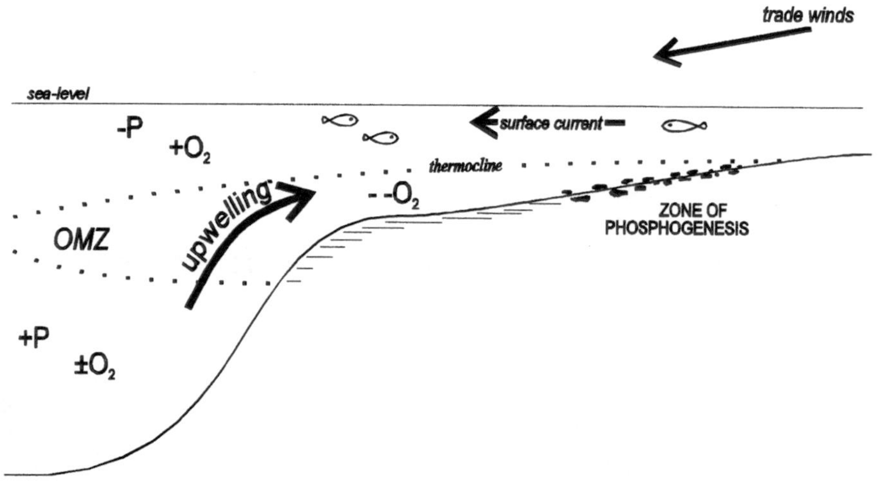

Fig. 5.1. The traditional upwelling concept which addresses the trade wind, west-coast related upwelling model. The supply of nutrients in shallow water enhances bioproduction which results also in enhanced organic matter deposition. The oxygen minimum zone increases in size and anoxic to suboxic conditions develop in outer shelf positions.

by KAZAKOV (1937), who investigated the P_2O_5-CaO-H_2O system and combined this theoretical view with field data. Phosphogenesis was thought to be triggered by a decrease of partial pressure of CO_2 and an increase of pH, which causes a decline in phosphate solubility and consequently results in apatite precipitation. The necessary high P-supply for the formation of significant deposits was explained by the oceanographic "upwelling model", which gave rise for the popular idea of upwelling derived formation of phosphogenic environments. The upwelling currents charge P-rich water masses from the relatively P-enriched deep ocean reservoir to the P-depleted surface water along the shelf margin. KAZAKOV thought that the decreasing partial CO_2 pressure in the elevated water is sufficient to cause supersaturation. Most of the following researchers favored the upwelling hypothesis as the main factor to create phosphogenic environments and initiate apatite precipitation. Later, these ideas have been intensively tested and furthermore developed by the works of McKELVEY et al. (1959) and SHELDON (1963). They replaced the inorganic precipitation of apatite by a biologic component for the deposition of phosphorus through organic matter. Increasing oceanographic knowledge has shown that the P-concentrations in the upwelled seawater are to low to start direct precipitation of apatite. In the modified upwelling model (Fig. 5.1), the P-supply through upwelling enhances bioproduction and consequently organic matter deposition.

Phosphorus which is liberated by the partial decomposition of organic matter is concentrated in the porewater until the system reaches supersaturation.

Besides the mainstream, BUSHINSKI (1966) and PEVEAR (1967) emphasized the fluvial/estuarine origin of the deposits. From a genetic interpretation of an evaporite-phosphate-Fe-rich sediment association, HITE (1978) developed a model for phosphate precipitation in an evaporite basin through mixing of P-rich continental solutions with saline brines. Final evidence for the model is pending. HENDERSON et al. (1979) proposed a model of diagenetic brine leaching mechanism for phosphatization of lime mud by phosphate-rich brines. AMES (1959) attributed apatite replacement of carbonate to purely diagenetic processes.

MANSFIELD (1940) was the first who recognized a spatial relationship between phosphogenesis and volcanism, but final proof of this relationship remained open. In the following decades, the temporal link between phosphogenetic intervals and volcanism was again proposed by SHATSKII (1955), STRAKHOV (1960), BRODSKAYA (1974), ILYINSKAYA (1968, 1970), ROONEY & KERR (1967, denied by SYNDER et al. 1984, RIGGS & MALETTE 1990). Especially, for the chert-phosphorite association, the relationship was discussed by BIDAUT (1953), LOWE (1972) and ZIMMERLE (1985).

For some deposits, a tectonic and paleotopographic control for the development of phosphogenic environments, respectively the local alteration of the current system by the paleotopography, was proposed (BENTOR 1953, REISS 1962, WÜRZBURGER 1968, WILCOX 1953, JARVIS 1980, RUSSELL & TRUEMAN 1971, RIGGS 1979b, 1984, FREAS & RIGGS 1965). Also in other environments, the water depth was in dispute. Whereas McKELVEY et al. (1959) postulated depth of 200-300 m for the black shale-chert-phosphorite association of the Permian Phosphoria Formation (USA), a peritidal to intertidal depositional environment was discussed for the Cambrian Georgina Basin phosphorites in Australia (DeKEYSER & COOK 1973, SOUTHGATE 1983, 1986a,b, 1988, SOUTHGATE & SHERGOLD 1991).

A weathering process with subsequent transportation of the residuum and reworked limestone was initially thought to be the origin of the enormous phosphate deposits in the southeastern USA which extend from Florida to North Carolina (CATHCART & McGREEVY 1959, ALTSCHULER et al. 1964). The primary sedimentary origin of these deposits was clearly demonstrated by RIGGS (1967, 1979a,b). True residuum phosphate enrichments are described by GERMANN et al. 1979, 1981, DE OLIVEIRA (1980) and SCHWAB & DE OLIVEIRA (1981).

The early researches assigned most phosphate particles to coprolites and fecal pellets (BUCKLAND 1843). In modern approaches, the fundamental role of organic phosphate coupling and passive or active role of microbes is unquestioned (e.g. LUCAS & PRÉVôT 1984, LAMBOY 1990, KRAJEWSKI et al. 1994).

Until the 1970s, several modern occurrences have been considered to be contemporary (e.g Namibia: BATURIN 1969, California: DIETZ et al. 1942, Peru/Chile: MANHEIM et al. 1975, VEEH et al. 1973, McARTHUR 1983, later also Mexico: JAHNKE et al. 1983), but more precise dating methods assigned most of them to ages of several hundred thousands years (KOLODNY 1969, KOLODNY & KAPLAN 1970, McARTHUR et al. 1988). Only the deposits off Peru/Chile, Namibia and E'Australia partly revealed a modern genesis (BATURIN et al. 1972, VEEH et al. 1973, 1974, BURNETT 1974, BURNETT & VEEH 1977, BURNETT et al. 1980, 1988, O'BRIEN et al. 1986). The possibility of direct observations in a phosphogenic environment gave phosphate research in the 1970s and 1980s a new perspective (FROELICH et al. 1988, GLENN & ARTHUR 1988, GLENN 1990, GLENN et al. 1988, 1994, GARRISON & KASTNER 1990).

By the mid-1970s, phosphate research significantly increased and was furthermore enhanced by the establishment of the IGCP project 156: "Phosphorites" (1978-1988, see also COOK et al. 1990) and the succeeding project 325: "Paleogeographic reconstructions with phosphorites and associated authigenic minerals" (1990-1996). Within this framework, numerous monographs and descriptions of single deposits have been published. The historical development of phosphate research and the progress which was achieved during the two IGCP projects is summarized also in COOK et al. (1990).

The different models proposed in the past reflect the broad spectrum of genetic pathways and depositional settings of phosphorite deposystems, although the conditions of apatite mineralization are very specific. In the past, the elegant upwelling model and the occurrence of the only modern deposit in an upwelling zone focused the interpretation of many ancient deposits very much on this idea regardless to a far more differentiated deposystem evolution.

5.3 The phosphogenic environment: the apatite precipitation window

5.3.1 Investigation techniques: methods and problems

The research on processes and environments of phosphogenesis has three main approaches which are (1) the observation and direct measurements in modern phosphogenic environments, (2) the interpretation of geological observations and geochemical data from ancient deposits, and (3) determination of parameters for the solution-apatite equilibrium through laboratory experiments and thermodynamic calculations. All approaches stimulated the phosphate research, but all methods are also yielding specific difficulties.

The modern occurrences, which are offshore deposits in several hundreds meters of water depth, differ significantly in facies and depositional setting from most ancient deposits (BENTOR 1980). However, fundamental geochemical processes have to be similar and may be representative for other deposits (GLENN et al. 1994). The investigations in modern environments require precise dating of the phosphate fabrics to prove contemporary phosphogenesis in the present porewater and to determine growth rates. Phosphorites incorporate substantial portions of uranium and the development of the U-disequilibrium (U-series dating) revolutionized the understanding of the modern phosphorites (BURNETT & VEEH 1977, BURNETT 1974, 1977, BURNETT et al. 1980, 1988).

Precise dating is also important in ancient deposits to distinguish primary fabrics from reworked grains. Phosphorites from the Neogene to Late Cretaceous, recently also some older Phanerozoic intervals, can now be correlated with the $^{87}Sr/^{86}Sr$ ratio and its secular variations (DEPAULO & INGRAM 1985, ELDERFIELD 1986, VEIZER 1989, HODELL et al. 1991, COMPTON et al. 1993, STILLE et al. 1992, 1994, McARTHUR et al. 1990, 1993, DENISON 1994a,b). But the geochemical investigation in ancient deposits yields even more uncertainties than the research in modern deposits. The geochemistry of trace elements, REE or isotopic compositions is very much overprinted by element fractionation during apatite precipitation and the primary composition is further altered during diagenesis and weathering (chapt. 3 and Case Study 3). Further difficulties rise from the common polygenetic composition of the phosphatic sediments with high organic matter contents which reflect a primary composition of the ocean surface water and furthermore clay minerals which may be weathering derived and have signatures from their precipitation site in the source sediment (PRÉVôT 1990, PIPER & ISAACS 1995, PIPER & MEDRANO 1995).

The problematics in precipitation experiments are the necessary simplification, and the only possible investigation of processes which are rapid compared to these observed in the geological record.

5.3.2
Modern and youngest phosphogenic environments

Very differentiated results reveal investigations in modern phosphogenic environments. In the meanwhile, very precise data were collected on the solid mineral phase and porewater profiles during several cruises to the occurrences off Peru and Chile. These provide a fairly precise picture from the environment of apatite precipitation and organic matter decomposition (BURNETT 1980a, BAKER & BURNETT 1988, BURNETT et al. 1988, FROELICH et al. 1988, GLENN & ARTHUR 1988, GLENN et al. 1988, KIM & BURNETT 1988, PIPER et al. 1988, GARRISON & KASTNER 1990, GLENN 1990b, McARTHUR et al. 1990, GLENN et al. 1994). Similar results came up also from the investigation of the Neogene to modern phosphorites on the E' Australian shelf (RIGGS et al. 1989, O'BRIEN et al. 1990, HEGGIE et al. 1990).

The uppermost sediment body (<20 cm) in both environments shows a very distinct stratification of the porewater and solid-phase composition as well as increasing oxygen depletion with depth. The stratification pattern varies with sediment accumulation rates and water depth. Phosphorus liberated from organic matter is rapidly concentrated in the porewater within the upper 20 cm soft

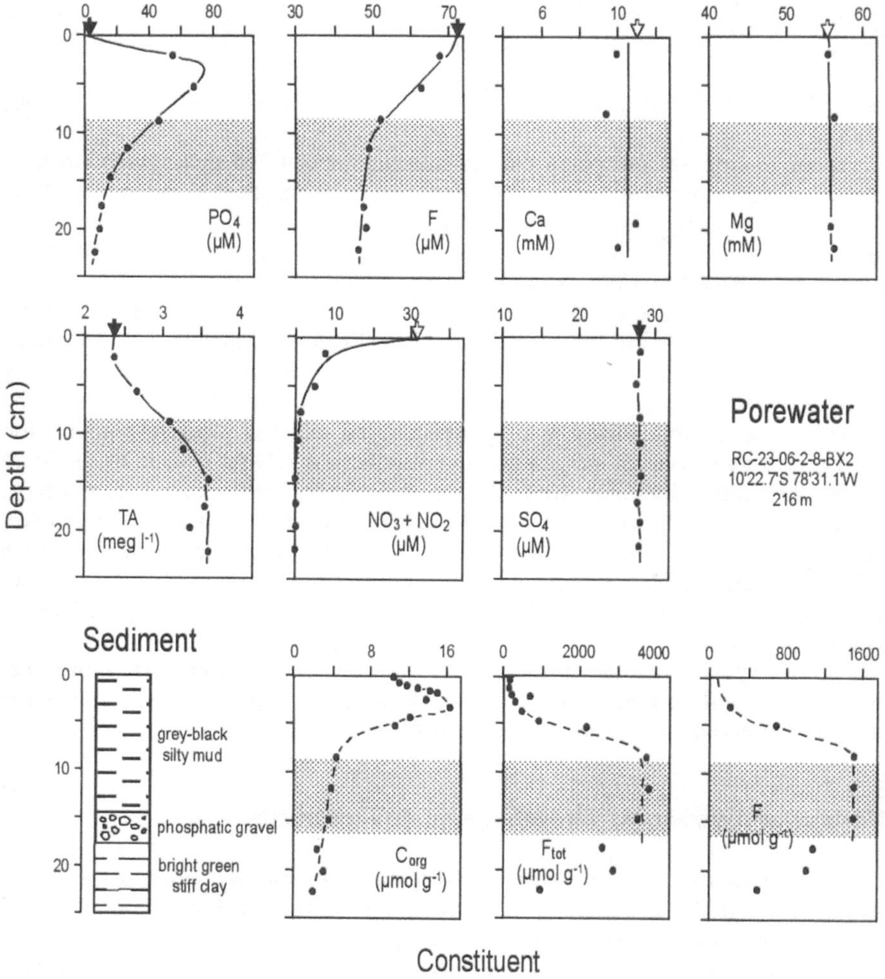

Fig. 5.2. Porewater and solid-phase geochemistry for slow accumulating sediments (0.017 mm/yr) in the modern upwelling area on the Peru continental margin, 230 km NW'Lima. Black arrows indicate measured seawater values (white arrows estimated concentrations). Stippled area shows depths interval of active phosphate precipitation with dissolved fluorine and phosphate porewater progressively decreasing in this layer in respect of increasing sediment P and F. This correlates with an increase in total alkaninity (TA) and low NO_3^- and NO_2^- contents which demonstrate mildly alkaline and suboxic conditions (no significant H_2S was detected). Organic matter is rapidly degraded in the phosphate precipitation zone, which covers both, the silty mud and the layer of older phosphate gravel (data from FROELICH et al. 1988, GLENN 1990b, modified from JARVIS et al. 1994).

sediment. In this sediment layer, increasing total sediment phosphorus which correlates with a drop of porewater-P marks the zone of active phosphate precipitation. This layer is located in the suboxic environment and the position in the sediment column depends on the O_2-stratification pattern (Fig. 5.2). In deeper sediment layers, phosphate precipitation is inhibited by rising total alkalinity.

FROELICH et al. (1988), and later O'BRIEN et al. (1990) and HEGGIE et al. (1990) demonstrated the importance of the iron for the phosphogenic processes. Their iron-pumping model (Fig. 5.3) provides a mechanism for the process of P enrichment within a few centimeter sediment depth. Phosphorus which is liberated from organic matter near the sediment surface must be scavenged from recycling into bio-coupling before being concentrated in the porewater. Fe-oxyhydroxy complexes which circulate through the oxic to sub-

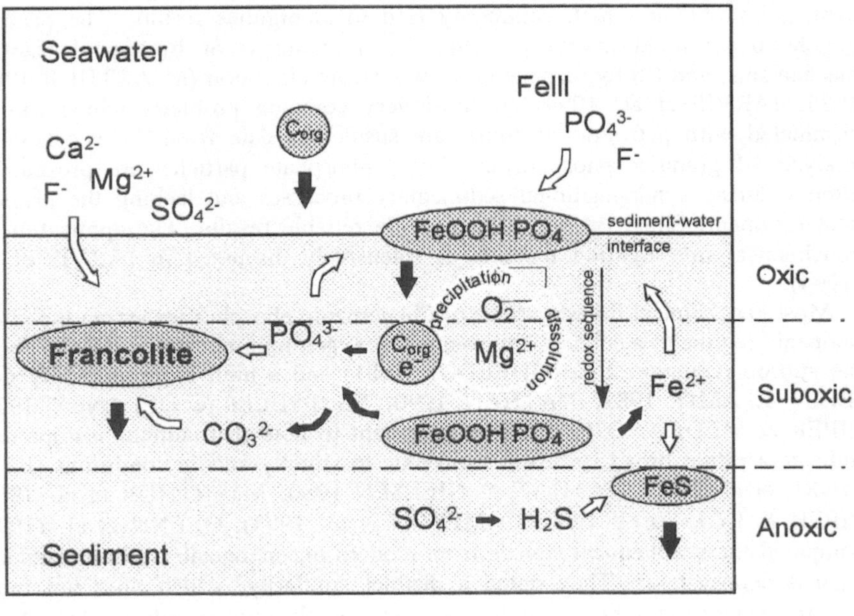

Fig. 5.3. Schematic diagram of the Fe-redox cycle. Light stippled areas represent solid phases, black arrows solid-phase fluxes. White-outlined black arrows indicate reactions, white arrows diffusion pathways. Liberated phosphate from the degradation of organic matter is scavange by Fe-oxyhydroxy complexes and shuttled into greater sediment depth where the Fe-complex decays. Phosphorus is enriched in the suboxic sediment layer until concentrations are reached that apatite or a precursor can start to precipitate. Excess phosphate diffuses upwards and is recycled (from JARVIS et al. 1994).

oxic sediment layers, driven by reduction and oxidation of the Fe "scavenge" liberated phosphorus in the upper layer. The complexes shuttle P in a deeper layer where it is protected from bio-coupling and concentrated in the porewater. The complex genetic relationship between Fe and P, respectively the resulting mineral phases glauconite, pyrite and apatite is well known from the investigation of ancient deposits (see chapt. 8).

5.3.3 Ancient phosphogenic environments

Various geochemical studies try to uncover the geochemical conditions and source of phosphorus of Phanerozoic phosphorites from the investigation of the trace element, REE, and isotope composition. The number of contributions is enormous, but systematic investigations are still rare (inorganic: e.g. COOK 1972a, MAUGHAN 1976, ALTSCHULER 1980, PRÉVôT & LUCAS 1980, 1985, BOCK 1987, PRÉVôT 1990, PIPER 1991, JARVIS 1992; organic: e.g. SANDSTORM 1986, NATHAN 1990). All these are dealing with two fundamental problems which commonly lead to ambiguous results. The primary geochemical signal is overprinted by (1) chemical or biological fraction mechanisms, and (2) by diagenetic or weathering alteration (McARTHUR 1980, 1985, JARVIS et al. 1994). A third very common problem, which can be eliminated with petrographic tools, are suspicious data from the geochemical analysis of granular phosphorites. These phosphate particles are presumable altered during syndepositional sedimentary processes and lacking the primary host sediment. Some isotopic data provide reliable results. The application of geochemical investigation methods is intensively discussed in JARVIS et al. (1994).

Most geochemical investigations of Phanerozoic phosphorites assign the phosphogenic regime to a suboxic environment (oxygen minimum zone, OMZ) above the sulfide reduction zone (SHELDON 1981) and a high bioproductivity area (BEIN & AMIT 1982, NATHAN 1990, FILIPPELLI & DELANEY 1994, PIPER & MEDRANO 1995). It is important to note that sulphur isotope data indicate a rather broad spectrum from oxic to slightly anoxic conditions (BENMORE et al. 1983, NATHAN & NIELSEN 1980, McARTHUR et al. 1986, PIPER & KOLODNY 1987, COMPTON et al. 1993). GLENN et al. (1994) compared the geochemical signals from modern phosphogenic environments and ancient deposystems. They noted a distinct similarity, which does not mean that the paleogeographic constellation consequently corresponds to that of ancient deposits. GERMANN et al. (1984) could link the trace element compositions to fluvial and open marine dominated environments in the Cretaceous deposits in Egypt. Recently, PIPER & MEDRANO (1995) discussed the major element, trace element and REE geochemistry from one section of the Phosphoria Rock Complex. They could relate the element composition to their detrital, biogenic and inorganic/hydrogeneous precipitation source. The variations of the geochemical signal from the host sediment and the phosphate component is also documented from the non-reworked Alaunschiefer phosphorites in Germany (see Case Study 3). These studies demonstrate the complexity of various overprinting geochemical signals which are obtained from phosphogenic sediment. These are only partly understood. In the present state knowledge, the fundamental geochemical conditions of ancient phosphogenic systems were similar to that in the modern occurrences. But the greater variability of depositional setting and biological processes (e.g. in stromatolites) which are not represented in the modern systems indicates a slightly broader spectrum of phosphogenic environments.

5.3.4 Physico-chemical parameters of apatite precipitation

Precipitation of the various apatite modifications can start when a geochemical system reaches supersaturation. These conditions are defined by steady-state parameters. Most of these are determined only for the modification Ca-fluorapatite (GULBRANDSEN 1969). In a more complex chemical system, such as the apatite system, mineralization from solution to a solid crystallized phase may pass a gelatinous to low order phase (MALONE & TOWE 1970, LUCAS & PRÉVôT 1985, PRÉVôT et al. 1989, FROELICH et al. 1988, VAN CAPPELLEN & BERNER 1991). A special, mainly undetermined role plays microbial activity by altering the microenvironment or affecting precipitation with organic substances (LUCAS & PRÉVôT 1985, PRÉVôT et al. 1989, KRAJEWSKI et al. 1994).

Inorganic phosphorus in water occurs in form of dissociations of phosphoric acid. In natural seawater three phases are present: H_2PO4^-, $HPO4^{2-}$, $PO4^{3-}$ (KESTER & PYTKOWICZ 1967). In general, seawater is supersaturated with phosphorus in the equilibrium to hydroxylapatite (0.02mg/l), that means three to four-fold (calculations of MIKHAYLOV, 1968 in BATURIN 1982). But hydroxylapatite precipitation in marine environments is not commonly observed. Phosphogenesis is rather a rare phenomenon than an abundant process. Precipitation experiments of MARTENS & HARRISS (1970) and NATHAN & LUCAS (1976) indicate that seawater is undersaturated in calcium phosphate. The phases hydroxylapatite and fluorapatite are less soluble than calcium-fluorapatite (KRAMER 1964, ATLAS & PYTKOWICZ 1977). BENTOR (1980) compiled the chemical characteristics of porewater from modern phosphogenic environments to determine the conditions of apatite precipitation and gave pH values of 7.2 to 7.5 and Eh-values of -210 mV. But the more recent studies revealed rapid chemical variations in the sediment column, although the generally increasing pH with depth is buffered by ammonia production and increase of total alkalinity. Experimental data are available from experiments of MARTENS & HARRIS (1970), ATLAS & PYTKOWICZ (1977), NATHAN & LUCAS (1976) and SAVENKO (1978). These experiments demonstrated also a clear Mg-inhibition effect on the precipitation of apatite substitutions at Mg-concentrations which are present in natural seawater and porewater. The studies of VAN CAPPELLEN & BERNER (1991) indicate that the effect is caused by hydrated Mg-complexes which block the Ca-sites (see also KIBALCZYC et al. 1990). A Mg-depletion in the porewater of modern phosphogenic environments compared to seawater was not reported (s. a. Fig. 5.2). In these environments, apatite precipitation increases with increasing carbonate alkalinity in greater sediment depth (>20 cm).

A common phenomenon is the phosphatization of calcium carbonate of shells, carbonate detritus, and carbonate rock surfaces. This type of phosphatization was early discovered by MURRAY & RENARD (1891) and later described by many authors. Experiments of AMES (1959, 1960) in which an alkaline solution of sodium phosphate was passed through a tube filled with grounded calcite obtained a carbonate apatite with about 10% CO_2. For the experiment, a solution of a pH of about 11 and a P-concentration of 1.5 g/l was used. The reaction follows the scheme:

$$NaOH + 3Na_3PO_4 + 5\ CaCO_3 \rightarrow Ca_5(PO_4)_3OH + 5Na_2CO_3$$

AMES calculated the threshold for the reaction at pH of 7-8 and a P-concentration of 0.1 mg/l. But the condition of the experiment (high P, low salt, high pH) are far from those in the modern sea.

The kinetic regime of fluorapatite (FAP) precipitation was investigated under simulated marine conditions by VAN CAPPELLEN (1991), VAN CAPPELLEN & BERNER (1991). The study could identify different pathways of precipitation: Depending on the degree of supersaturation FAP is precipitated via metastable calcium phosphate precursor phases (ACP or OCP_p) or at low degree of supersaturation, FAP nucleates directly from solution. Direct nucleation of FAP is a very slow process and, hence, not very effective to initiate apatite crystal growth. At high degree of supersaturation, the precipitation of a precursor phase is kinetically favored over FAP nucleation. Two precursors are forming (1) an octacalcium phosphate-like precursor (OCP_p) and (2) amorphous calcium phosphate (ACP). In the presence of dissolved magnesium, Mg^{2+} incorporation results in the precipitation of amorphous calcium magnesium phosphate (ACMP), but the fluor contents remain low in all precursor phases. The precipitation pathways of the various phases are depending on the total phosphate concentration in the solution and are compiled in Fig. 5.4. In natural environments only concentrations are reached for direct FAP nucleation. Only some topmost sediment layers off Peru and Chile provide a potential environment for precursor precipitation (JAHNKE et al. 1983, FROELICH et al. 1988, VAN CAPPELLEN & BERNER 1988). These conditions may have occurred during other intervals of the geological record.

RUTTENBERG & BERNER (1993) showed evidence for direct nucleation of apatite in form of "dispersed phosphate precipitation" in coastal non-upwelling sediments.

Once apatite was nucleated, precipitation of an in-situ phosphate fabric takes place very rapidly so that initiation of phosphogenesis appears as an event-like phenomenon. JAHNKE et al. (1983) and VAN CAPPELLEN & BERNER (1988) presented data from upwelling areas on the western continental margin off Mexico which indicate the formation of several percent of authigenic apatite

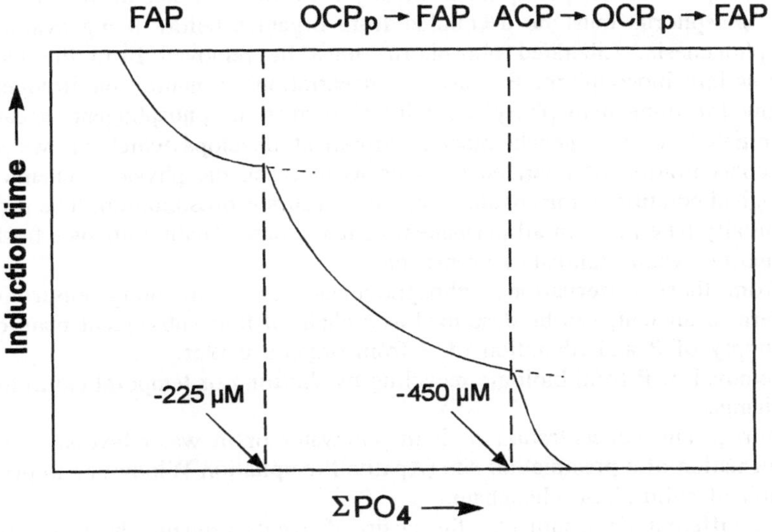

Fig. 5.4. Induction time (time between moment of supersaturation and precipitation) for calcium phosphate precipitation as a function of supersaturation (Mg-free seawater, pH = 8, t = 25°C) . The diagram shows that at low phosphate concentrations, FAP is precipitated directly without any amorphous precursor phase (from KRAJEWSKI et al. 1994).

within a few hundred years. LAMBOY (1990b) calculated from microscopic studies of a phosphatic crust from the continental margin off Peru the number of phosphate particles of $10^9/cm^{-3}$. The precipitation of such a large number of authigenic apatite particles rapidly depletes the surrounding porewaters of dissolved phosphate and fluoride. The apatite formation ceases and appears to be a short lived phenomenon.

5.4 The process of phosphogenesis: a 4-step model

All phosphogenic systems regardless their broad spectrum of environments, scales, processes, and resulting sedimentary fabrics are based on a fundamental chain of events. Any sedimentary system in which phosphate minerals, respectivly apatite, are forming requires an inceased supply of phosphorus, because the availability of the element in natural environments is very limited. Most of the phosphorus is bound in organisms and in dead organic matter, or is buried dispersed in shallow marine sediments (see chapt. 1). In contrast to KAZAKOV (1937), apatite is not precipitated directly from the seawater in most cases. The bulk phosphorus must be decoupled from organics before being available for phosphogenesis. Liberated phosphorus must be removed from the recycling process into biocoupling. Because P-concentrations in natural environments are to low for immediate phosphate mineral formation, phosphogenesis can only be initiated when a geochemical environment develops which allows storage and concentration of liberated phosphorus to reach the physical, chemical and biological conditions for apatite or apatite precursor precipitation. The processes potentially take place in all permanent or temporary aquatic milieus - freshwater (lacustrine, vadose/phreatic) or marine.

From these observations, phosphogenesis in sedimentary environments, modern or ancient, can be described as a chain of four subsequent main events:

1. Supply of P and liberation of P from organic matter.
2. Removal of P from biologic recycling by shuttling or temporal environmental changes.
3. Storage and concentration of P in porewater or in water layers.
4. Formation of a precipiation site (**A**patite **P**recipitation **W**indow) and precipitation of solid-phase phosphate.

In the different environments, the chain of events remains the same, but the processes and mechanisms initiating these events vary widely, e.g. phosphorus enrichment can be achieved rapidly via high flux rates or in a stable geochemical milieu through time. The interrelationsships within the chain of fundamental events and the processes controlling these are best described in all their complexity in a box model. An early three box model of phosphogenesis and subsequent phosphorite formation was introduced by LUCAS & PRÉVôT (1993). On the basis of this early concept, a new significanly extended model was now developed. The single evolutionary stages of the phosphogenic process which are displayed by the boxes are initiated by various mechanisms. These mechanisms, their preference, and their interelationship are strictly controlled by the sedimentary and biological environment (Fig. 5.5).

Stage 1: P supply - degradation of organic matter

The global phosphorus cycle and quantitative data from seawater and freshwater reflect the deficiency of liberated phosphorus in most natural environments and its coupling to organic matter. The strict bio-coupling of phosphorus in natural environments confines P-supply in phosphogenic environments to organic matter deposition (see chapt. 6). Organic matter deposition is controlled by two major factors which are the primary bioproductivity, respectively the nutrient availability (bioproductivity models), and the development of oxygen-deficient conditions to prevent complete oxidation before deposition (preservation models). In the sediment, the strict biocoupling requires a liberation of P via microbial oxidation from organic matter to retard dispersed P burial. The decomposition of organic matter follows several stages which each reveal a typical isotopic signature (IRWIN et al. 1977) (Fig. 5.6). Current isotope data

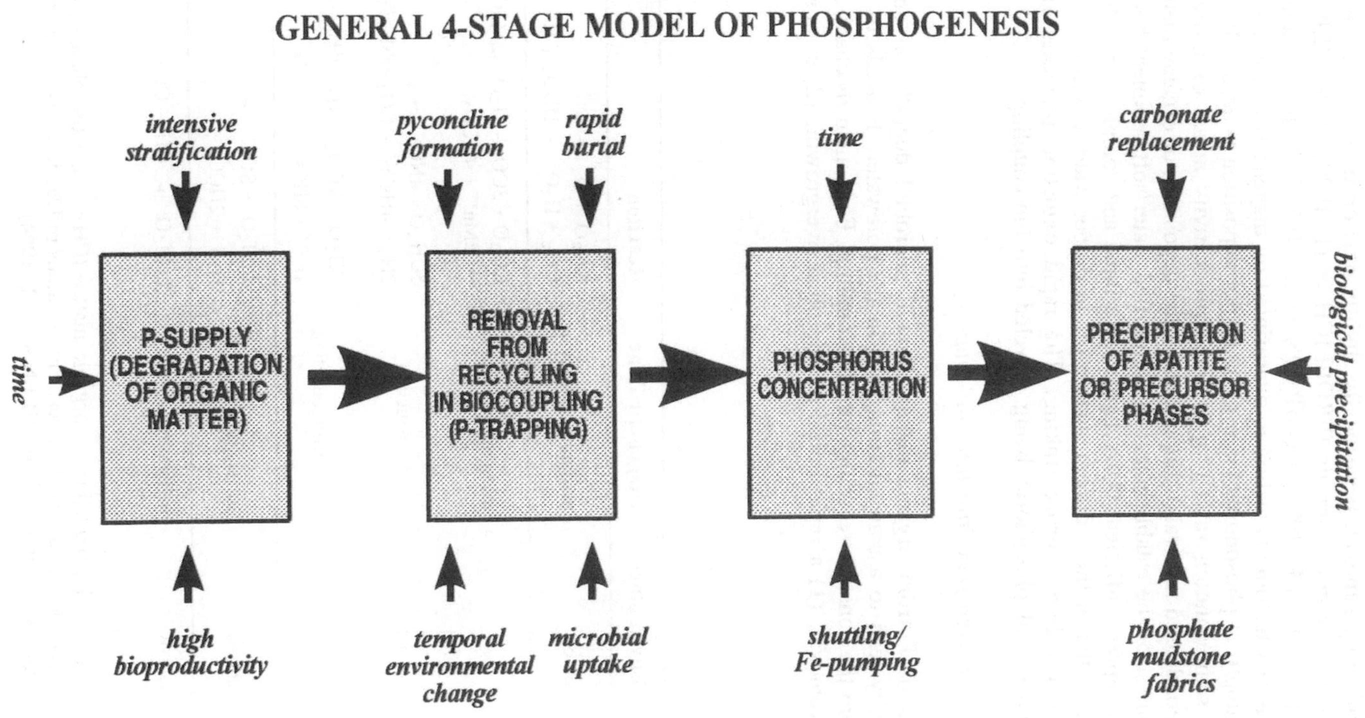

Fig. 5.5. The concept of a general model for phosphogenesis. The phenomenon is understood as different chains of events which have 4 major "bottlenecks" in common. Various genetic processes control a broad spectrum of phosphogenic pathways. Their selection is purely environmentally driven.

indicate a dominantly suboxic, occasionally slightly oxic or anoxic oxygenation stage in phosphogenic environments (FROELICH et al. 1979, 1988, GLENN et al. 1988, 1994, PIPER 1991, COMPTON et al. 1993). In many ancient deposits, the initial isotopic signature is overprinted by diagenesis and weathering. A high supply of phosphorus via organic matter deposition is accomplished through high sedimentation rates (e.g. in high productivity areas) or through time (e.g. in condensed intervals). Important for the deposition of organic matter is the rapid burial or the establishment of a stratified water/soft sediment body. The processes supress sufficient O_2-recharge for a rapid and complete oxidation of the organic matter at the sediment surface or, in deeper water environments, within the water column during sinking. The rapid oxidation increases significatly the chance of phosphorus being recycled into bio-coupling.

Stage 2: Removal from recycling into bio-coupling

The liberation of P from organic matter through microbial degradation does not consequently lead to a concentration of P in the porewater. P needs to be scavenged from the constant recycling into biocoupling. Four main mechanisms generate a P-removal: (1) a rapid burial or microbial overgrowth, (2) seasonal

TDC - $\delta^{13}C$	Environment	Diagenetic Zone	Reaction
+/-0.5‰ (botton water)	oxic	aerobic oxidation	$CH_2O + O_2 \rightarrow$ $CO_2 + H_2O \rightarrow HCO_3^- + H^+$
↓	suboxic	manganese reduction	$CH_2O + 3CO_2$ - $H_2O + 2MnO_2$ $\rightarrow 2Mn^{++} + 4HCO_3^-$
		nitrate reduction	$5CH_2O + 4NO_3^- \rightarrow$ $2N_2 + 4HCo_3^-$ - $CO_2 + 3H_2O$
		ferric iron reduction	$CH_2O + 7CO_2$ - $4Fe(OH)_3 \rightarrow$ $4Fe^{++} + 8HCO_3^- + 3H_2O$
-25‰	anoxic	sulfate reduction	$2CH_2O + SO_4^{--} \rightarrow$ $H_2S + 2HCO_3^-$
TO: +25‰	anoxic	methanogenisis	$2CH_2O \rightarrow CH_4$ - CO_2

Fig. 5.6. Stages of microbial oxidation of organic matter (CH_2O) and the changes of carbon isotope composition of marine porewater. The reactions reflect downward decreasing metabolic free energy yields (based on GLENN et al. 1994).

changes which affect the rate of bioproductivity and P-consumption, (3) shuttling of P into a deeper sediment layer, or (4) P-fixation within microbial mats.

(1) The rapid burial and the establishment of a P-enrichment zone is reported from some sites off Peru (FROELICH et al. 1988, JARVIS et al. 1994). In areas with low accumulation rates, the concentration of phosphorus occurs where microbial mats form a seal at the sediment surface.

(2) Temporal changes in bioproductivity induced by seasonal variation of water temperature, oxygenation stage or salinity should affect the P-uptake and the P-surplus.

(3) A spatial removal requires a well-defined oxic to anoxic stratification in the water/sediment interface and the presence of high rate of Fe in the porewater. A P-flux into deeper sediment layers, where P is protected from recycling into bio-coupling, is induced by a shuttle system for phosphorus via Fe-pumping (HEGGIE et al. 1990, O'BRIEN et al. 1990).

(4) The biologic uptake of phosphorus via microbial mats transfers phosphorus from the sea-water into the chemical micro-environment of the microbial sediment fabrics.

Stage 3: Concentration of P in porewater systems or at the sediment/water interface

The concentration of P to reach the supersaturation for apatite precipitation, starting at about 100M (VAN CAPPELLEN & BERNER 1988), requires a substantial storage potential near the sediment/water interface. The time frame for the concentration process is uncertain, but seems to be in the scale of $x10^3$ years or more in condensed sediments. All sediments are too drastically depleted in terms of phosphorus to initiate broadly phosphogenesis (see chapt. 1). In all cases, the concentration to the required level of supersaturation is always a slow process. There are no indications for rapid increase. Consequently, the phosphogenesis demands a storage environment for P in which the concentration increases slowly through permanent net P supply. As well as the removal of phosphorus from the biocoupling, the storage environment must provide low benthic bioproductivity, which is only present under suboxic to anoxic conditions. As a result, a potential phosphogenic environment has to provide in spatial, questionably also in temporal coexistence, an environment of intense microbial activity for liberation of P and a second environment of low bioproductivity where P is stored until concentration reaches supersaturation. An exception may be biomineralization within soft tissues.

The formation of a storage environment for phosphorus can only by achieved through a microbial decomposition rate which exceeds the O_2 recharge. This situation leads consequently to oxygen depletion, decreasing new bioproduction, and reduced P-consumption. Factors initiating the process are high organic matter deposition and stratification to reduce oxygen recharge. Supersaturation is reached relative rapidly, if these two driving processes are intensively developed. Under weakly developed conditions, supersaturation state is reached through time. From the geological record, both phenomenons are common.

Concentration through time is documented in the abundant occurrence of phosphatization associated with condensation and omission (FÖLLMI 1990, 1996, JARVIS 1992). More rapid concentration can be reached by an enhanced rain of organics. The mechanism is a self-supporting process. Intensified organic matter deposition results in an increased demand of oxygen for decomposition and affects oxygen depletion if the water column is not permanently mixed.

In Fe-rich, potentially phosphogenic sediments, the process of Fe-pumping was identified to be a sufficient mechanism to concentrate phosphorus in a soft sediment porewater system. Even in environments with low organic matter deposition rates, Fe-pumping was efficient to concentrate phosphorus to the state of supersaturation (O'BRIEN et al. 1990, HEGGIE et al. 1990).

Stage 4: Precipitation of a solid-phase phosphate in the sediments

Apatite precipitation results form three basic processes: (1) Precipitation of cryptocrystalline apatite aggregates. (2) phosphatization and/or impregnation of carbonate sediments, and (3) biomineralization. Microbial mediation may also be an important factor in the first two processes. The various forms of apatite precipitation and resulting products depend on the geochemical environments, but also on the dominating mineral phases present in the phosphogenic sediment. From the observations of ancient phosphorite depositional systems different trends of phosphate mineralization are emphazised. The presence of lithified carbonate rock or soft carbonate muds shifts mineralization to phosphatization of carbonates. The process of replacement or impregnation is not well understood. Siliciclastic sediments do not have any replacement potential. Apatite mineralization occurs in form of concretionary growth, or various forms of apatite cements are precipitated in the pore space. A third form of apatite precipitation is biomineralization within soft tissues.

5.5 Biogenic phosphogenesis: the biological approach

The biologic component in phosphogenic processes influences the phosphate research from its early beginning. The discussion focuses on three issues:

(1) Most early phosphate researchers attributed phosphate particles to coprolites and fecal pellets (BUCKLAND 1829, PORTER & ROBINSON 1981, LAMBOY 1982a). These assumptions could only be verified in rare cases (CAYEUX 1950, SLANSKY 1986, GLENN et al. 1994). Evidence was also not found for the hypothesis that a major portion of the pelletal phosphorite grains are kidney stones of mollusks (DOYLE et al. 1978). The fecal pellet origin is still in dispute for some deposits (LAMBOY 1982b, SCHRÖTER 1986).

(2) Many invertebrate and vertebrate organisms construct parts or the entire skeleton in phosphate mineralogy (RHODES & BLOXAM 1971, McCONNELL 1971, 1973, McCONNELL et al. 1961, LOWENSTAM

1972, LOWENSTAM & WEINER 1989). Primary phosphatic bioclasts in form of shell fragments, teeth and bones, contribute considerable amounts of phosphate material to some phosphate deposits, e.g. in the Ordovician deposits in Estonia (HEINSALU 1990, ILYIN & HEINSALU 1990) and the Miocene phosphate deposits in Florida (RIGGS 1979a,b).

(3) Undoubtedly, a significant role in phosphogenesis plays microbial activity. LUCAS & PRÉVôT (1981, 1984a,b, 1985), El FALEH (1988), HIRSCHLER (1990), HIRSCHLER et al. (1990a,b) could demonstrate the important role of bacteria for the precipitation of phosphatic aggregates. Another eminent factor is the passive mediation of the microenvironment in the porewater system of sediments or microbial mats through the living activity of bacteria. This process is nicely documented in the formation of concretions around fossil remains (SLANSKY 1986). The pathway of phosphatization in phosphatic stromatolites remained uncertain so far (KRAJEWSKI et al. 1994).

5.5.1 Microbial mediation

Microbial fabrics and phosphatic stromatolites have long been recognized in numerous Precambrian and Phanerozoic phosphate occurrences (e. g. CAYEUX 1936, OPPENHEIMER 1958, BUSHINSKI 1964, 1969).

Phosphatic stromatolites as discrete biosedimentary structures demonstrate a potential phosphogenic environment which is directly bound to microbial mats. Autogenic apatite occurrence in most of these fabrics is strictly limited to the microbial structure. Variations in mineralized phosphate contents are correlated with patterns linked to biolamination (KRAJEWSKI 1981, KUMAR & MULLER 1988, TEWARI 1993, MARTIN-ALGARRA & VERA 1994). The processes of apatite precipitation in microbial mats remain poorly understood, because no modern equivalents were found so far. Detailed isotopic and biologic studies may uncover the mechanisms in the future. The occurrence of autogenic minerals (glauconite, pyrite), which is linked to late stages of organic matter degradation, indicates a potential development of oxygen-deficient micro-environments in these structures. Three main processes contribute to the development of phosphogenesis in stromatolite building communities which are (1) transitory fixation of phosphorus, (2) regulation of dissolved phosphorus and (3) modification of the chemical environment of phosphogenesis (KRAJEWSKI et al. 1994). The accumulation rates of phosphate in these microbial communities indicate an external source of phosphorus for apatite autogenesis.

The occurrence of phosphatic stromatolites shows a spatial maximum in the Precambrian and Cambrian, with emphasize on the Asian-Pacific province. The decline correlates with the development of dominantly metazoa communities in the Ordovician. KRAJEWSKI et al. (1994) postulate a second maximum in the Mesozoic. The majority of the phosphate literature anyway focuses on this time interval and a casual evidence remains uncertain.

The growth forms and the internal fabrics of phosphatic stromatolites correspond to their carbonate equivalents. Most phosphate stromatolites display a calcium carbonate and apatite mixture, indicating alternation of calcification and phosphatization processes (BANERJEE et al. 1986, KRAJEWSKI et al. 1994, ROUGERIE et al. 1997). Mineral composition is additionally affected by the amount of detrital trapping by the mats. Microbial phosphorites associated with chamositic iron ore deposits reveal shifts between apatite and Fe-mineral precipitation within the microbial community (TRAPPE & ELLENBERG 1994). Phosphatization discordant to the microbial lamination in some occurrences documents apatite replacement processes in calcium carbonate stromatolite fabrics (e. g. BANERJEE 1971, VALDIYA 1972, CHAUDAN 1979, SOUTHGATE 1986a,b, 1988, EGANOV 1988). Phanerozoic phosphate stromatolite occurrences are commonly related to starved sedimentation intervals during high sea-level stages or platform drowning (KRAJEWSKI 1981, 1984, FÖLLMI 1989, 1990, MARTIN-ALGARRA & SANCHEZ-NAVAS 1995). The depositional environment of phosphatic stromatolites ranges from pertidal/intertidal to pelagic (SCHMITT & SOUTHGATE 1982, FÖLLMI 1990, KRAJEWSKI et al. 1994, MARTIN-ALGARRA & SANCHEZ-NAVAS 1995).

Phosphatic microbial fabrics have been recognized in many deposits (e.g. BREMMER 1980, O'BRIEN et al. 1981, ZIMMERLE 1982, 1994, BATURIN 1983, SOUDRY & CHAMPETIER 1983, KRAJEWSKI 1984, MULLINS & RASCH 1985, ZANIN et al. 1985, SOUDRY 1987, 1992, RAO & NAIR 1988, SOUDRY & LEWY 1988, 1990, ABED & FAKHOURI 1990, GARRISON & KASTNER 1990, LAMBOY 1990a,b). The fabrics are mostly molds which reveal a morphology of various cyanobacteria, other bacteria, and fungal filaments. The open pore space or the cement filling of the molds are thought to represent decomposed organic matter. Wall-like structures and capsules of apatite mineralogy are rarely preserved. Any evidence is lacking that the microbes have played an active role of apatite precipitation. Their passive role as outlined in the discussion of phosphatic stromatolites is fairly unknown. Small crystal size of most phosphate mineralizations give microbial structures an excellent chance of preservation compared to other sediments.This fact may be one reason for the common occurrence of microbial structures in sedimentary phosphate fabrics.

LUCAS & PRÉVôT (1981, 1984a,b, 1985), VAILLANT (1987), EL FALEH (1988) and HIRSCHLER (1990) investigated the influence of biological processes of bacteria in the formation of apatite in laboratory experiments. They could find random extracellular apatite mineralizations around cell bodies, but the cells seem to act only as substrate. The membraneous and wall macromolecules provide binding sites for calcium ions and promote the nucleation of calcium minerals on cellular structures (ELIOTT 1985).

Ultrastructure studies of phosphate particles demonstrated that many phosphate fabrics are composed of apatite crystalites resembling bacteria (e.g. O'BRIEN et al 1981, ZIMMERLE 1982, BATURIN 1983, LAMBOY 1990a,b, GARRISON & KASTNER 1990). These particles are micro-sized rods, globules or spindles of massive apatite which reminded some authors of replaced bacteria cells. But no evidence was found for an intracellular phosphate concentration and apatite precipitation. The laboratory experiments reveal a broad spectrum of inorganic bacteria-like apatite crystals (VAILLANT 1987, El FALEH 1988, HIRSCHLER 1990).

In summary, microbes commonly occur in phosphate rocks in which preservation chances are potential high due to low crystallite size and rapid mineral precipitation. Evidence for an active role in phosphogenesis is lacking. The metabolic activity in accumulation and decomposition of organic matter and the subsequent effects on the porewater chemistry may support the development of phosphogenic micro-environments (BATURIN 1982).

5.5.2 Skeletal phosphogenesis

In contrast to microbial activity which seems to affect only a biologic mediation during phosphogenesis, a direct apatite biomineralization occurs in vertebrate organisms and some invertebrate groups. These organisms precipitate phosphate minerals to stabilize the elements of their endo- or exoskeletons.

(1) Vertebrate bones and teeth are composed of calcium carbonate and contain up to 35% P_2O_5 mostly in form of the apatite variety fluorapatite. In the fossil record, the conodonts as an early vertebrate group segregated their apparatus in apatite mineralogy.

(2) Inarticulate brachiopod shells are composed of the apatite variety dahllite and francolite, some organisms from six other modern classes form hard tissues completely or partially of dahllite, francolite, phosphatic amorphous hydrogels and other amorphous and crystalline phosphate precipitates (LOWENSTAM 1972, LOWENSTAM & WEINER 1989).

Only bone material and shells of inarticulate brachiopods form sedimentary concentrates which have been recognized in the geological record. Bone beds as marker horizons and fossil sites are well known (e.g. REIF 1982). In most younger phosphate deposits, vertebrates and teeth are one of the minor constituents of the phosphate particle spectrum, e.g. Florida (RIGGS 1979a,b), Morocco (ARAMBOURG 1952, TRAPPE 1989, PRÉVôT 1990). Complete shells and shell fragments of inarticulate brachiopods are the major constituent of the Ordovician economic phosphate deposits in Estonia (see Case Study 1). Other occurrences of phosphatic shell concentrates are reported from Ordovician rocks in other regions.

Very little and only fragments are known on the processes of phosphate biomineralization. For the precipitation of vertebrate bones and teeth material, nucleation by proteins and macromolecule matrices provide molds which control crystal growth and cessation. Similar processes presumably control also the phosphatic shell segregation of inarticulate brachiopods and other invertebrates, but data are even more spare than for vertebrate organisms (LOWENSTAM & WEIMER 1989).

CASE STUDY 1
Shelly phosphorites in Estonia and westernmost Russia, Early Ordovician

Anatomy of the deposit

Location: A narrow strip along the Finnish Bay/Baltic Sea, between Tallinn in Estonia and St. Peterburg in Russia; 59°30'N, 25°E to 59°10'N, 29°E.
Stratigraphy, age: Lithostratigraphic system: Kallavere Formation subdivided in Maardu (former "*Obolus* conglomerate"), Suurjöse & Katela Members or equivalents to the east (KALJO et al. 1986, HEINSALU & RAUDSEP 1993), upper and basal contact unconformable (Fig. 5.7). Age: (?Late Cambrian) Lower Tremadoc (KALJO et al. 1988, PUURA & HOLMER 1993, pers. comm. ERDTMANN 1995), Late Cambrian age in dispute, probably derived from reworked fauna.
General paleogeography: Southern margin of a narrow SE-NW trending epicontinental gulf on the SW' edge of Baltica, near the NE end of the gulf (Fig. 5.8), Paleolatitude: ~ 30°S, cold climate (GOLONKA et al. 1994); nearshore: peritidal to intertidal; low relief hinterland.
Paleotectonic setting: Epicontinental sag basin.
Depositional system, facies architecture: Siliciclastic system, depositional prims along a shoreline, irregular stacking of bar sediment bodies, conglomerates at the base from reworked underlying Cambrian strata (Tiskre & Ülgase/Tsitre Fms.), weakly cemented (soft sediment), horizontally bedded, locally intensive crossbedding, trend to more fine-grained, clay-rich sediments to the south, thickness decreases to the east.
Sea-level control, sequence stratigraphic interpretation: Transgressive systems tract (2. or 3. order cyclicity), above well defined transgressive surface (superimposing directly sequence boundary). Overlying C_{org}-rich shales may represent maximum flooding sediments or a newly rapid transgressive pulse.
Type of phosphorite particles: Bioclasts of primary phosphatic inarticulate brachiopods, common abrasional features (Fig. 5.9).
Model for the phosphogenic processes: Biological skeletal phosphate precipitation.
Phosphorite formation: Frequent reworking and redeposition during a transgressive interval in pertidal to intertidal environments, formation of bars and channels.
Reserves, resource quality: Total resources are 600-700 million t P_2O_5 (KRASILINIKOVA & ILYIN 1989), more recent estimates: 450 million t P_2O_5 (RAUDSEP 1993), depocenters at Maardu near Tallinn and Rakvere, mining abandoned.

The Estonian phosphate deposits represent a unique example of biologic phosphate deposits of economic importance. These deposit demonstrate a major biologically driven mineralization pathways in the box model. But in contrast to sedimentary pathways, major parts of the chain of events and processes for the formation of a suitable environment for large scale biologic apatite precipitation are still widely enigmatic (see chapt 5.5). In the Estonian deposits, enormous amounts of phosphatic shell fragments are accumulated with relative mature sands in nearshore agitated water.

Phosphogenesis is purely restricted to biologic processes. The phosphatic shells, dominantly of the genera *Schmidtites* and *Ungula*, contain 35-37% P_2O_5 (KRASILINIKOVA & ILYIN 1989, ILYIN & HEINSALU 1990, PUURA & HOLMER 1988). The average phosphate contents of the mineable beds are 6-15% P_2O_5. The source of these enormous quantities of phosphate which was precipitated from seawater by these brachiopods is unknown. The sea area was a narrow epicontinental basin, which neither provides sufficient P-reservoir nor a typical upwelling situation (Fig. 5.8). Rare phosphate cementation in shell fractures or around quartz grains is thought to originate from post-depositional phosphate mobilization (ILYIN & HEINSALU 1990).

The existing sediments are entirely clastic and are composed of a mature quartz component, phosphatic bioclasts and glauconite (Fig. 5.9). The phosphate-hosting sediment suite, which reaches a maximum thickness of 15 m, rests with a basal conglomerate on an unconformity surface (Fig. 5.7). The conglomerate consists of phosphate cemented sandstone pebbles which are derived from reworking of the underlying Cambrian sandstone. Their phosphatization documents the presence of a phosphogenic environment during the initial transgressive stage. The Kallavere sediments are commonly cross-bedded, including herringbone cross-stratification, and show intense lateral variation in facies and thickness. The phosphate bioclasts and the quartz detritus are commonly polished and show abrasion. The abrasional features and the maturity of the quartz sand and the sediment structures support a nearshore to intertidal origin. The grade of shell abrasion and sediment structure from the peritidal to intertidal realm indicate an environment of frequent reworking and redeposition. Intensive lateral facies changes support a depositional environment with rapidly accumulating and decaying bars and channels. The depositional setting caused rapid lateral variations in thickness and facies, respectively of the phosphate contents. The random pattern of stacked bar and channel sediment bodies affects an unpredictable facies architecture. Intensive drilling revealed some depocenters with slightly high subsidence rates, which increased the accommodation space (Fig. 5.7, 5.10). In some localities, organic matter-rich shales are intercalated in the suite and represent low energy areas between shoals. HEINSALU et al. (in press) correlated the facies variations with paleotopographic elements and presented lithofacies maps of the Rakvere area, where the topography is documented during the transgressive onlap phase.

ILYIN & HEINSALU (1990) discuss the earlier proposed relationship between the biological phosphogenesis, availability of phosphorus, and black shale deposition. They relate the growth of brachiopod colonies on shoals to times of low sedimentary P-fixation from the sea water in black shales. During regressions, the colonies were destroyed and the shell fragments are deposited in depressions. HILLER (1993) pointed out the similarity of the Estonian deposits

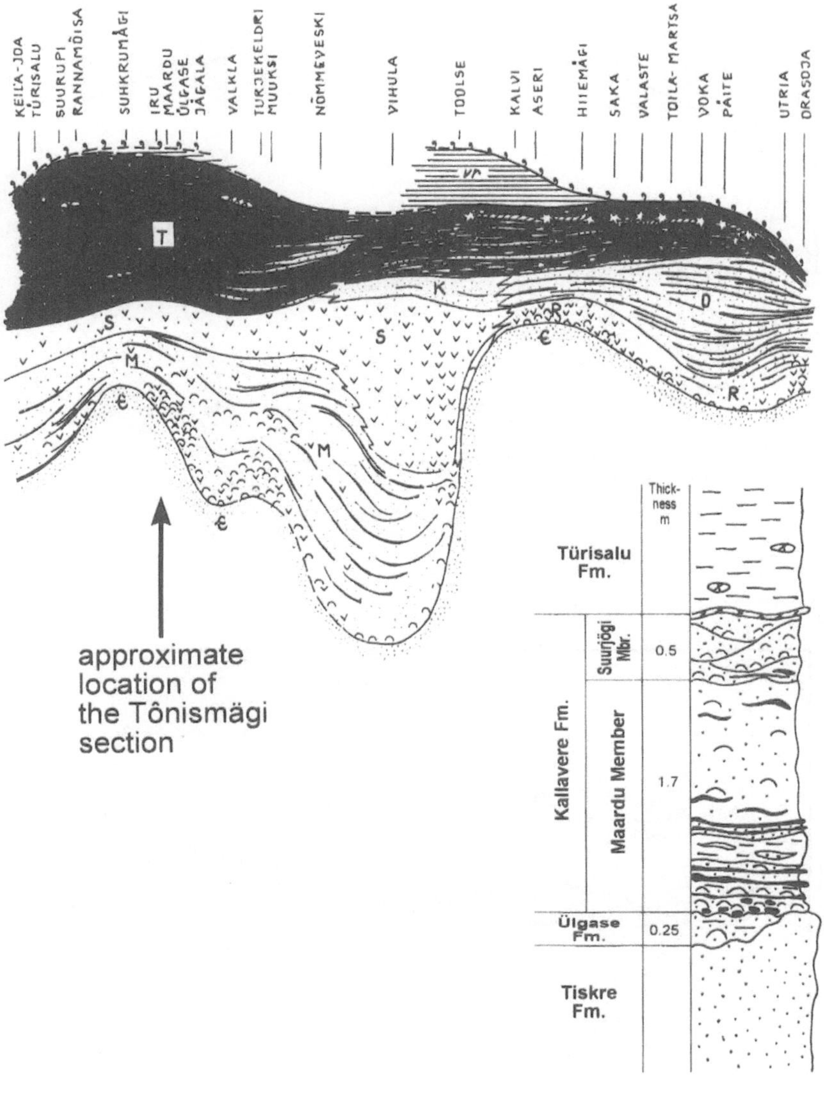

Fig. 5.7. Stratigraphic column of the phosphate bearing sediment suite in a typical example in Tallinn (Tônismägi section) based on KALJO et al. (1988) and the lateral facies development along a W-E cross-section through northern Estonia (modified from HEINSALU 1990). The phosphate bearing sediment suite rests on a distinct erosional surface above Cambrian Ülgase sandstones. The erosional surface is paved with phosphatized sandstone pebbles. Thickness in m. vr = Varangu Fm., T = Türisalu Fm., K = Katela Mbr., S = Suurjögi Mbr., M = Maardu Mbr., O = Orasoja Mbr., Rannu Mbr., E = Cambrian.

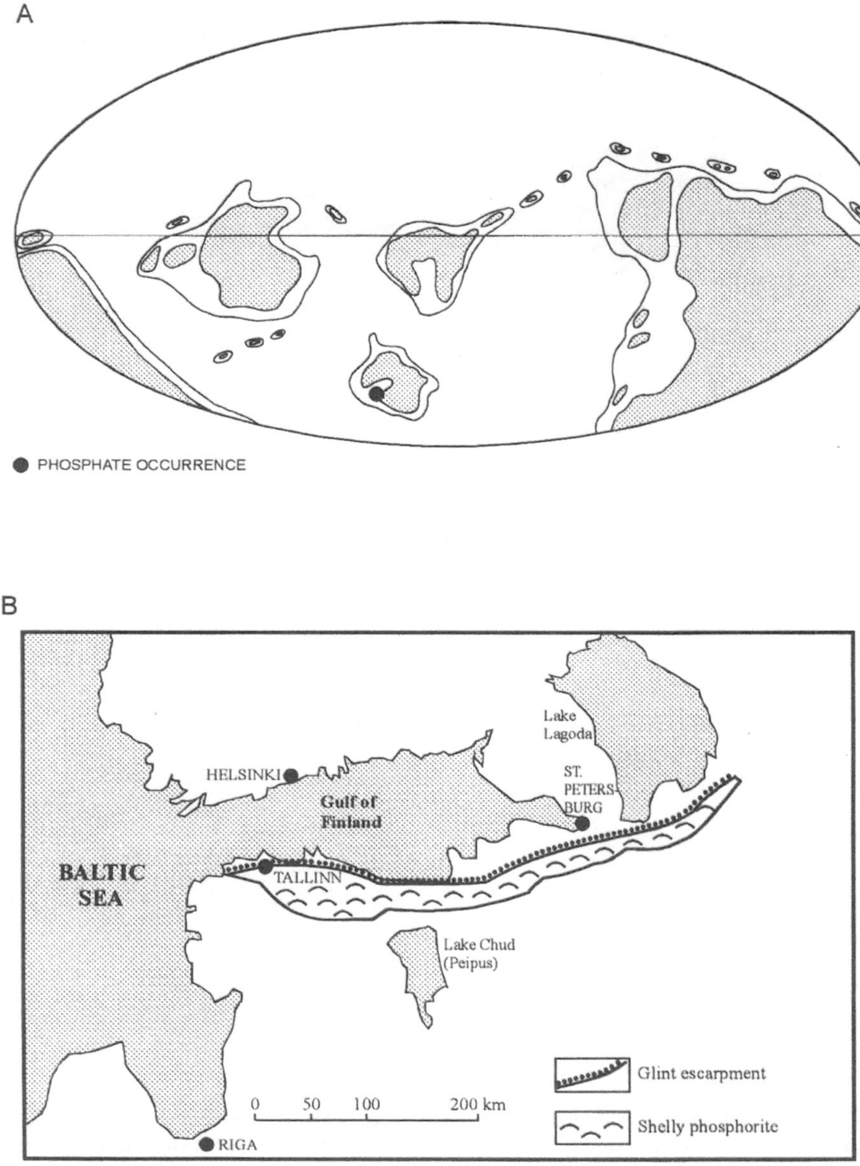

Fig. 5.8. Paleogeography of the Estonian phosphorite deposits: (A) general paleogeographic position based on paleogeographic reconstructions of GOLONKA et al. (1994), (B) distribution of Early Ordovician phosphate-bearing sediments in the Baltic region (modified from ILYIN & HEINSALU 1990).

to *Discinisca* shell accumulations along the Namibian coast. The conclusion, that the potential upwelling source of phosphorus in Namibia may indicate upwelling in the Ordovician gulf in Estonia denies the modern paleogeographic reconstructions. More plausible is a selective concentration of stable sediment constituents during transgressive blanket sand formation and subsequent reworking and redeposition during following minor sea-level changes. In this model, the phosphate sediment component is a concentrate derived from reworking of a larger sediment pile and a significant larger area.

Fig. 5.9. (next page) (A) Outcrop of phosphate bearing strata in a road construction area in Tallinn. (B) Outcrop appearance of the crossbedded shelly phosphorite beds of the Maardu Member. Same section. (C) Microfacies of slightly lithified horizons in the Maardu Member. Rounded phosphatic shell fragments of centimeter size are the major phosphate constituent. Additionally, some lithoclasts of phosphatized sandstone occur which are derived from the basal erosional surface of the Ordovician succession.

Fig. 5.9.

A

B

C

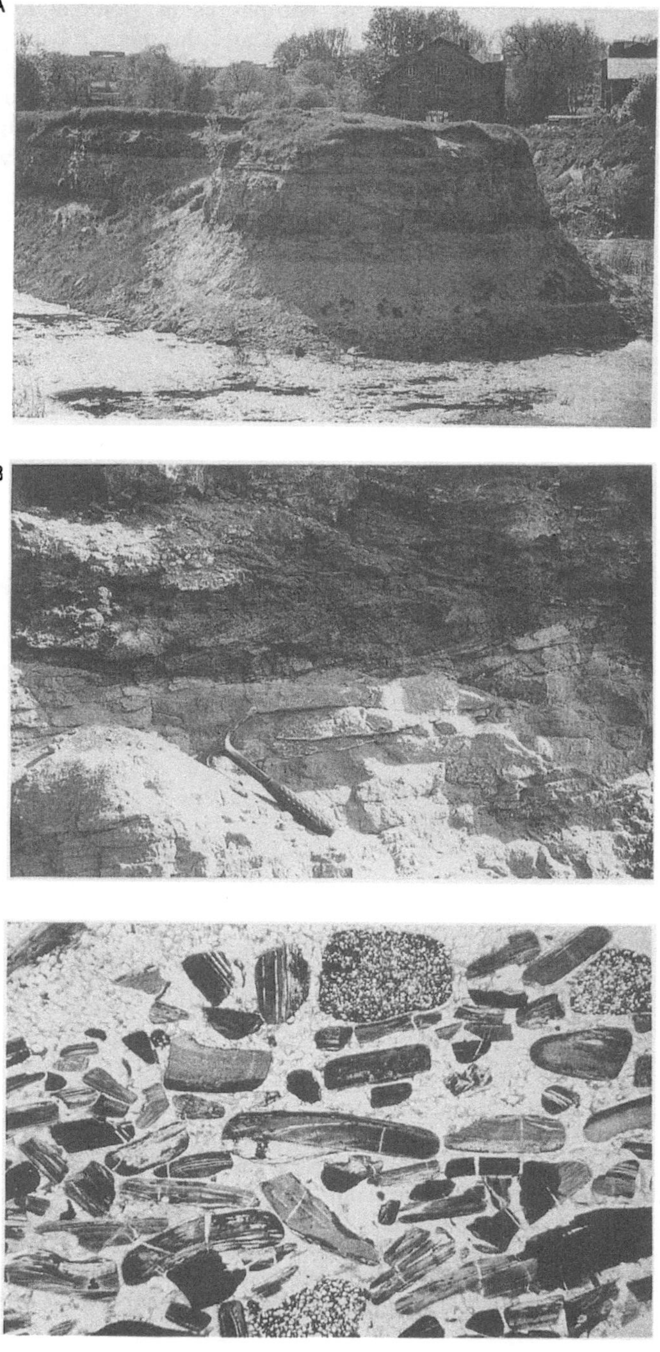

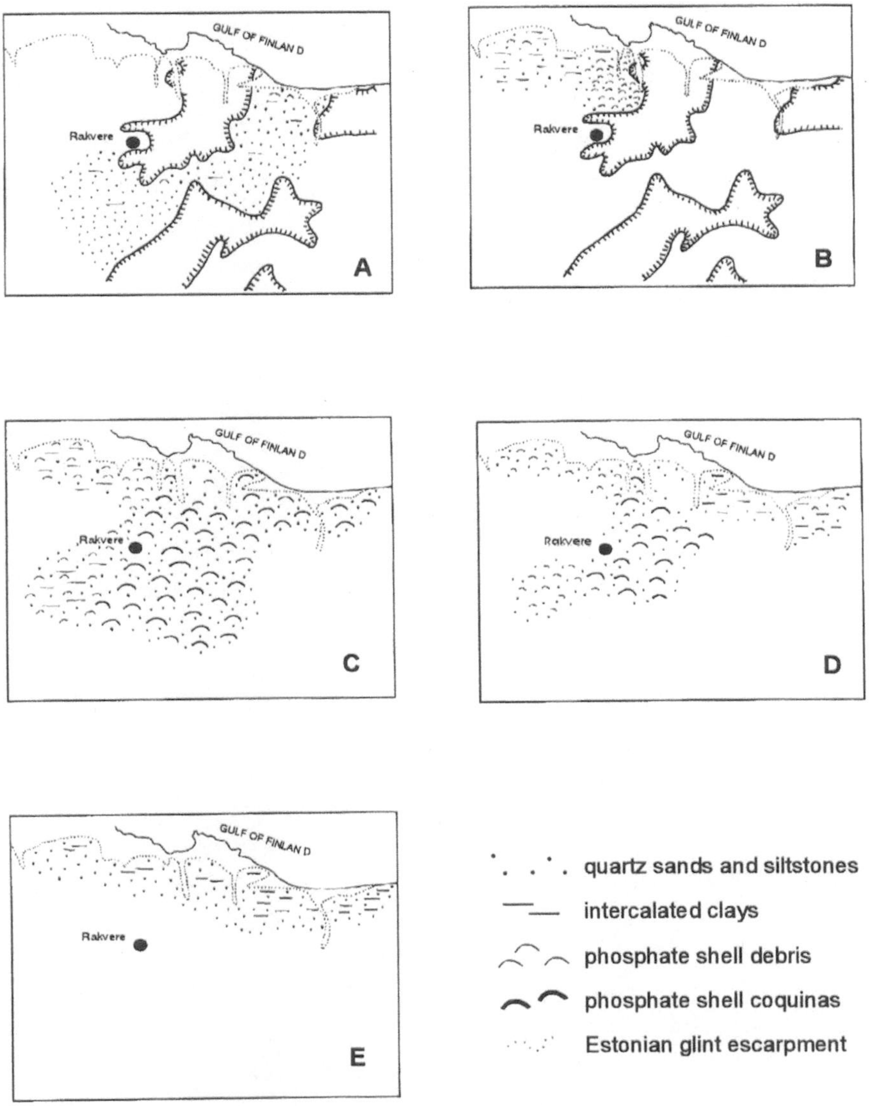

Fig. 5.10. Spatial and temporal facies development of the Late Cambrian to early Tremadocian time (modified from HEINSALU in press). A = Tiskre and Ülgase Fms, B = lower Maardu Member, C = upper Maardu Member and lower Suurjögi Member, D = upper Suurjögi Member, E = Katalena Member.

6 Pathways to phosphogenic environments - from a steady state to an environmental approach to phosphogenesis.

After determining the fundamental processes and parameters of phosphogenesis, the question raises how and in which natural environments phosphogenesis took place during the Phanerozoic. The investigation of ancient economic phosphate deposits reveals a wide range of facies and paleogeographic constellations, very much contrasting the modern occurrences. The ancient deposits further demonstrate that phosphogenic environments during the Phanerozoic developed in depositional settings which are not represented in the modern world. The spectrum of phosphogenic environments during the Phanerozoic was much wider than today, and the modern occurrences seem to display more an exceptional situation. Which are typical depositional systems? Which environmental conditions are required? Which pathways of phosphogenesis developed in adaptation to the environment during the Phanerozoic? And which are the resulting phosphate fabrics and lithofacies associations?

The 4-step model of phosphogenesis demonstrates which minimum requirements a depositional system has to accomplish for being potentially phosphogenic. These are a sufficient phosphorus supply, a P-storage and concentration system and finally a geochemical regime which passes the conditions for the precipitation of apatite (apatite precipitation window) during its evolution. The comparison of paleogeographic setting, facies, and evolution of ancient and modern phosphogenic deposystems indicates significant genetic differences by the completion of these minimum requirements for phosphogenesis. The development of phosphogenic environments follows a broad spectrum of genetic pathways and cannot be reduced to a single genetic model.

Whereas the parameters for apatite precipitation in sediments are static, the depositional environments and consequently also the evolution of phosphogenic environments are changing in the dynamic system Earth. The design of depositional systems is always significantly affected by the paleogeographic position further by global and regional environmental changes. The broad facies variation in phosphate sediments and their spatial and temporal occurrence in different sedimentary host systems indicate that every phosphogenic environment is a unique adaptation to the specific depositional history of its host sediment. Consequently, pathways to a phosphogenic environment are numerous and are strictly determined by the evolution of the various depositional systems. But all genetic pathways which finally lead to phosphogenesis in a depositional system have in common to pass "gateways" or "bottlenecks". These are defined by the physico-chemical requirements of phosphogenesis. The pathways, processes and mechanisms should be similar in related depositional systems if the latter are understood as an image of the environmental conditions.

The spatial dimensions of phosphogenic environments range from the enclosed environment within a shell to a region within an epicontinental basin or shelf of several thousands of square kilometers. A wide range is also evident in the temporal existence of phosphogenic environments. The formation of the Meade Peak or Retort Phosphatic Members, the two major phosphogenic intervals in the Phosphoria Formation, was achieved within a single conodont zone (see Case Study 4). Phosphogenic systems on omission surfaces may have been stable for time intervals documented by several biozones.

6.1 Major supporting factors for the development of phosphogenic environments

The knowledge of major supporting factors of phosphogenesis helps to identify vivid phosphogenic settings in ancient paleogeographic constellations. From the investigation of the depositional setting of modern and ancient phosphate deposits, a number of environmental phenomenons are recognized being significant for the genesis of phosphate deposits. Most depositional settings combine several driving factors for phosphogenesis.

Bioproduction: Enhanced bioproduction due to nutrification from a fluvial or oceanic source in shallow marine sea areas or lakes supports organic matter deposition rates (SOUTHAM et al. 1982, PARRISH 1982, GLENN & ARTHUR 1985, TYSON & PEARSON 1991, SMITH 1992, ARTHUR & SAGEMAN 1994). The increased organic matter deposition consequently increases also the potential flux rates of bio-coupled phosphorus in the sediments. Enhanced decomposition rates by elevated bioproduction are also responsible for rapid O_2-depletion when the O_2-recharge by mixing is reduced. Recently, INGALL & JAHNKE (1997) could find enhanced P-regeneration rates in O_2-depleted water-columns which has a positive feedback on the bioproduction. SUESS (1981) proposed that the high sedimentation rate of fish bone in upwelling areas is a major source of phosphorus.

Stratification: Phosphogenic systems can only develop in partially or completely oxygen-deficient depositional systems where P-recycling into biocoupling is suppressed. The formation of a O_2-deficient water or sediment body is the result of microbial O_2-depletion and reduced O_2-recharge. For the latter, the formation of boundary layers and a stratified water/sediment body has fundamental importance (DEGENS & STOFFERS 1976, SOUTHAM et al. 1982, TYSON & PEARSON 1991). GLENN & ARTHUR (1985) demonstrated that bottom anoxia by density-stratification in closed basins leads to organic carbon and phosphorus accumulation. Boundary layers (pycnoclines) to various degree of intensity and temporal persistence commonly develop in freshwater and marine environments. Permanent or seasonal stratification develops due to gradients of temperature, salinity, alkalinity, or is current induced.

Upwelling: The most intensively discussed supporting phenomenon is oceanic upwelling. The oceanic influence on the development of phosphate deposits is sustained by faunistic relationships, biomarkers, inorganic geochemistry, and paleogeographic modeling (e.g. KAZAKOV 1937, SHELDON 1980, 1981, ARTHUR & JENKYNS 1981, FRAZIER 1981, BATURIN 1982, PARRISH 1982, RIGGS 1984, BARRON & MOORE 1994). The second argument is the occurrence of most modern phosphate sediments in upwelling regions. A single oceanic phenomenon supports all main requirements for the development of a phosphogenic system which are enhanced bioproduction through slightly P-enriched water and stratification through the formation of boundary layers. Due to rapidly decreasing bioproduction at the base of the photic zone, the net nutrient supply by the rain of organic matter exceeds the demand. High decomposition rates in the water column cause further O_2-depletion and limit new bioproduction. If bottom currents run against a submarine rise (dynamic upwelling) or are induced by a water deficit along the shore by dominantly seaward directed wind (equatorial or monsoonal upwelling), deeper water will be forced to flow upwards (PICKARD & EMERY 1990, POND & PICKARD 1991). These oxygen-poor, but very nutrient-rich waters reach the photic zone where the oxygen content is recharged by frequent mixing. The water masses now generate an enormous primary bioproduction, respectively an intense rain of dead organic matter. The different direction of bottom and surface currents, especially in equatorial upwelling systems, also affects a distinct stratification.

Fluvial P-supply: The fluvial influx of P as a supporting factor was discussed (BUSHINSKI 1966, PEAVER 1967). The flux is only locally important. GERMANN (1984) and BOCK (1987) proposed river input for a part of the Egyptian phosphorites. The source is probably significant for lacustrine systems. Slow mixing of sea and river water in an estuarine setting may contribute to the development of stratification.

Low net sedimentation rates: Intensive carbonate production or siliciclastic sediment supply would dilute the organic deposition. Field evidence shows definitely a correspondence of phosphorite formation and relative low sedimentation rates (GLENN et al. 1994, FILIPPELLI 1997). Modern sediments off Peru/Chile display a broad variation of accumulation rates between 3 mm yr^{-1} and 0.02 mm yr^{-1} (FROELICH et al. 1988), which has major effect on the porewater profile and apatite precipitation site. High sedimentation rates may also influence negatively the development of microbial communities through rapid burial (FÖLLMI 1990).

Condensation and omission: The commonly recognized phosphatization of hardgrounds explains the significance of condensation for phosphogenesis (e. g. PEDLEY & BENNNETT 1985, FÖLLMI 1990, JARVIS 1992, CARSON

& CROWLEY 1993). Under extreme low or absent carbonate or siliciclastic sedimentation over longer periods, but a constant rain of organic matter, P-contents in the porewater at or close to the hardground surface reach concentrations at which apatite precipitation is initiated. Slow or absent burial affects also the establishment of stable microbial communities which support liberation of P from organic matter.

Fe-pumping: The iron pumping model from SCHAFFER (1986), HEGGIE et al. (1990) and O'BRIEN et al. (1990) provides a striking model for the shuttling process of microbial liberated P into deeper soft sediment layers to prevent recycling into bio-coupling. Sufficient amount of Fe is present in many depositional environments. The existence of Fe in phosphatic sediments is documented by coexisting iron-crusts on hardground surfaces or by the association of phosphate with glauconite.

Cold water systems: Many phosphorites are associated with non-tropical carbonates. Corals are commonly absent or extremely rare. The biofacies assemblage commonly form the brymol-association, which is adapted to cold or moderate water regimes. The carbonate production in colder water is commonly lower than in warm water areas. Low output of the carbonate factory again prevents dilution of organic matter deposition and P-porewater concentrations.

Shallow shelfal and subtidal epicontinental settings: The global phosphate cycles identify these environments as a major phosphorus reservoir resulting from high P-flux rates into the sediment. In moderate water depths, balanced organic matter deposition and degradation provides best chances for the development of phosphogenic systems. In deeper water environments, sinking organic particles are completely decomposed within the water column. The microbial degradation generates only little new production. In paleogeographic constellations where the marine water body is not permanently mixed, dysaerobic conditions through microbial oxygen consumption develop in the water depth of maximum decomposition rates. Surface sediments at the contact zone of the dysaerobic water layer with the slope or ramp are consequently also characterized by low oxygenation levels at the sediment surface. Biologic activity is suppressed and P-recycling into biocoupling is reduced. In moderate water depths, a sufficient rain of organic material reaches the sediment surface instead of being degraded in the water column. Decomposition of this material liberates larger quantities of phosphorus. In shallow water, rapid degradation and recycling is enhanced by oxygen recharge. The situation in deeper water depends on the paleogeography. In closed basins e.g. Black Sea, the deeper water body is anoxic. In shelfal setting, undercurrents may supply normal oceanic waters amd are oxic. The deep oceanic water is not O_2-depleted because most organic matter is oxidized in smaller water depths (e.g. Peru margin)(GLENN & ARTHUR 1985, FROELICH et al. 1988, ARTHUR & SAGEMAN 1994).

6.2 Phosphogenic trends in siliciclastic versus carbonate systems

The different physical and chemical properties of carbonate and siliciclastic sediments have fundamental influence on the genesis of various phosphate fabrics. The controlling factors on the specific phosphogenic mechanism are the chemical reactivity and the porosity/permeability, which both affect the porewater regime. Clay minerals in marine muds exchange fluid phases with the porewater through incorporation of alkalines into the crystal structure.

Carbonate muds and mudstones in natural porewater composition are characterized by small crystal size, high internal surface, and a large potential of chemical reactivity, which leads to common apatite replacement or impregnation behavior (LAMBOY 1993).

These fabrics are absent in siliciclastics and shales. The large crystal size of detrital quartz and stability under most porewater compositions allows only authigenic or biologic growth of phosphate fabrics within these sediments. The phosphorus uptake of clay minerals in marine sediments is poorly known (ZANIN et al. 1990, COOK 1969). The presence of phosphatic shales is a commonly recognized phenomenon, but phosphorus in these sediments is usually associated with its source, the organic matter, or represents dispersed authigenic apatite mineralization (SHELDON 1963, COOK 1969, 1970, RUTTENBERG & BERNER 1993). Visible phosphogenic fabrics in siliciclastic and pelitic sediments are consequently related to inorganic concretionary growth, cementation, phosphate mud deposition (microsphorite), and microbial mediation.

The reactivity behavior affects also the lithification. Whereas siliciclastics and pelitic muds can be preserved as soft sediment over hundreds million years, most shallow marine carbonates lithify within a few thousands years. Phosphatization of erosional surfaces is consequently a common phenomenon in rapidly lithifing carbonates, but more exceptional in siliciclastic sediments (JARVIS 1992, FÖLLMI 1990). Long maintained permeability in siliciclastics and shales, resulting porewater flow, and late lithification support phosphatic concretion growth or phosphate cementation phenomenons.

Special interest has to be drawn to the Mg-bearing minerals. Mg-inhibition effects on phosphogenesis is in discussion since the experimental work of MARTENS & HARRIS (1970), KIBALZCYC et al. (1990), SISODIA & CHAUHAN (1990). Empirical observation of the common association of phosphorites with dolomites, palygorskite or other Mg-rich minerals are well established (BURNETT 1977, PRÉVôT 1990, JARVIS et al. 1994). Porewater profiles and kinetic studies neglect a direct relationship, but outline the importance of passive effects (VAN CAPPELLEN & BERNER 1991, GLENN et al. 1994).

6.3 Depositional systems and common pathways to phosphogenic environments

Phosphogenic environments can develop in most continental and marine depositional systems, but the majority of all ancient and modern phosphate occurrences is found in the marine realm (Fig. 6.1). Due to the fact that the development of phosphogenic environments requires balanced rates of organic matter degradation and deposition, which is mostly bound to oxygen-deficient conditions and poorly mixed water bodies, these systems are absent in all zones of permanently turbulent water.

This general environmental requirement addresses the initiation of phosphogenic processes, but does not exclude the formation of phosphorites in these environments by reworking, transportation, and redeposition mechanisms.

6.3.1 The continental and insular realm

Phosphorites which originate from precipitation in the continental regime are extremely rare, but major deposits developed on oceanic islands, many of them of economic importance. In the continental realm, a potential P-reservoir is generally present (see chapt. 1), but in most continental sediments, the O_2-recharge exceeds demand resulting in complete decomposition and P-recycling. Consequently, the conditions for phosphogenesis are generally given only in peculiar and very exceptional settings (Fig. 6.3).

The terrestrial and fluvial regime:These environments comprise four general types of phosphate rock occurrences. The first three of them are principally small-sized and are not reaching the scale of economic deposits. The terrestrial occurrences comprise deposits of very different origin and sedimentary fabrics:

(1) Phosphatization of coprolites in terrestrial sediments is known from Triasssic strata of the Ural (CHALYCHEV 1968). Coprolites primary rich in organic matter, respectively phosphorus, tend to phosphatize in poorly ventilated sediments. Phosphogenesis is initiated through decomposition of the high contents of organic matter and development of a geochemical microenvironment within the coprolites. The latter may be supported by organic mucus coatings which inhibit porewater exchange.

(2) The formation of modern phoscretes on old land surfaces above phosphorite exposures or supratidal paleo-surfaces has been reported from Australia and Asia (BLISKOVSKII 1967, DeKEYSER & COOK 1972, SOUTHGATE 1986a,b). These irregularly laminated crusts, which are phosphatic equivalents to calcretes or silcretes, are described in detail from the Georgina basin in Australia by SOUTHGATE (1986a,b). The development of modern

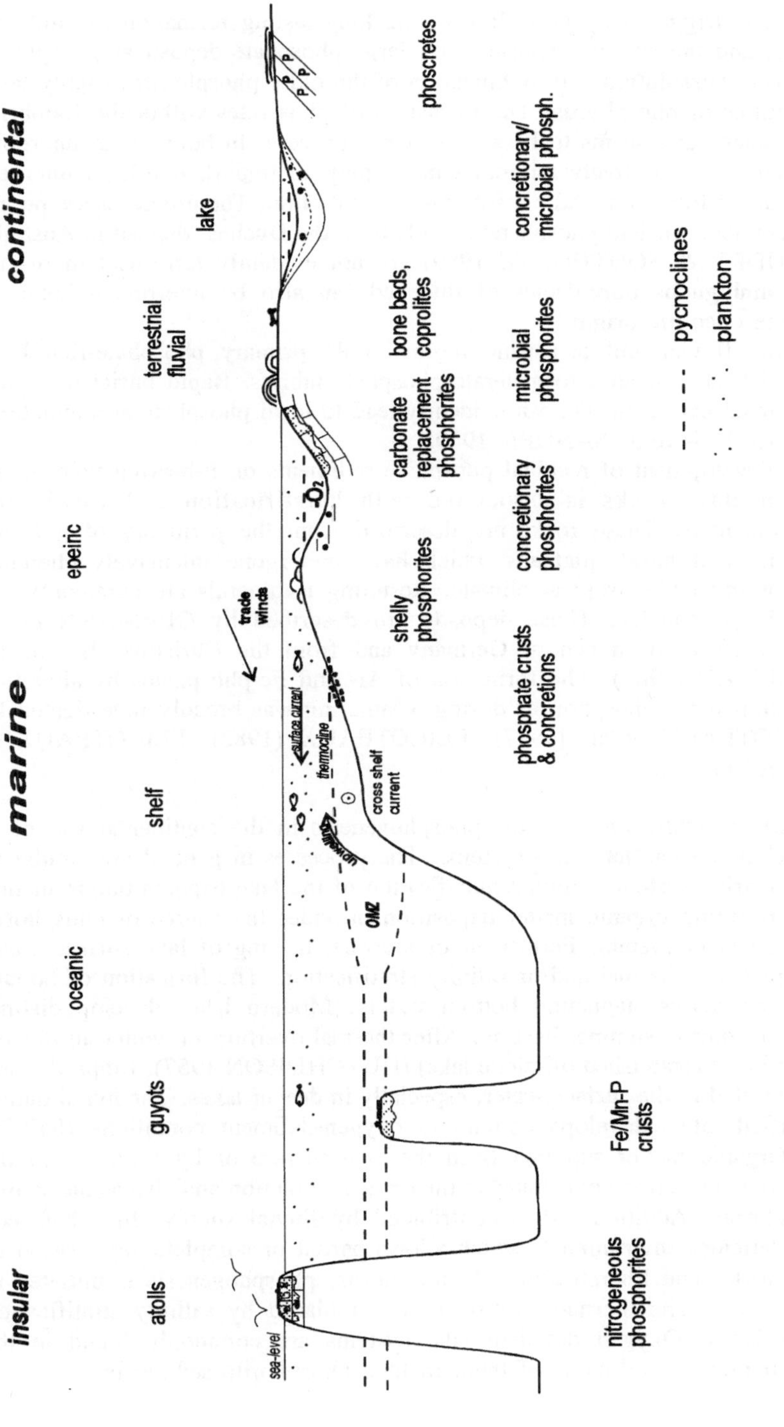

Fig. 6.1. The location of phosphogenic environments and major precipitation products in an idealized ocean to continent profile.

crusts is triggered by arid climate, the long lasting formation of land surfaces, and the surface exposure of a large phosphate deposit as phosphorus sources. Dissolution and mobilization of the older phosphorites supply large quantities of phosphorus. The formation of phoscretes within the Cambrian phosphate suite seems to be a more rapid process. In both cases, microbial mediation, respectively microbial mats, play an important role in microenvironment formation and phosphorus concentration. The processes are poorly understood. Ancient phosphatic pisoids from the Duchess deposit in Australia (SOUDRY & SOUTHGATE 1989) are not evidently terrestrial in origin. The ambiguous mircofacies of this bed can also be interpreted being of marine oncolitic origin.

(3) In the fluvial and lacustrine regime, only primary phosphate-rich bone material has a chance to generate phosphate fabrics. Rapid burial in poorly ventilated bar sediments can evidently lead to local phosphate accumulations in bone beds (e.g. SANDER 1989).

(4) The development of residual phosphate sediments on P-bearing volcanic or sedimentary rocks is associated with karstification and weathering phenomenons. These rocks are described from the periphery of volcanic massifs and basalt plateaus which have undergone intensively chemical weathering under tropical climate. Resulting P-minerals are commonly Al- and Fe-phosphates. These deposits are described by GERMANN et al. (1979, 1981) from central Germany and from the Christmas Islands by TRUEMAN (1965). The formation of Al- and Fe-phosphates by alteration of sedimentray phosphorites during weathering was broadly investigated by FLICOTEAUX et al. (1977), FLICOTEAUX (1982), FLICOTEAUX & LUCAS (1984).

Lacustrine systems: Larger scale phosphogenesis in the continental regime is bound to dysaerobic lake deposystems. The processes in general are similar to these in marine systems. Fluvial nutrification of the lake triggers bioproduction and the resulting organic matter deposition provides the source of phosphorus in the porewater system. Permanent or seasonal heating of lake surface water affects intensive thermal and/or salinity stratification. The formation of booundary layers causes stagnating bottom waters. Modern lakes develop distinct pycnoclines during summer heating. After thermal overturn in winter an inverse stratification is established (dimictic lake) (HUTCHINSON 1957). Bioproduction is concentrated to the surface water, especially in deeper lakes. The hypolimnion in stratified lakes develops commonly oxygen-deficient conditions (KELTS 1988). Organic matter supplied from the lake surface or by leaves from the surrounding vegetation contributes to the oxygen depletion and also as the source of phosphorus. Additional P is contributed by fluvial supply. In a balanced oxygen deficient environment, which allows partial or complete degradation of organic matter and liberalization of phosphorus, phosphogenesis is initiated in lake sapropels. The thermal stratification is replaced by salinity stratification in playa lakes. Oxygen deficient lake systems are commonly found in the geological record, and some of them include phosphorite sediments.

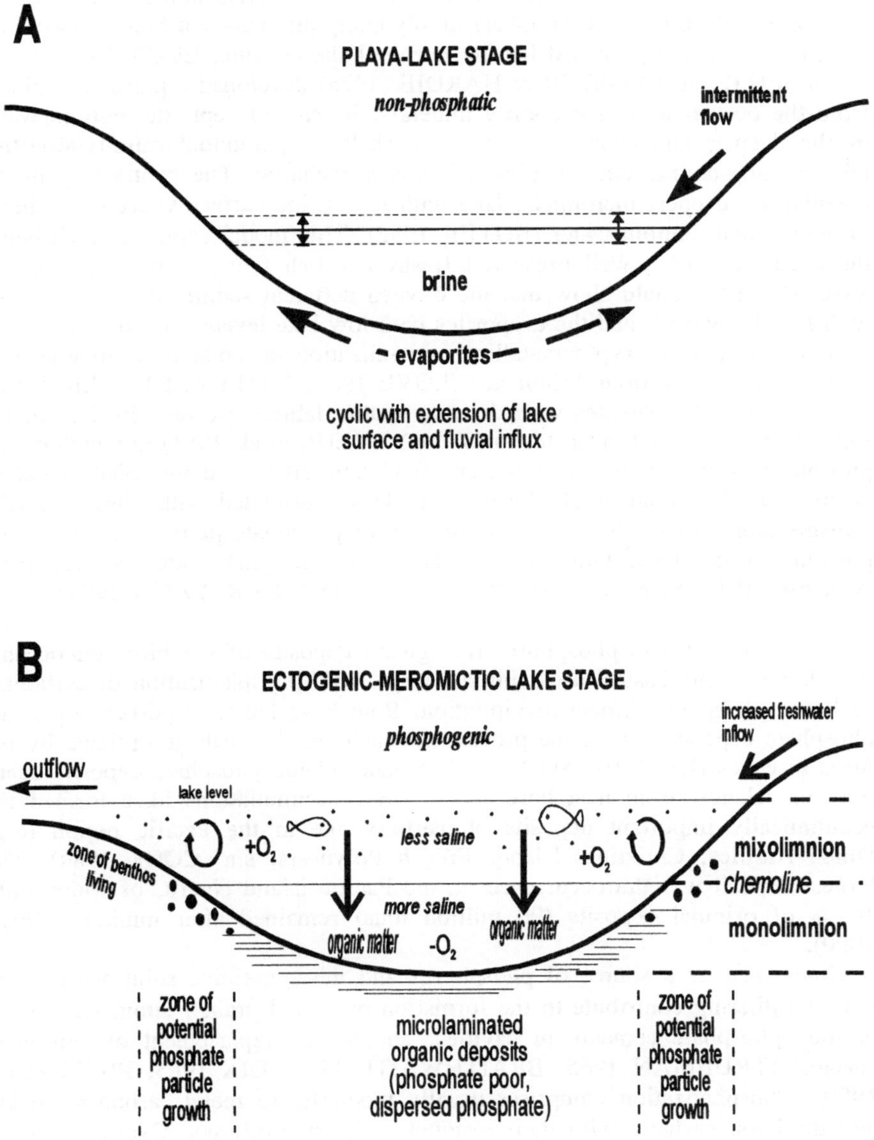

Fig. 6.2. The development of the phosphogenic facies in the lacustrine Green River Formation (Eocene, Wyoming and adjacent States) based on the playa to meromictic lake model of BOYER (1982). During lake level highstands, organic matter is deposited in the deeper portion of the lake, where oxygen deficient conditions provide a P-flux into the sediment. In areas of suboxic conditions, phosphogenesis is initiated.

The best investigated ancient phosphogenic lake system is the Eocene Green River Fm. (Wilkins Peak Member) in Wyoming and adjacent States. Two contrasting models are proposed for the Eocene Lake Gosiute. EUGSTER & SURDAM (1973) and EUGSTER & HARDIE (1975) developed a playa lake model from the occurrences of evaporite minerals. In this concept, the bottom water of the deepest and central portion of a shallow, perenntial salinity-stratified lake became anoxic due to inhibited oxygen-recharge. The contrasting model postulates a deeper, meromictic lake with less saline surface water overlying a brinely anoxic bottom water (BOYER 1982). This model explains much better the occurrence of a well preserved freshwater fish fauna in these sediments. SMOOT (1983) could show that the oxygen deficient sediments are associated with lake highstands and the evaporites with low lake levels (Fig. 6.2.). Apatite occurs as dispersed cryptocrystalline mineralization in moderately organic matter-rich rocks of various lithofacies (LOVE 1964, MOTT & DREVER 1983).

Lacustrine phosphorites with discrete apatite fabrics are described from the Pliocene Glen Ferry Fm. in Idaho (SWIRYDCZUK et al. 1981) in which ooidic phosphate grains, carbonate replacement phosphorites, and phosphate cements occur. The development of phosphate rocks is correlated with phases of lake transgression and mechanical concentration of phosphate particles in nearshore portions of the lake. Other ancient lacustrine phosphates are reported from Argentina (LEANZA et al. 1989), and China (YANG & YANG 1994).

Insular settings: The phosphorus-rich guano deposits of sea birds on oceanic islands and some coastal areas initiate large scale phosphatization of carbonate rock and phosphate cement precipitation. Both have led to important economic phosphate deposits. The same process in much smaller scale is initiated by bat feces in caves (HUTCHINSON 1950). Guano-related phosphate deposits occur only on islands or in nearshore areas. These accumulations have formed the economically important deposits of Tertiary age in the Pacific region (e.g. Dutch Antilles, Christmas Island, French Polynesia, s.a. COOK 1984). The largest deposit was the occurrence on the Pacific island Nauru, of which only 10 % of original deposits (90 million tons) remained after mining (PIPER 1990).

Guano acts as a source of phosphorus and the aggressive solutions derived from weathering contribute to the formation of coated grains, laminated crusts, various phosphate cements in cavities, and to the replacement of carbonate surfaces (TRUEMAN 1965, BRAITHWAITE 1980, DIX 1983, PIPER et al. 1990). Phosphatization comprises usually bioclastic to reefal carbonate rocks, and the karst surfaces with their residual sediment infillings. Cycles of emergence and submergence caused reworking and formation of clastic phosphate grain concentrates. These deposits are characterized by a carbonate fluorapatite, and a distinct mineral assemblage of ammonium phosphates, mostly struvite (CULLEN 1988). PIPER (1990) proposed a process of rainwater reaction and apatite precipitation in the groundwater zone for the Nauru deposit supported by evidences from the isotope and minor element compositions. The presence of an alumosilicate substrate leads to the development of Fe- and Al-phosphates

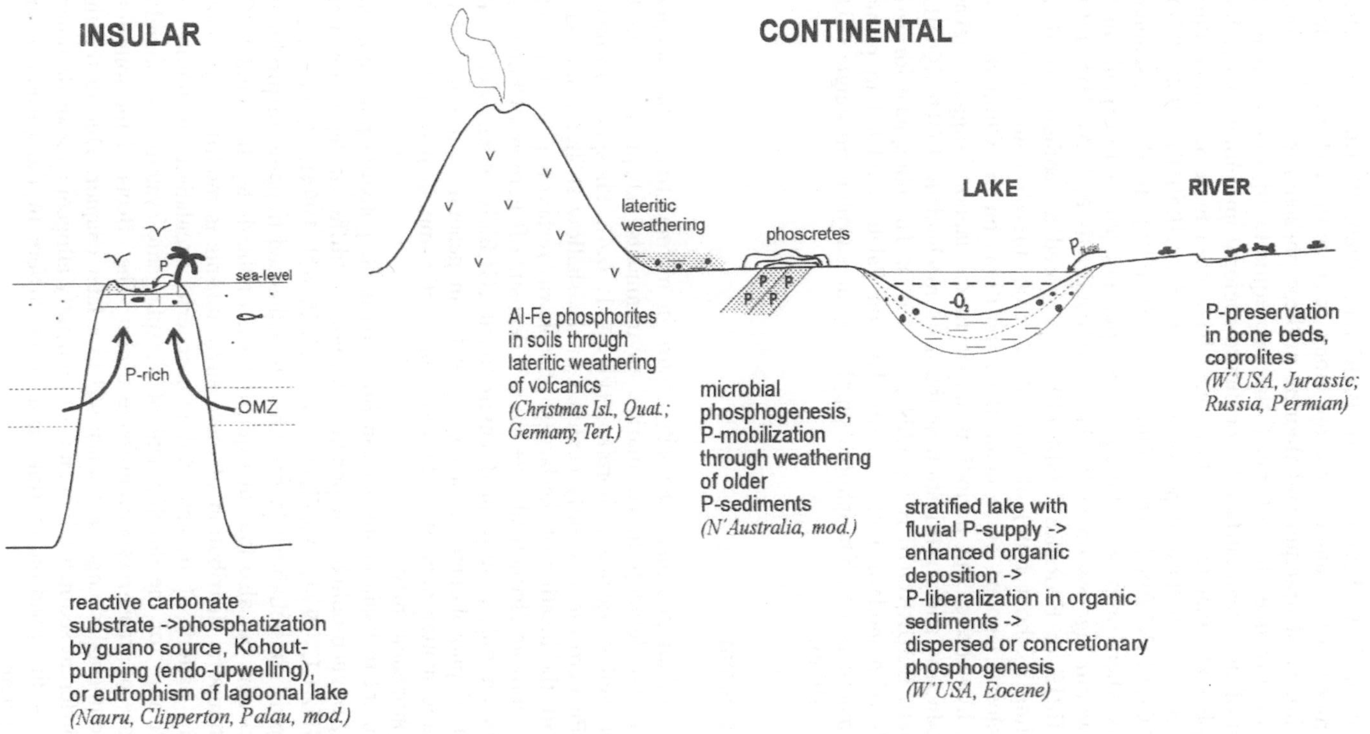

Fig. 6.3. The location of phosphogenic environments, processes and major precipitation products of the insular and continental realm including lakes. Examples in italic.

(TRUEMAN 1965, BURNETT 1980, BARRETT 1989, PIPER et al. 1990). The guano-model leaves unanswered questions, such as the lack of bird fossils or the enrichment of uranium and fluorine in these phosphorites. Most insular phosphate occurrence do not meet the prerequisite of dry climate and euthrophism of the ocean surface water due to regional upwelling today. Some authors emphasized that the conditions may have been more favorable during the Pleistocene, when these deposits were formed (BURNETT 1980, STODDART & SCOFFIN 1983, AHARON & VEEH 1984). In the consequence, other models where proposed which all relate phosphogenesis to organic matter deposition within lagoons on atoll islands. ROUGERIE & WAUTHY (1989) and ROUGERIE & FARERSTROM (1994) proposed a geothermally driven endo-upwelling (Kohout flow) of nutrients through the structure of the atoll into the enclosed lagoon. Trace element compositions on the Clipperton atoll support the hypothesis, but periodical anoxia in the lagoons suggest elevated primary production and stratification as the main mechanism (BOURROUILH-LE JAN 1980, BOURROUILH-LE JAN et al. 1985). The phosphate formation in sediments of a modern anoxic marine lake on Palau is related to primary production maintained by phosphorus supply by the surrounding jungle (BURNETT et al. 1989).

6.3.2
The marine realm

The majority of all phosphate deposits occurs in marine strata. Phosphogenic deposystems can develop in all marine environments except in the highly agitated and well oxygenated shoreface (Fig. 6.1, 6.4). Phosphate formation during the Phanerozoic is clearly centered in the shallow shelfal zone, which correlates with the location of the largest phosphorus reservoir (Fig. 1.1). This distribution contrasts distinctively the modern deposits off Peru and Chile which are found in an outer, deep shelf environment (GLENN et al. 1994). The development of phosphogenic systems correlates in general with zones of enhanced organic matter deposition to meet the first common perquisite of the 4-step phosphogeneic model.

In a continent to basin center (ocean floor) transect, potential phosphogenic, organic-rich, oxygen-deficient deposition develops in different depositional settings (WIGNALL 1994, ARTHUR & SAGEMAN 1994). A number of paleogeographic constellations favor either the enhanced trapping or production of organic matter, but also the subsequent stages of the 4-step box model. The spatial and temporal distribution of these environments is modified by eustasy and changing pattern of oceanic and atmospheric circulation. A number of supporting factors for the development of phosphogenic systems are bound to specific phases of transgressive/regressive cycles. The effects of the long-term global environmental changes are discussed in a later chapter. Due to the fundamentally different sediment dynamics - dominantly allogenic versus dominantly autogenic - the position in time and space differs in the carbonate and siliciclastic system.

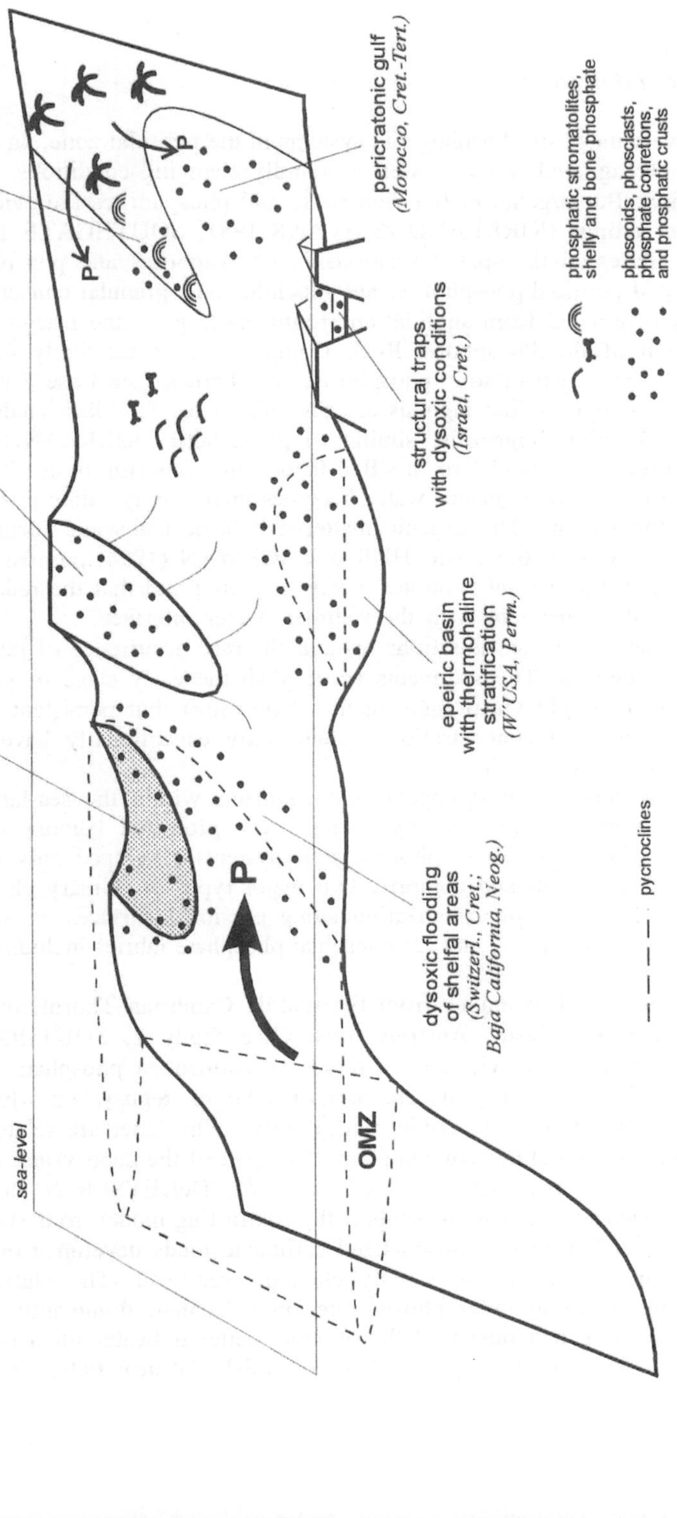

Fig. 6.4. Phosphogenic environments in the continental margin to epeiric realm. Examples for these settings in italic.

6.3.2.1
The peritidal zone

The development of phosphogenic systems in the peritidal zone, an environment of broadly agitated water as well as rapidly changing conditions, seems to be a paradox. But organic matter rich rocks and phosphorites are evidently found in these settings (KIRKLAND & EVANS 1981, SOUTHGATE 1988). HITE (1978) addressed the spatial relationship of evaporites and phosphorites. The majority of peritidal phosphorites are allochthonous, granular concentrates which initially originated from subtidal environments (e.g. in the marginal Park City Formation of the Phosphoria Rock Complex, see Case Study 4). From the peritidal zone, also pristine phosphorites are known (see Case Study 2,3).

Meso- to hypersaline lagoons are favorable sites for black shale formation, respectively phosphogenesis, similar to playa lakes. KIRKLAND & EVANS (1981) proposed a model for a silled lagoon or evaporitic basin. The influx of plankton-rich marine surface water becomes increasingly saline due to evaporation in the lagoon. The organic matter-rich brine sinks and forms an anoxic bottom water (Fig. 6.5). ARTHUR & SAGEMAN (1994) pointed out that the sediments in the coastal zone are mostly reducing and that the redox boundary is commonly coincident with the sediment/water interface.

These sediment configurations explain the rare occurrence of peritidal phosphogenic regimes. The sediments accomplish the early stage of phosphogenic pathways (P-supply via organic matter deposition), but persistent P-liberalization, storage, and concentration systems only exceptionally have chances to develop within the sediment.

Ancient peritidal phosphogenic environments within the sea-land transition zone are bound to low energy coasts, very protected lagoons or emergent carbonate platforms. These phosphogenic systems developed only in carbonate dominated sediments and comprise two major types of primary phosphatic features which are (1) phosphatization along erosional surfaces on soft to semi-durated carbonate muds, and (2) microbial phosphate fabrics including phosphate stromatolites.

Well preserved examples from the middle Cambrian Thorntonia Formation in the Georgina Basin, Australia (see Case Study 2, SOUTHGATE 1983, 1986a,b, 1988) show evidence for rapid synsedimentary phosphatization of carbonate muds. The sediments are part of a larger deposystem which includes non-carbonate granular to ooidic phosphorites. The latter are related to an organic matter-rich pelitic sedimentation. This part of the deposystem was initially attributed to the peritidal facies (COOK 1972, DeKEYSER & COOK 1972). But microfacies investigations support the contrasting model from SOUTHGATE & SHERGOLD (1991). Phosphatized carbonate muds developed on a peritidal platform in the context of rapidly changing sea-level. The platform margin facies consists of granular phosphorites in a basinal, dominantly pelitic host sediment. The geochemistry of the organic matter indicates intensive supply of marine plankton to these peritidal to intertidal platform flats. SANDSTORM

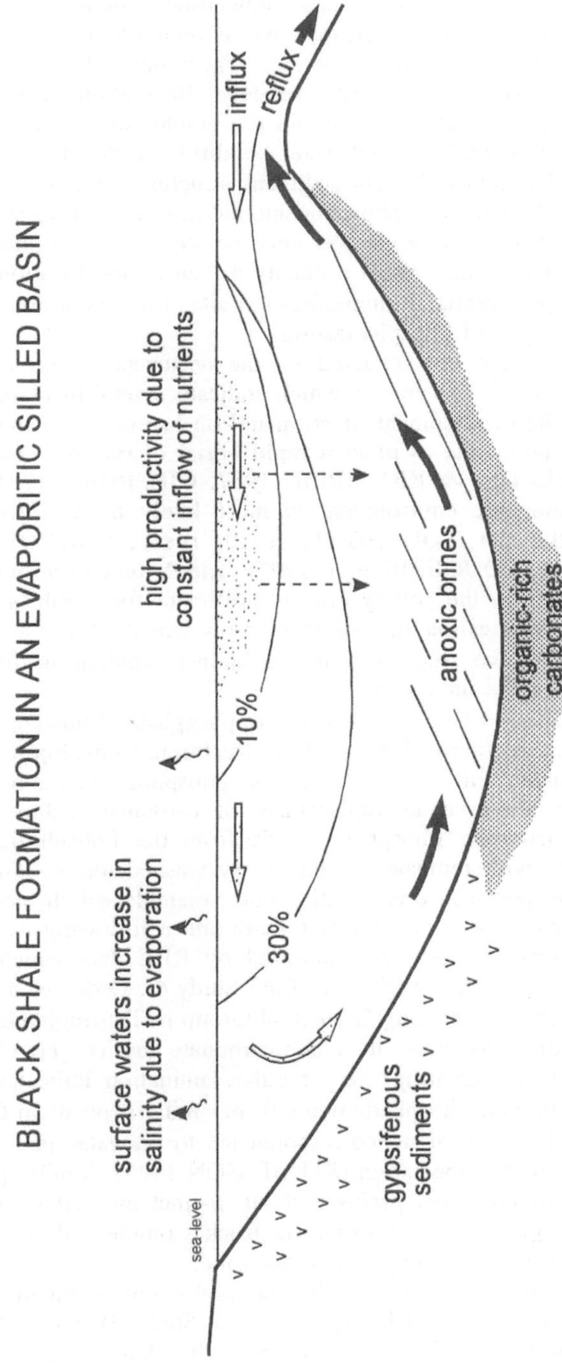

Fig. 6.5. Model for the formation of organic-rich sediments in a silled basin or mesosaline lagoon by KIRKLAND & EVANS (1981). From WIGNALL (1994).

(1986) proposed a continues P-recharge via an upwelling system. Phosphatization in the carbonate mud dominated peritidal flats is bound to intervals during which erosion exceeds carbonate deposition rates. These intervals are bound to the earliest transgressive stages of 5- order sea-level fluctuation. Sedimentary structures, multiple hardground stacking and biostratigraphic data indicate short-termed formation with temporal agitated water conditions rather than a hiatus derived condensation. By evidence from sediment structures, favorable conditions for the initiation of a complete phosphogenic pathway leading to carbonate mud replacement must have been developed very rapidly. The phosphatization zone follows the erosional surface, most probably driven by the location of the redox boundary as the preferential P-mineralization site. Enigmatic remains the P-budget for this rapid type of phosphatization.

The intertidal zone is also characterized by the occurrences of microbial/stromatolitic phosphate fabrics, some of which indicate synsedimentary phosphate mineralization. The development of stromatolitic layers is also bound to a relative low sedimentation rate to prevent rapid burial of the microbial mats (KRUMBEIN 1983, GERDES & KRUMBEIN 1987, GINSBURG 1991). The formation of a phosphogenic environment is most likely derived from the microenvironment in the mat (SOUTHGATE 1980, 1986c, EGANOV 1988, KRAJEWSKI et al. 1991, ROUGERIE et al. 1997). The P-contents accumulated in the organic matter within the colony are not sufficient for a substantial or complete phosphate mineralization or carbonate replacement of the structure. The necessary external source arises a similar P-budget problem as discussed above. Modern examples are unknown.

More specifically related to this environment are phosphatic ooids and coated grains. Continuously and regularily laminated or structureless envelopes require frequent motion by current or wave action. The phosphogenic pathways of direct microbial precipitation versus replacement of carbonate sediments are uncertain in most occurrences. Phosphatic ooids from the Phosphoria Rock Complex are phosphatic with remnants of the partly washed matrix. Granular cement in the remaining pore space is calcitic. This assemblage indicates more an early replacement process of the primary texture and a following carbonate cementation. This microfacies data are supported by REE data which show similar patterns to other phosphate fabrics (see Case Study 4). Ordovician oo/oncolitic fabrics from the iron ore hosting Gräfenthal Group in Thuringia/Germany are alternating chamositic phosphatic in a non-carbonate matrix (TRAPPE & ELLENBERG 1994). The alternating, very regular lamination indicates more a primary apatite precipitation. Phosphatic ooids from similar deposits in Estonia show a REE pattern which reveal no correspondence to seawater pattern and indicates more likely microbial mediation (STURESSON 1995). Similar processes have to be applied to nucleated grains without distinct internal lamination of the phosphatic envelopes. In the Phosphoria Rock Complex, these fabrics are also related to the peritidal to upper subtidal zone.

Precipitation of phosphate aggregates in the enclosed environment of moldic bioclasts is also found in the peritidal zone (see Case Study 4). Characteristic for the nearshore environments is an association with glauconite, in single fabrics or within the same bioclast. This mineral association originates from the continental influx of Fe, Al and Mg in the coastal waters. The conditions for glauconite precipitation are close to sedimentary apatite (ODIN & LETOILLE 1980, ODIN & MATTER 1981, SEIBERTZ 1992).

Thermohaline stratification and fluvial supply of phosphorus should generate potential phosphogenic environments in estuarine settings, and was proposed by BUSHINSKY (1935, 1964), PEVEAR (1967), and CULLEN et al. (1990). Evidence for precipitation of phosphate fabrics above the dispersed apatite mineralization of RUTTENBERG & BERNER (1993) was not found.

Endo-upwelling in lagoons of oceanic atolls causing enhanced primary bioproductivity and the formation of phosphogenic environments was discussed in chapt. 6.3.1.

6.3.2.2
Subtidal shelfal or epicontinental environments

The subtidal shallow marine sediments of the continental shelves cover only 5 % of the todays Earth's surface, respectively 7% of the marine environments. According to the data from MACKENZIE et al. (1993)(see also chapt. 1), these environments host roughly 50 % of the total phosphorus in the biosphere, and concentrate also 20% of the marine primary bioproduction in this region. Higher eustatic sea-level stages during most Phanerozoic periods affected a significant increase of the surface covered by these productive shallow marine waters (VAIL et al. 1977). Most ancient deposits evidently originate from this zone, very much contrasting the modern outer shelf deposits.

The first prerequisite for the formation of phosphogenic environments in the box model is the ***deposition of organic matter***, thus encircling the formation of phosphogenic environments to areas of organic matter deposition. High net organic sedimentation may be derived by various processes, e.g. by increased bioproductivity, reduced background sedimentation, or by enhanced preservation conditions. The modern shelfal and epeiric marine environments identify two major configurations as favorable for these conditions, which are the deep enclosed basins, e.g. Black Sea, and the zones of trade wind upwelling along the west coasts of the continents. In the latter configuration, phosphorites are forming today near to the outer shelf margin. But the occurrence of modern phosphorites off E'Australia proves that elevated primary bioproduction is not absolutely necessary if the following prerequisites of the box model are accomplished. The absence of phosphorites in many black shales and modern organic-rich sediments emphasizes the importance of subsequent prerequisite which are liberalization of phosphorus during coinciding removal from biocoupling. These complex conditions are bound to exceptional deposystem developments.

The modern occurrence and even more the investigation of ancient systems raises the controversial question whether primary production or the formation of favorable conditions, e.g. through thermohaline stratification, is responsible for initiation of phosphogenic pathways.

The occurrence of organic-rich sediments and black shales in non-actualistic deposystem configurations (e.g. the Cretaceous Western Interior Seaway, USA, or the Jurassic epeiric black shales in Central Europe) and the formation of phosphorites in condensed deposits, where the time factor limits the modern approach, draws the attention to geologic settings of phosphogenesis which have to be explained by additional geologic models. These are intervals of evidently wider and more frequent occurrence of oxygen-deficient water bodies in shallow marine areas (WRIGHT et al. 1984, 1987, WIGNALL 1994, WIGNALL & HALLAM 1992). The climax of these phenomenons are oceanic anoxic events (OAEs). It is important to note, that OAEs are not directly associated with enhanced phosphogenesis (see chapt. 11), but less intense stages of regional or global oceanic oxygen depletion have to be considered. Extensive epeiric organic-rich deposition during the Jurassic, which was not related to OAEs, requires the development of new non-actualistic models (TYSON et al. 1979, OSCHMANN 1988, 1990, WIGNALL 1991). These models are dominantly based on various thermohaline stratification effects. Modifications of these models in respect to less intense oxygen-depletion, which allows more rapid P-liberalization, have to be applied to epeiric phosphate basins. The depositional pathways in these approaches are totally different from P-burial and enrichment via organic deposition on omission or condensation surfaces. In the latter, time and sediment starvation are the major factors.

All potential models come down to three causal model types, which are enhanced primary productivity models, stratified environment models, and condensation models.

Productivity models: The upwelling model versus the geological problems

The interpretation of subtidal phosphogenic systems in the past emphasized the upwelling productivity model. The upwelling phenomenon explains the supply of large quantities of phosphorus as well as the formation of a suitable environment. Further evidence for the model is the occurrence of modern phosphorites off Peru within such a setting (see chapt. 5.3).

The history of the upwelling model in phosphate research was outlined earlier in chapt. 5.2. Upwelling in oceanographic sense is based on the theory of EKMAN (1905) on the relationship of wind and currents. Coastal upwelling requires a slope and main winds in seaward direction. These winds induce a surface current which drives surface water toward the open ocean. The water deficit is compensated by an undercurrent of deeper water. This water body is usually colder, relative poorly oxygenated, less saline, but rich in nutrients. It is important to mention that upwelled water in modern oceans is not deep oceanic water. The water masses arrive from depths of 50-300 m with a maximum of 350 m in patchy areas. The upwelling zones, e.g. off Peru, are about 7-50 km wide and several hundreds kilometers long. The intensity of upwelling varies with the seasonal climatic changes (FRAZIER 1981, BATURIN 1982, PICKARD & EMERY 1990, POND & PICKARD 1991). Zones of permanent seaward winddrift are the trade wind belts, where atmospheric circulation

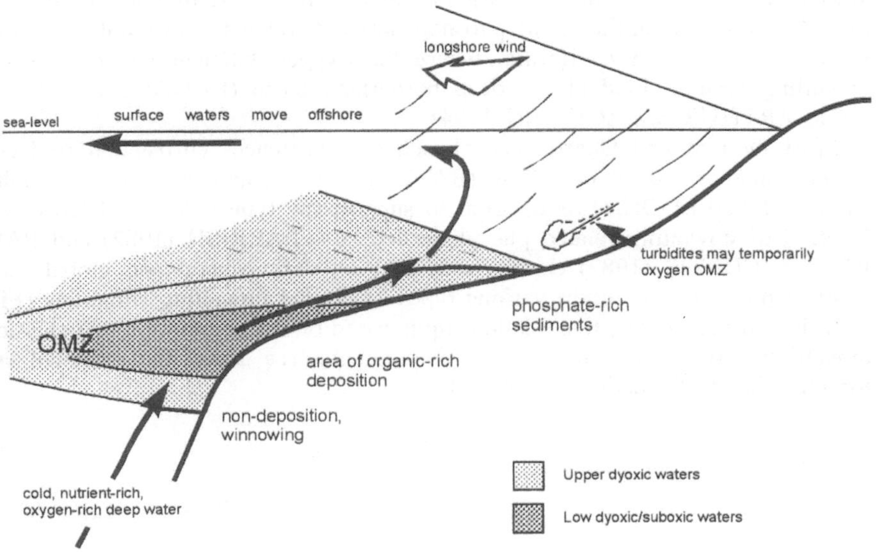

Fig. 6.6. Generalized configuration of bioproductivity, water masses and sedimentation on the outer shelf off Peru and Chile. Phosphorites are dominantly forming at the roof of the oxygen minimum zone (OMZ). Variations in detrital sedimentation rates cause different phosphate facies related to rapid burial or condensation. Based on WIGNALL (1994).

generates winds in direction to the equator. The Coriolis force drives the winds into westerly orientation. The relationship of upwelling intensity and broadness of the shelf is contradictory. The trade winds occur with seasonal variations in zones between 5°- 30° on the northern and southern hemisphere. Modern upwelling is a phenomenon of the outer shelf to upper slope. In shallow water less than 10 to 20 m depending on the water energy, the EKMAN stratification collapses due to wave mixing.

The supply of nutrients to the surface water increases the primary bioproduction, and consequently also the organic matter sink. The high amount of organic matter sink causes oxygen depletion in water depths of 100-500 m, where organic matter decomposition rates reach a maximum (REIMERS & SUESS 1981). As a consequence, a large oxygen-poor water body is forming within this oxygen minimum zone (OMZ). The contact of this water body with the seafloor at the upper slope and outer shelf generates favorable conditions for organic deposition (Fig. 6.6). In the modern system off Peru, the central portion of the OMZ is characterized by dysaerobic to suboxic conditions. Organic matter deposition exceeds decomposition rate. Below this dysaerobic mass, sediment starvation and current systems cause non-deposition. In the roof of the OMZ water body, oxygenation stage tends to less dysaerobic conditions. This allows partial degradation of organic matter in the sediment, phosphorus liberalization, and initiation of phoshogenesis. A related configuration is represented in Southern California, where the oxygen-deficient waters from an upwelling system flood also a deep borderland basin (RHOADS & MORSE 1971, ARTHUR & SAGEMAN 1994).

Most modern and Quarternary phosphate occurrences correspond perfectly to the zones of coastal upwelling (BATURIN 1982), and the paleogeographic studies of SHELDON (1964b) seem to support the typical low-mid latitudinal location of upwelling related phosphate deposits. PARRISH (1982) and PARRISH & CURTIS (1982) could show, that on very general and global scale many deposits occur within various types of coastal upwelling situations (Fig. 6.7). But in most cases, this relationship is based on the assumption that classical upwelling systems operate also in epicontinental sea areas, the principal site for phosphorite formation in the past.

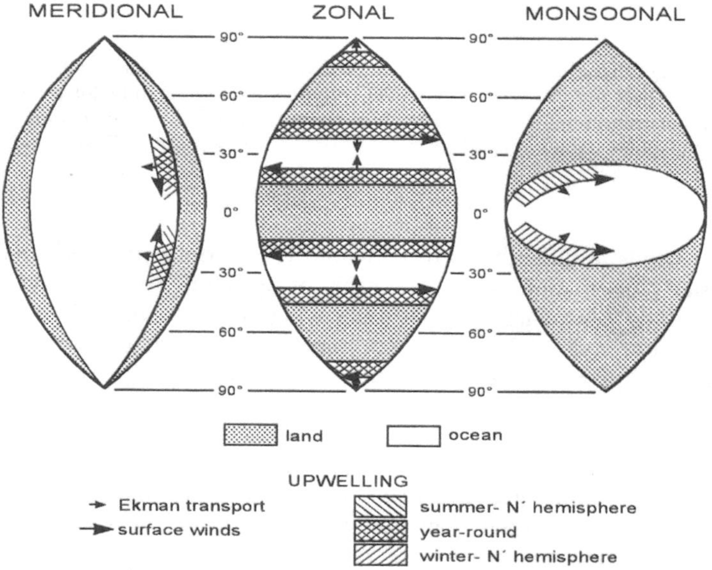

Fig. 6.7. Schematic diagram of various types of coastal upwelling situations. Meridional (trade wind) upwelling is related to the subtropic high pressure cells. Zonal upwelling is generated by zonal winds, and monsoonal upwelling is driven by cross-equatorial winds during the monsoons (Modified from J.T. PARRSIH © 1982, reprinted by permission of the American Association of Petroleum Geologists).

Forms of upwelling are multifarious and upwelling currents occur also in epeiric seas (FRAZIER 1981, PICKARD & EMERY 1990, POND & PICKARD 1991, HAY & BROCK 1992). But it remains questionable whether these potential upwelling situations in epicontinental seas can supply coastal areas with waters from a significantly different water body. Upwelling currents can be created on all slopes in moderate water depths (>2 m) with friction forced surface circulation, thus in shallow basins as postulated for many occurrences. But the effects on the nutrient supply by this upwelling currents are only hypothetical. It is uncertain if a deeper water reservoir even in a stratified shallow epeiric basin can provide a significant P-flux, and if the intensity of these upwelling currents can serve as a sufficient P-pump. The P-supply through these currents depends on the size of the potential reservoir. In epicontinental basins or marginal sea areas, the upwelled water masses have to be replaced by downwelled water in other places. The budget of these marginal basins depends very much on the river supply and ocean current water supply, but data are sparse. The differences of nutrient contents of deeper water and surfaces water should also be minor in most seas due to frequent mixing. Only where stratification is intense and stable, an effect is expected (see below). BRASS et al. (1982) postulated that during non-glacial Phanerozoic times the extent of upwelling derived OMZs may have been larger than in the modern configuration.

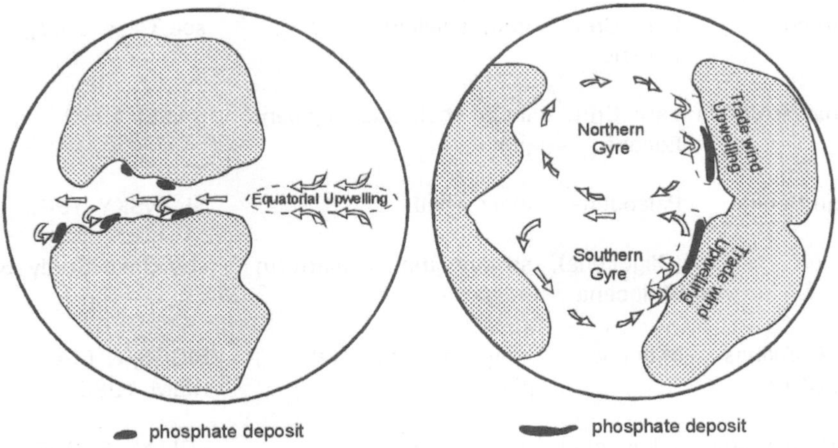

Fig. 6.8. The model of major phosphate occurrences in relation to meridional (trade wind) upwelling or equarorial upwelling by SHELDON (1980).

Table 6.1. Paleogeographic positions of major and well investigated Phanerozoic phosphate deposits

Location	Age	Depositional environment	Reference
Kazakhstan	Cambrian	gulf, shallow	EGANOV et al.1986
China (Yangzte Platform	Cambrian	inner to mid shelf	SIEGMUND 1994
N'Australia	Cambrian	pertitidal flats to mid-shelf	see Case Study 2
Estonia	Ordovizian	pertitidal of epeiric gulf	see Case Study 1
W'USA	Permian	open, silled gulf, shallow	see Case Study 4
W'Russia	Late Cret.	epeiric sea, carb. platforms	ILLYIN 1994
France	Late Cret.	epeiric sea	JARVIS 1994
Israel/Jordan	Late Cret.	inner shelfal platform with structural depressions	BENTOR 1953
Egypt	Late Cret.	inner shelf to deltaic	GERMANN et al. 1987,GLENN 1990a
Morocco	Late Cret. - Eocene	gulf, shallow	see Case Study 6
Tunesia/Algeria	Late Cret. - Eocene	inner shelf and adjacent gulf	SASSI 1980
Togo/Benin	Paleogene	inner shelf	SLANSKY 1986
Florida	(Oligocene), Miocene	shallow marine platform	see Case Study 5
off Carolinas (SE-USA)	Miocene	inner to mid-shelfal platform	RIGGS & MAN-HEIM 1988
California, Baja California	(Oligocene), Miocene	outer shelf	FÖLLMI & GAR-RISON 1991, SCHWENNICKE 1994

The numerous Mesozoic/Cenozoic phosphorite deposits from the Tethyian realm were formed along generally W-E directed coasts in low latitudes. McKELVEY (1967) combined EKAMN's theory of coastal upwelling with dynamic upwelling by westerly currents to address these low latitude occurrences. SHELDON (1980) reintroduced the idea of HUTCHINSON (1950) of dynamic upwelling in the Pacific to find support for the equatorial upwelling model (Fig. 6.8). Dynamic upwelling as occurring along the Atlantic coast of SE-USA is current derived upwelling at any kind of submarine rise or slope (RIGGS 1984). RIGGS & SHELDON (1990) in their global model for the Neogene tried to link the global climatic configuration with these types of upwelling.

The upwelling model in all variations very elegantly explains phosphogenesis in the theoretical approach for the young and modern systems. The advantage of that model is based on the number of requirements for phosphogenesis such as P-supply, stratification, organic deposition, modern equivalents, which are all covered by a single phenomenon. Several authors (summarized in GLENN et al. 1994) noticed that the paleogeography and the facies of ancient deposits poorly correspond to the modern analogs and their upwelling setting. Most of them are found in very shallow inner shelfal settings, gulfs or shallow epicontinental basins (Tabl. 6.1). Also modern phosphorites occur in both, upwelling environments and in non-upwelling settings, such as the E-Australian continental margin (RIGGS et al. 1989).

So far known, none of the major ancient phosphate deposits occur in a paleotectonic position similar to the modern upwelling related occurrences on the continental margin off Peru. GLENN et al. (1994) tried to draw the attention to the similarities of the modern Peru margin, upwelling derived phosphorites and the ancient deposits, but the analogies are mostly facies and general process related.

Stratification models: silled basins, anoxic puddles, and transgressive oxygen-deficiency

In contrast to phosphate deposits, a number of different models have been developed for the occurrence of black shales in non-upwelling settings. None of these concepts have been applied to phosphorite basins so far, although similarities to phosphate basins are striking. The major difference to these organic matter preservation models in their application to phosphate basins is the necessarly higher degree of organic matter degradation to liberalize sufficient amounts of phosphorus for the initiation of apatite formation. All these models have in common to operate with thermohaline stratification. Intense organic deposition is achieved by the suitable preservation environment instead of increased primary bioproductivity.

Dysoxia to anoxia develop in some modern silled basins, e.g. dysoxia in parts of the Baltic Sea in moderate depth or anoxia in the deep Black Sea basin (DEMAISON & MOORE 1980, GLENN & ARTHUR 1985). Stratification in both cases is achieved by a less saline upper water body and a higher saline lower water reservoir which is supplied by an underflow of normal saline marine water. Phosphorites are not forming today in these examples, but these scenarios with slightly modified conditions and a suitable position of the redox boundary have to be discussed as potential phosphogenic sites to explain the ancient, non-actualistic epeiric configurations. Most epeiric phosphate basins, however, are situated in warm and arid to tropical climates and are associated with evaporites and red beds in their marginal portions, e.g. the mid-Permian Phosphoria Sea or the late Cretaceous to Paleogene Mediterranean occurrences (SHELDON 1964b, HITE 1978). These configurations contrast the modern examples which are situated in moderate humid zones. Critiques on the application of these models fo the explanation of large scale epeiric oxygen-deficiencies were expressed by TYSON & PEASON (1991), who argued that

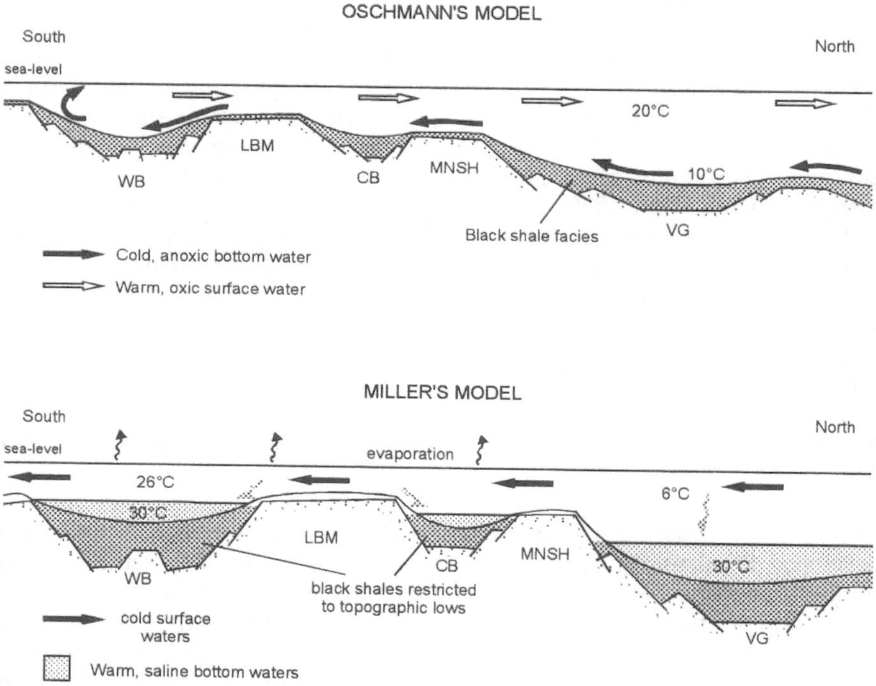

Fig. 6.9. Two contrasting stratification models for the Jurassic North Atlantic water passages and the deposition of the organic matter-rich Kimmeridge Clay. MILLER's model proposes stratification by "warm, saline bottom water" formation. The OSCHMANN model postulates an undercurrent of cold and poorly oxygenated polar water. See also text for discussion. WB = Wessex Basin, LBM = London-Brabant Massif, CB = Central Graben, MNSH = Mid North Sea High, VG = Viking Graben (from WIGNALL 1994).

"estuarine" salinity stratification is not stable in large basins, that ancient examples evidently formed in more shallow basins, and that the temporal pattern of elevated freshwater influx contrasts the dominantly transgressive position of oxygen-deficiency in the ancient occurrences.

A number of other non-actualistic models address the paleogeographic situation of ancient oxygen-deficient sediments more precisely. They consider two different phenomenons as responsible: (1) rising chemoclines during OAEs or (2) paleogeographic configurations which induce temperature or higher saline water stratification. Rising chemoclines in sluggish oceans in combination with a drastically rising sea-level theoretically produce an oxygen-deficient flooding of continental areas (SCHLANGER & JENKYNS 1976). This model is closely ralated to productivity models. The expansion of the OMZs was initially attributed to the extension of the shallow marine areas and the nutrient recycling from the continental areas, or later to the formation of "warm saline bottom water" by increased low latitude evaporation, finally resulting in more intense

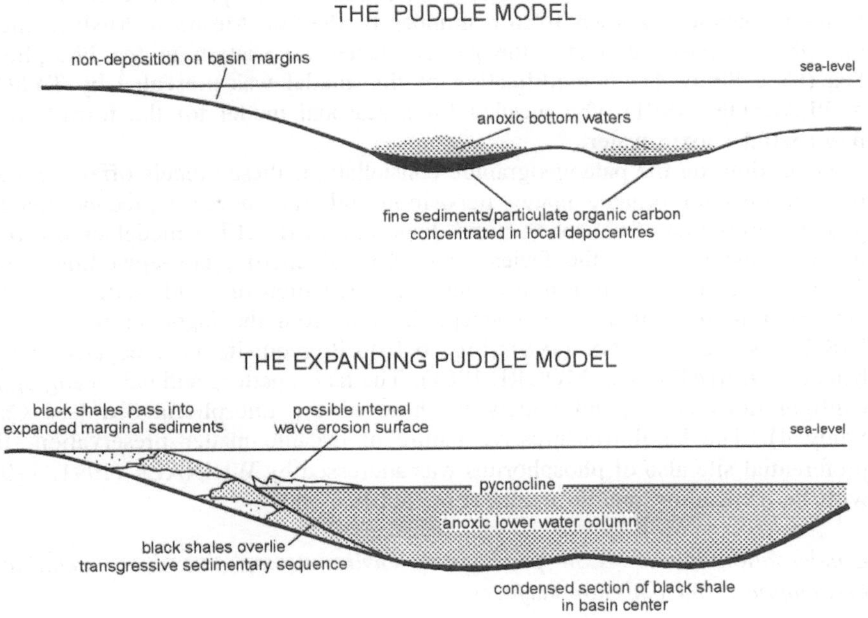

Fig. 6.10. WIGNALL's puddle model. During lowstand and early transgressive stage poorly oxygenated marine bottom water is restricted to submarine depressions, which are characterized by black shale deposition. During the following transgressive flooding phase, this water body spreads over most of the basin and establishes extended black shale deposition. Following regression renewed dominantly detrital sedimentation (From WIGNALL 1994).

upwelling and elevated productivity (SOUTHAM et al. 1982, BRASS et al. 1982). These models are widely accepted, though earning also critiques (TYSON & PEARSON 1991). A temporal relationship of intervals of large scale phosphate deposition and OAEs is absent (see also discussion in chapt. 10), although being striking for black shale formation.

Non-actualistic stratification models have been developed to explain organic-rich sedimentation of large extent in the Jurassic (Fig. 6.9, 6.10). So far, these concepts have not been introduced to phosphogenic systems although paleogeographic constellations are comparable for some phosphate basins with a dominance of the black shale-phosphorite lithofacies association. These models combine silled basin concepts with thermohaline stratification mechanisms and a current system within the Kimmeridgian North Atlantic water passage. OSCHMANN (1988) deployed an anoxic undercurrent of cold polar water for the flooding of subbasins, but the model cannot explain why the polar water is oxygen-depleted. A contrasting model was proposed by MILLER (1990) on basis of ideas from TYSON et al. (1979) in which warm saline bottom waters in depositional depressions are the cause of the oxygen-deficiency. These "puddles" of unmixed bottom water become anoxic by organic matter degradation and suppressed O_2-recharge. Evaporation and anoxic puddles correspond to facies associations known from a number of the late Mesozoic Mediterranean phosphate deposits and also the general facies association in the Phosphoria Sea (Case Study 4). A modification of this model was presented by TYSON & PEARSON (1991) who pleaded for a seasonal model for the formation of mid-shelfal anoxic water.

Depending on the paleogeographic constellation, these models offer contrasting concepts for organic matter deposition and trapping as a precondition for potential initiation of phosphogenesis. Especially, MILLER's model shows striking correspondence to the facies association of epeiric, non-upwelling areas. Similar facies patterns in a more land locked situation or a gulf setting are also generated in the silled basin concept derived from the lagoonal model from KIRKLAND & EVANS (1981)(Fig. 6.16). Prerequisite is a negative water balance (DEMAISON & MOORE 1981). The facies pattern and paleogeographic configurations correspond widely to those of the Phosphoria Sea (see Case Study 4). The basal transgressive nature of organic matter preservation, the preferential site also of phosphorites was addressed by WIGNALL (1991, 1994) with his expanding puddle model (Fig. 6.10).

Condensation models: Low productivity environments, starved sedimentation, Fe-pumping, and omission surfaces

Another mechanism for the accumulation of phosphorus from organic matter sources is not coupled to favorable environments or elevated productivity, but operates with time and intense shuttling processes. Many phosphorites occur in spatial association with hardgrounds, stratigraphic hiatus contacts, and stratigraphic condensation horizons (see chapt. 9). The origin of these beds or surfaces is related to low and ultra-low detrital or carbonate sedimentation rates, but a persistent rain of organic matter to various degrees from the overlying water body. These evident conditions from field observations, in the past, stand in contrast to an enigmatic mechanism of PO_4^{3-} concentration. The introduction of the Fe-pumping and shuttling model by SCHAFFER (1986) offered a concept for the understanding of these processes. Similar to most phosphogenic environments, a further precondition is the formation of a redox boundary near the sediment/water interface. FÖLLMI (1989a,b, 1990, 1996) investigated in detail the phosphogenic processes in condensation horizons. The sedimentary record supports three different mechanisms:

- Fe-shuttling of liberated phosphate from degradation of organic matter at the surface into the sediment. Important is a sediment mixing by burrowing organisms.
- rapid, catastrophic burial of organic mats by sand sheets.
- current-induced convection-diffusion processes which transfer dissolved phosphate from bottom water into the interstitial waters.

The Fe-shuttling process is evidently a mechanism for phosphate concentration in low-productivity areas (HEGGIE et al. 1990, O'BRIEN et al. 1990).

Further development of the phosphogenic pathways ***(P-liberalization, P-concentration and apatite precipitation)***, respectively the specific mechanisms to accomplish the subsequent stages of the box model, are controlled by the characteristics of the various environments in the shallow subtidal realm and the sediment dynamics of the substrate - carbonate versus siliciclastics:

Phosphogenic pathways in subtidal carbonate systems

Carbonate deposystems of the shallow subtidal are characterized by benthic biological or microbial carbonate production. In oxygenated environments, enhanced primary bioproductivity is directly linked to elevated carbonate production, dilution of the organic matter deposition, and phosphorus concentration. Truly planktonic carbonates are restricted to the pelagic realm (see chapt. 6.3.4). The correlation between benthic carbonate production and oxygenation of the seawater usually causes rapid decomposition of the organic matter and recycling into bio-coupling. If accommodation space is available, potential high sedimentation rates counteract processes of P-enrichment in the sediment. Consequently, the subsequent prerequisites of the box model, phosphate storage and concentration, have to be restricted to abnormal sedimentation stages or specific microenvironments. Stratification and bottom water poisoning through oxygen depletion lead to a shut down of the so far dominating carbonate factory and subsequently to a shift from carbonate to non-carbonate deposystems (see below). Potential phosphogenic pathways in carbonate systems have to include interruption or extreme reduction in the productivity of the carbonate factory. These stages open way for initiation of condensation related P-concentration and storage when the redox boundary is becoming situated close to the sediment surface and Fe-shuttling is initiated (see chapt. 5). After sufficient P-enrichment, phosphogenesis of carbonate sediment surfaces or relict grains is generated. Additionally, low sedimentation rates create an environment for the growth of microbial colonies, which evidently become phosphatized if the necessary supply of phosphorusis available (see chapt. 5.5.1).

Subtidal shelfal or marginal basin environments of carbonate dominated systems are best described by models for ramps or for carbonate platforms (READ 1982, 1985, BURCHETTE & WRIGHT 1992). In all cases, carbonate deposition in the shallow subtidal is dominated by allochemical sedimentation or is related to organogenic buildups. Primary sedimentation rates are strictly controlled by ecologic factors affecting the carbonate factory (e.g. water temperature, salinity, clarity, etc.). The sediment dynamics, which cause the formation of potential phosphogenic low-deposition intervals by erosion, are very much driven by the water depth of the specific environment, inclination of the ramp or platform margin, and distinctively by sediment remobilization due to base level drop during eustatic sea-level fall (SARG 1988);(see also chapt. 9.3). Reduced carbonate production is further coupled to drowning stages, when the carbonate ecosystem is disturbed by relative rapid sea-level rise or siliciclastic poisoning. Important for the development of a suitable environment for potential P-enrichment is further a hinterland which supplies only a minimum of detrital sediment and would not be able to compensate the effect of reduced carbonate deposition.

The most important phenomenon is the formation of phosphatic hardgrounds due to erosion or drowning. Phosphatization of erosional surfaces is an abundantly described phenomenon (e. g. JARVIS 1980, 1992, PEDLEY & BENNETT 1985, ILYIN 1994). Erosional surfaces and subsequent condensation in carbonate depositional systems dominantly occur on platforms and in shelfal settings during sea-level drop. The soft sediment body is removed by increasing water energy until the lithified lower level is exposed. New faunal assemblages get established on these hardgrounds. Bioerosion and karstification cause new mechanisms of rock destruction - a hardground is formed (BROMLEY 1975, FÜRSICH 1979). Sediment starvation as a major process of hardground formation in siliciclastic systems occurs in carbonate systems only during rapid platform drowning (FÖLLMI et al. 1994). Micritic rocks show a dark phosphatization zone along the irregular surface. The grade of phosphatization decreases with depth of the penetration zone. Erosional surfaces (omission surfaces) and condensed sediments evidently represent a sufficient time interval to accomplish a P-concentration to reach supersaturation for apatite or a precursor. But the conditions to maintain a long lasting prevention from P-recycling into bio-coupling remain widely uncertain. Organic matter breakdown on the surface continuously adds phosphorus to the porewater systems in the micritic rock. Phosphatization occurs as calcium fluor apatite precipitation in the interparticle space or as true replacement of carbonate. Thermal and saline stratification, O_2-depletion through organic sedimentation and decomposition, and especially Fe-pumping provide potential mechanisms, but the process itself is not clear.

This type of carbonate replacement was also found on the surface of carbonate intraclasts e.g. in the Park City facies of the Phosphoria Rock Complex (Case Study 4), or Hawthorn Group Florida (Case Study 6). It is beyond stratigraphic

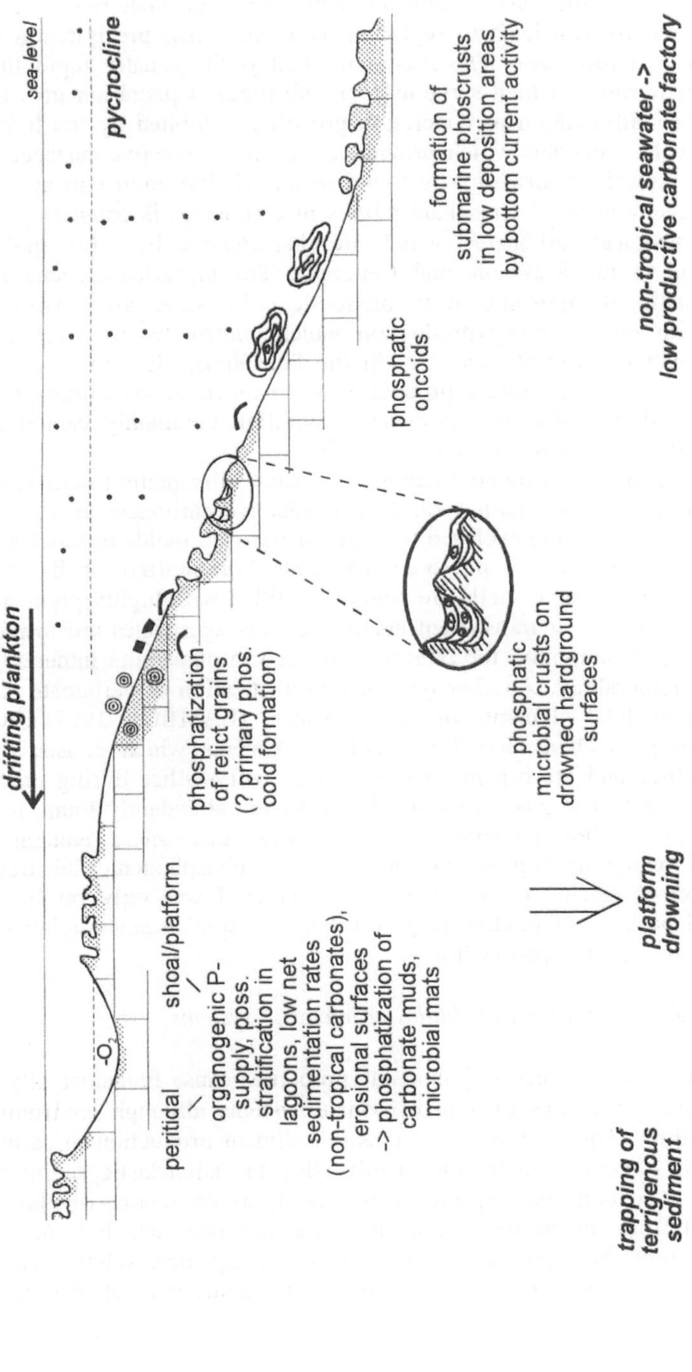

Fig. 6.11. Summary sketch of subtidal phosphogenic environments, processes and control factors in carbonate deposystems.

resolution in these deposits to prove if the occurrences of these grains as part of the allochem spectrum are exclusively derived from erosional surfaces.

In contrast to siliciclastic sediments, authigenic phosphate fabrics in carbonate deposystems are restricted to replacement mechanisms, precipitation in molds, and microbial processes. The reason for that is the usually rapid lithification of carbonate muds, which serve as rock substrates. Concretion growth may be slower than lithification or concretion growth is inhibited by the large number of small calcite crystals which provide an enormous reactive surface. The maximum of phosphate formation is found in mid-shelfal environments.

The development of phosphate fabrics in carbonates is commonly associated with non-tropical carbonates, which are characterized by a bryomol biofacies association in the Mesozoic and Cenozoic. The biofacies association assigns the depositional environment to moderate cold water areas with generally reduced carbonate factory production which contributes to a relative higher organic matter sedimentation, e.g. in the Phosphoria Sea. In tropical shallow subtidal areas, the carbonate production is so much overwhelming that due to the high sedimentation rate porewater condition commonly cannot reach the required P-concentration near the surface.

Phosphogenesis during continuous carbonate sedimentation rates is dominantly intraclast- and bioclast-hosted. In a productive carbonate factory with high oxygenation stages, the enclosed, organic matter-rich molds in bioclasts provide a potential phosphogenic micro-environment. Foraminifera shells, zooecia of bryozoans, and bivalve shells are commonly filled with highly pigmented phosphatized sediment, or transparent apatite cements aggregates are formed within the skeleton. Commonly, the fabrics occur also in glauconite mineralogy or are of mixed mineralogy. Another type of phosphatization of carbonate particles is bound to skeletal elements of echinodermata (LAMBOY 1987a). Only this faunal group is characterized by skeletal elements which consist of a very regular lattice and are organized in a way to recrystallize during diagenesis to a single crystal. The gussets within the lattice are abundantly found to be filled with cryptocrystalline phosphate. The small interlattice space reducing the fluid flow and remaining organic matter provide a phosphogenic substrate. Phosphatization of calcitic or aragonitic shell material was only rarely observed. The solid package of relative large crystals resulting in small relative reactive surface prevents phosphatization.

Phosphogenic pathways in subtidal siliciclastic systems

Other sediment dynamics and substrate properties cause fundamentally different phosphogenic pathways than in carbonate systems although environments are corresponding. The dominantly autogenic sediment production in carbonate environments is contrasted by the mainly allogenic siliciclastic sediment supply in clastic environments. Organic matter supply as the source of phosphorus is uncoupled from the benthic sediment production rate, which is mostly determined by hinterland paleogeography. As a consequence, relative elevated organic deposition is mainly deployed by the allogenic control of sediment flux and accommodation space on the shelf or within the basins. In the framework of the above discussed models of organic matter deposition, subsidence and paleogeographic configuration are becoming more important for organic matter trapping, the first prerequisite of the box model. In normal shelf profiles or within epeiric basins, lowest detrital sediment flux characterizes the transgressive stage when shorelines are retreating, base level is elevated, and the sea area is expanded. In the presence of a stable redox boundary at the sediment surface, the relative elevated organic matter deposition causes organic matter deposition and associated phosphogenesis. Evidence for this common relationship is widely known from the geological record (HECKEL 1977, TRAPPE 1993). This pure sea-level controlled process is locally superimposed by factors such as subsidence and broadness of the depositional basin (accumulation space), climate, weathering style, and current activity.

The prerequisites of the box model following organic matter trapping, which are decomposition behavior for phosphorus liberalization as well as P-concentration and storage, are very much controlled by the sediment properties. In coarse clastic systems, intense porewater to seawater interchange and porewater fluxes due to high porosity and permeability inhibit widely the development of oxygen-deficiency in those sediments. The conditions in pelitic sediments are more favorable. All known phosphogenic systems in siliciclastic systems are related to sediment starvation and condensation which emphasized the organic matter deposition (KENNEDY & GARRION 1975, FÖLLMI 1990). The permanent rain of organic matter over longer time periods during drowning or oxygen deficient flooding is able to establish a phosphogenic system in a porous medium with the support of microbial mat growth. FÖLLMI (1990) described a good example from an inner shelfal environment in the mid-Createceous of the Helvetian zone of the Alps. Authigenic phosphate precipitation is associated with glauconite. The developing fabrics are concretions and crusts which result from phosphate cementation of the pore space or the mineralization of organic mats. Additionally, phosphate mineralization occurs within enclosed microenvironment of shells. In these environments rarely occurring structureless apatite aggregates of sand size have their Quaternary equivalents in shelves of the Niger, Namibia and E'Australia (e.g. BIRCH 1980, RIGGS et al. 1989) and may result from authigenic growth. The phosphatic Shedhorn Sandstone of the Phosphoria Rock Complex hosts foraminifera-nucleated, structureless phosphate clasts (see Case Study 4).

The potential for the initiation of phosphogenic pathways in pelitic sediments is much higher. A number of these systems are characterized by sediment starvation which is achieved if the accommodation space exceeds sedimentation rate and the carbonate factory is turned off. The detrital fine-clastic sedimentation rates are usually much lower and relatively elevating the organic matter component. The lower permeability further supports the development of a P-storage and concentration system.

But only a very few of these systems develop phosphogenic conditions, most of them lead to black shale deposition. The decisive factor is the grade of oxygen depletion which ranges from dysoxic to anoxic (TYSON & PEARSON

1991). The development of phosphogenesis requires additionally a rapid partial or complete degradation of the organic matter to liberalize sufficient quantities of phosphorus. But recycling into bio-coupling has to be inhibited by rapid burial, microbial overgrowth, or the formation of a stable redox-boundary. Especially transgressive oxygen deficiencies which are characterized by temporal mixing during storm events were identified being favorable for phosphogenesis. Temporary mixing which is indicated by phosphate grain winnowing, prevents complete oxygen depletion and the preservation of dysoxic condition over longer periods (HECKEL 1977, WIGNALL 1993, TRAPPE 1993, GLENN et al. 1994). Suppressed or slow decomposition leads to organic matter preservation and subsequent dispersed phosphorus preservation.

The lack of replacement minerals for phosphates creates completely different phosphate fabrics in pelitic sediments than in carbonate systems. The variety of processes is very much reduced to nucleated and non-nucleated concretionary growth or dispersed phosphate precipitation (phosphatic shales). Replacement of fecal pellets is an again and again discussed process, but proof is difficult because concretionary and coprolitic origin result in structureless particles (e.g. LAMBOY 1982, SCHRÖTER 1986, see also discussion in GLENN et al. 1994). Nucleation for concretionary growth was described from several occurrences (e.g. CRESSMAN & SWANSON 1964, PRÉVôT 1990, SCWENNICKE 1992), but most particles lack any internal fabrics. From modern deposits off Peru, microbial oncoid structures are described which grow on the sediment surface (BURNETT 1977). Similar structures from a more shallow Ordovician deposystem are reported by TRAPPE & ELLENBERG (1994). Any kind of concretionary growth remains poorly understood. The concretion forming process may be diffusion or flow controlled. Most concretions are stratiform elliptical in shape, which is thought to be derived from the porewater flow. Sediment structures from pristine deposits and fossil preservation prove a very early diagenetic growth. BURNETT et al. (1988) could show that micro-concretions grow in a very short time span of several years. Whether the process of concretion growth is continuous or only seasonal is still uncertain. The synsedimentary growth is also supported by the fact that phosphate concretions are known for the precious preservation of fossils. In Lower Carboniferous deposits in Germany, radiolarians are well preserved in the concretions, but not in the enclosing host sediment. The in-situ nature is indicated by sediment structures (see Case Study 3). Most concretions are structureless enclosing different amounts of organic matter which influence the transparency of the fabric. Larger concretions from some Phanerozoic black shales show more specific internal structures (KIDDER 1985, SIEGMUND 1995, see also Case Study 3). In these, aggregates of organic matter are enclosed by various generations of apatite cements. The concretions from various localities document all stages of initial, patchy cementation to complete structureless concretions.

The above discussed subtidal deposystems are composed of only two main components which are carbonate and organic matter or silicilastics and organic matter. But a sufficient number of deposits are multi-component systems. Most of the Cretaceous and Tertiary deposits are marl dominated sequences with a

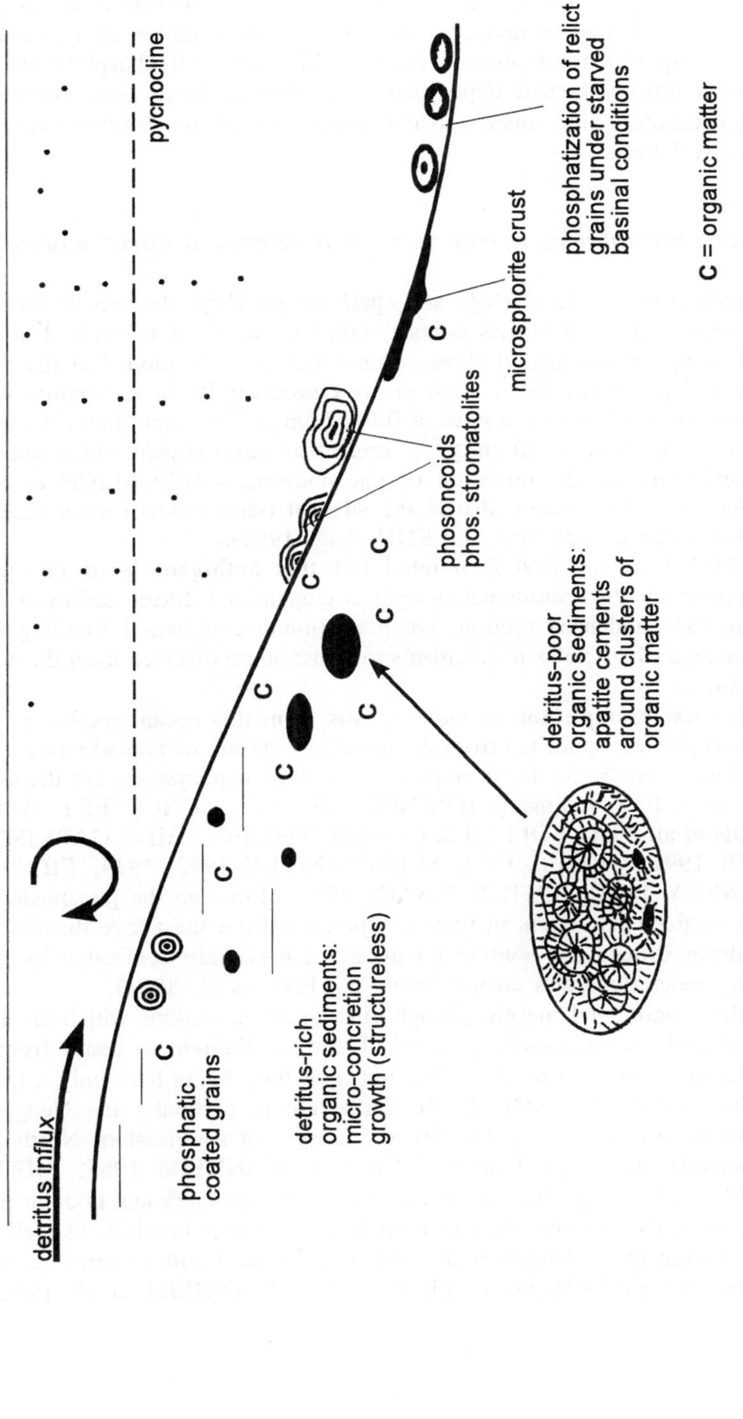

Fig. 6.12. Summary sketch of subtidal phosphogenic environments, processes and control factors in siliciclastic deposystems.

dominance of carbonates in marginal basin areas (see also chapt. 8.1). The probably most complex system is the Phosphoria Rock Complex with black shales, cherts, carbonate, and sandstones. The composition of allogenic and autogenic sediment components leads to a more complex environment/process relationship and a complicated facies architecture. All phosphogenic processes in the multi-component deposystems are derived from these outlined in the earlier chapters but superimposing processes and mechanism cause a larger facies variation.

6.3.2.3
Continental margin, ocean floor, and seamount/guyot environments

Depending on sea-level stage and shelf morphology, the genetically important productivity-derived OMZs contact with the sea floor is situated at the outer shelf to upper continental slope. This situation is developed at the site of the modern deposits off Peru. High primary productivity in association with both, relative low sedimentation rates of 0.0017 cm yr^{-1} or rapid burial by a sedimentation rate of about 0.30 cm yr^{-1}, creates an environment which satisfies most preconditions for the initiation of phosphogenesis (FROELICH et al. 1988). Similar conditions occur also at the sites of other modern outer shelf to continental slope phosphorites (GLENN et al. 1994).

LOUTIT et al. (1988) pointed out that authigenesis of phosphate and glauconite on the continental margin is emphasized during sediment starvation within the condensed section, when maximum continental flooding shifts the source area of siliciclastic sediments in a maximum distance from the continental margin.

The best example for ancient deposits from this oceanographic position are the Neogene phosphorites from the Monterey system or related strata in California. Characteristic for these slope or near-slope deposystems are the association with mass flow sediments (D'ANGELAN 1967, SCHUFFERT 1988, GARRISON et al. 1990, FÖLLMI & GRIMM 1990, FÖLLMI & GARRISON 1991, PIPER 1991, GRIMM 1992, SCHWENNICKE 1992, 1994, FILIPPELLI & DELANEY 1994, PIPER & ISAACS 1995). But also the phosphate facies of all these deposits differs in their composition from the above discussed epeiric sea phosphorites. The continental margin deposits are dominated by phosphate crusts, concretions and coated grains (GLENN et al. 1994).

Other continental margin phosphorites occur in regions which are related to typical trade wind upwelling zones. These are thought to result from upwelling-derived enhanced productivity, but upwelling being here only a local effect (RIGGS 1984a, POENOE 1990). These are in particular the Neogene phosphorite occurrences along the Atlantic margin of southeastern North America, respectively the Blake Plateau (RIGGS & MANHEIM 1988, RIGGS 1989, POPPE et al. 1992). They consist of micro-concretionary and nodular phosphate particles, carbonate replacement particles, phosphatic bioclasts as well as phosphate pavements or hardgrounds, and are associated with ferromanganese crusts and nodules (MANHEIM & PRATT 1967, MANHEIM et al. 1980, 1982).

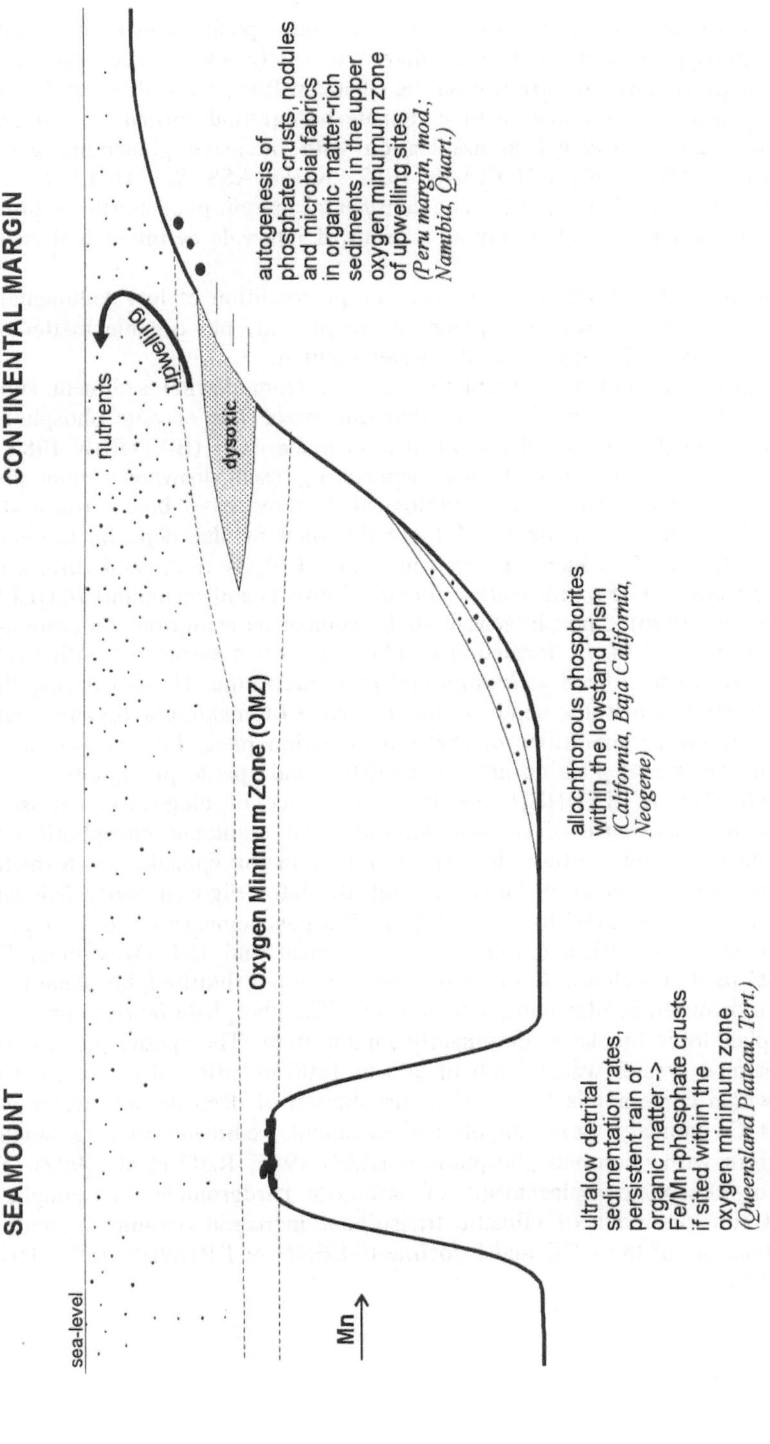

Fig. 6.13. Phosphogenic environments and phosphorite occurrences in the oceanic realm.

Their facies shows more affinities to the Florida deposits than to the modern Peru margin-type phosphorites. A similar, intensively investigated deposit represents the phosphorite occurrence on the Chatham Rise, west off New Zealand. Phosphogenesis was initiated with carbonate hardground formation and phosphatization, and proceeded in association with intensive glauconite genesis (CULLEN 1980, 1989, KUDRASS 1982, KUDRASS & CULLEN 1982, KUDRASS & VON RAD 1984). The continental margin phosphorites represent lag deposits and pavements today as a result of intervals of intensified current activity.

Ocean floor phosphorites are absent. The precondition of low sedimentation rates and low oxygen recharge potential are present, but organic matter sink is commonly degraded higher in the water column.

A peculiar environment, completely cut off from detrital sediment supply are seamounts, guyots and isolated submarine rises. The oceanic phosphorites differ in sedimentological and geochemical characteristics (BATURIN 1982 for summary report). A number of these deposits represent drowned insular phosphorite occurrences. Their sedimentological features have been discussed in chapt. 6.3.1. The geochemistry of these drowned insular deposits is characterized by low REE contents as well as a low F/P_2O_5 ratio, and differs significantly from that of permanently submerged guyots and seamounts (CULLEN & BURNETT 1986). Phosphogenesis under submarine conditions on seamounts and guyots has led to the formation of phosphate crust associated with Fe/Mn nodules and crusts as well as hardground phosphatization. If sited during their history in moderate water depth within the zone of maximum organic matter degradation, the preconditions of low sedimentation rates, low oxygen supply and organic matter deposition are accomplished and apatite precipitation is initiated. GLENN & KRONEN (1993) see evidence of diagenesis within the OMZ in the occurrence of ferroan dolomite. All seamount phosphorites are Late Cretaceous and Tertiary in age with two major episodes of formation during the late Eocene/early Oligocene and the late Oligocene/early Miocene, and two minor events (HEIN et al. 1993). The geochemistry of these deposits is characterized by REE enrichment, low pyrite and U/P_2O_5 values. The dominant apatite is calcium fluor apatite causing higher fluorine/phosphate ratios than from drowned insular phosphate deposits. The phosphate fabrics commonly enclose planktonic bioclasts, dominantly foraminifera. The apatite precipitation in an assumed greater water depth of 500 to 1600 m, affected a discrete REE pattern which corresponds to the REE distribution of deep oceanic water with a distinct Ce-anomaly. Some guyots and seamounts sediments indicate shallow water origin with microbial phosphorites (RAO 1986, RAO et al. 1992). The periods of phosphate replacements of carbonate hardgrounds are thought to represent time intervals of climatic transitions, increased oceanic circulation, and fluctuations of the CCD and lysocline (GLENN & KRONEN 1993, HEIN et al. 1993).

CASE STUDY 2

Peritidal to platform margin phosphogenesis: Thorntonia Formation and equivalents, northern Australia, Middle Cambrian

Anatomy of the deposit

Location: Isolated erosional remnants in the Barkly Tablelands in northwestern Queensland and directly adjacent northeastern Northern Territories; 137°30'E to 140°30'E and 18° to 22° S.

Stratigraphy, Age: Phosphatic sediments are incorporated in a number of different lithostratigraphic units and biostratigraphic niveaus (ÖPIK et al. 1957, SMITH 1972, SHERGOLD & DRUCE 1980)(Fig. 6.14). Age: Middle Cambrian, Ordian to late Templetonian (SHERGOLD & DRUCE 1980, SOUTHGATE & SHERGOLD 1991).

General paleogeography: Part of a larger shelfal sea area at the northern margin Gondwana (Fig. 6.15) between Tasman continental margin and terrestrial Westaustralia. Paleolatitude: ~ 20°N, north of potential equatorial humid zone, warm climate (GOLONKA et al. 1994); shallow shelfal to peritidal.

Paleotectonic setting: Cratonic sag basin or inner shelfal sag basin, subdivided in subbasins and stable tectonic platforms (TUCKER et al. 1979); gulfal sea area, partly nearshore (De KEYSER & COOK 1972, HOWARD & HOUGH 1979) reinterpreted as a carbonate platform (SOUTHGATE & SHERGOLD 1991).

Depositional system, facies architecture: Carbonate deposystem which interfingers with a siliciclastic system, initially interpreted to resemble a coastal siliciclastic belt and a basinal off-shore carbonate facies (DE KEYSER & COOK 1972); contrasting model assigns siliciclastic phosphate-hosting facies to the transgessive suite of a platform and platform margin deposystem, on the inner stable platform transition of shelfal to peritidal phosphorites (SOUTHGATE 1988), non-phosphatic carbonates represent highstand suite of platform to outer ramp origin (SOUTHGATE & SHERGOLD 1991).

Sea level control, sequence stratigraphic interpretation: Phosphorites are part of the transgressive systems tract of third order cyclicity, individual phosphate occurrences related to the retrograde portion of parasequences. The individual deposits are time transgressive with the sequential transgression (SOUTHGATE & SHERGOLD 1991).

Type of phosphorite particles: Carbonate system: phosphatized erosional surfaces on semidurated carbonate mud, phosphatic intraclasts derived from erosion of the hardgrounds, phosphatic stromatolites and phoscrusts, phosphatic bioclasts, phosooids and phosoncoids (initially reported as phospisoides); siliciclastic system: phosphatic lithoclasts (winnowed micro-concretions, bioclasts); weathering profiles: phoscretes.

Model for the phosphogenic processes: Carbonate system: carbonate mud re-

placement along erosional surfaces, formation of phosphatic microbial mats, ooids and oncoids (SOUTHGATE 1983), impregnation of bioclasts, biological skeletal; siliciclastic system: micro-concretionary growth, phosphatic mudstones (?intense dispersed apatite precipitation), biological skeletal, phosooid formation.
Phosphorite formation: Carbonate system: Rapid reworking during parasequences, tempestites; siliciclastic system: winnowing in shallow water, formation of ripples and submarine dunes.
Reserves, resource quality: 1.8 million tons with average of 16.9% P_2O_5 and roughly 1 million tons inferred (BRITISH SULFUR CORPORATION 1987). Mining momentarily abandoned.

The Georgina basin phosphorites document in an unique way the development of different phosphogenic pathways in adaptation to depositional environment, substrate, and sea-level, thus demonstrating and explaining the dynamic processes which have been outlined in course of chapter 6. The deposit is also extraordinary in terms of preservation of the carbonate-phosphate sedimentary suite and biostratigraphic resolution, which both offer the possibility to observe phosphogenic processes in carbonate deposystems and the relationship to sea-level changes in great detail.

In the Georgina basin, a phosphogenic carbonate depositional system interfingers in a complicate time space relationship with a siliciclastic phosphogenic system (DE KEYSER & COOK 1972, COOK 1976, 1989, SHERGOLD & DRUCE 1980, HOWARD 1986, 1989a,b, 1990, SOUTHGATE & SHERGOLD 1991). This black shale-chert-phosphate lithofacies type was intensively mined and therefore accessible for geologic investigation in trenches and mine pits in a generally outcrop-poor landscape. The carbonate-phosphate suite is present in natural outcrops. Additional data were obtained in the past by stratigraphic drilling projects (see results in SOUTHGATE & SHERGOLD 1991).

The siliciclastic-dominated sequence reveals a well bedded suite of deeply weathered black shales with all transitions to bedded chert (DE KEYSER 1969, RUSSELL & TRUEMAN 1971, COOK 1972, DE KEYSER & COOK 1972, FLEMMING 1977, COOK & ELGUETA 1986). Phosphoclastic beds are intercalated in the entire suite and are partly crossbedded. In response to the pelitic/siliciclastic substrate, the granular phosphate facies is dominated by structureless phosintraclasts, phosooids, phosphoncoids and phosphatic and non-phosphatic bioclasts (Fig. 6.18). Important is the occurrence of phosphatic mudstones which form thin (several mm) beds of structureless phosphate layers with a porcellanite appearance (DE KEYSER & COOK 1972, ROGER & CRASE 1980, SOUTHGATE 1983). Reworking fabrics have not been recognized from these beds, but the beds remain a potential source of the phosphoclasts. These are structureless lithoclasts which are identical to micro-concretionary derived clasts in other deposits, e.g. the Phosphoria Rock Complex. These phosphorites are black or brown in drilling core and whitish in outcrop. These phosphorites as well as the associated shales are widely silicified (FLEMMING 1974). Equivalent phosphorite facies occurs also within a carbonate matrix.

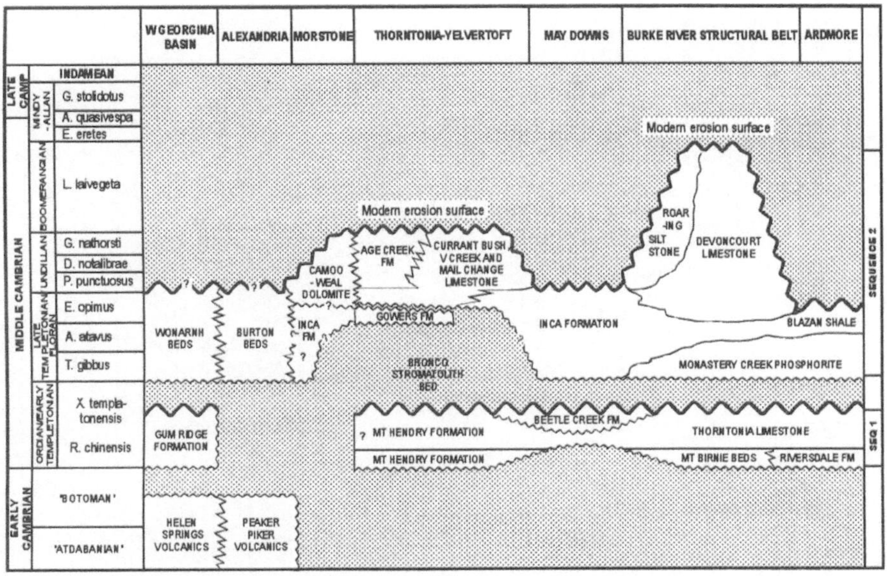

Fig. 6.14. Lithostratigraphic chart for the Middle Cambrian phosphate-bearing strata in the Georgina Basin based on correlations of SOUTHGATE & SHERGOLD (1991).

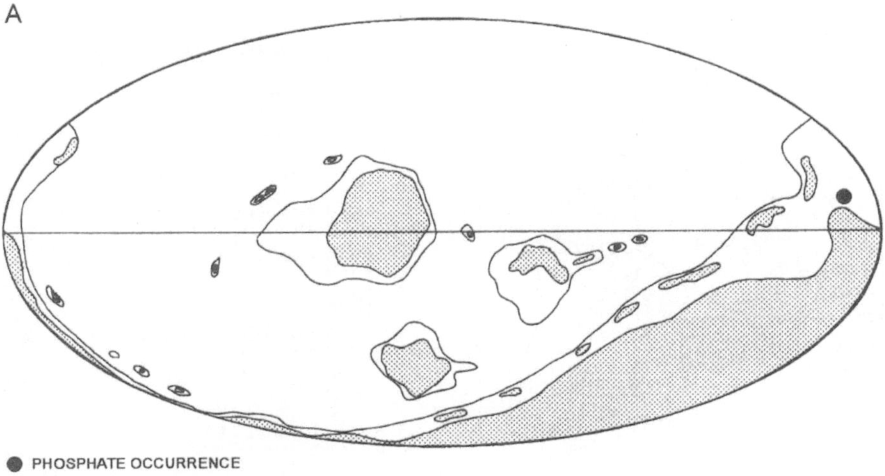

Fig. 6-15. Paleogeographic setting of the Georgina basin deposits: A: global position based on GOLONKA et al. (1994), B: paleogeographic gulf model of DE KEYSER & COOK (1972), C: alternative platform model of SOUTHGATE & SHERGOLD (1991).

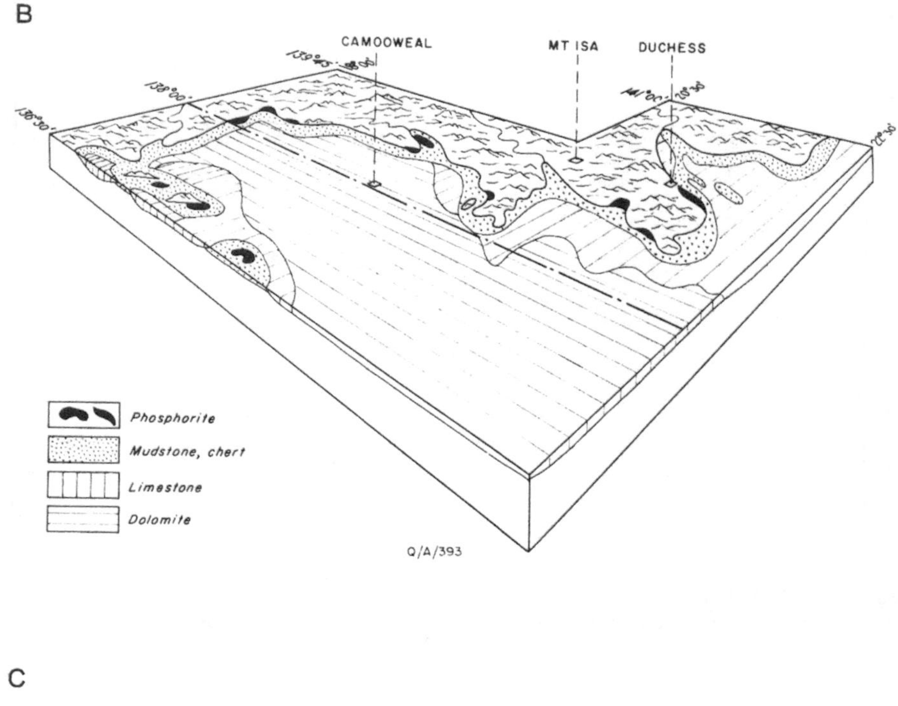

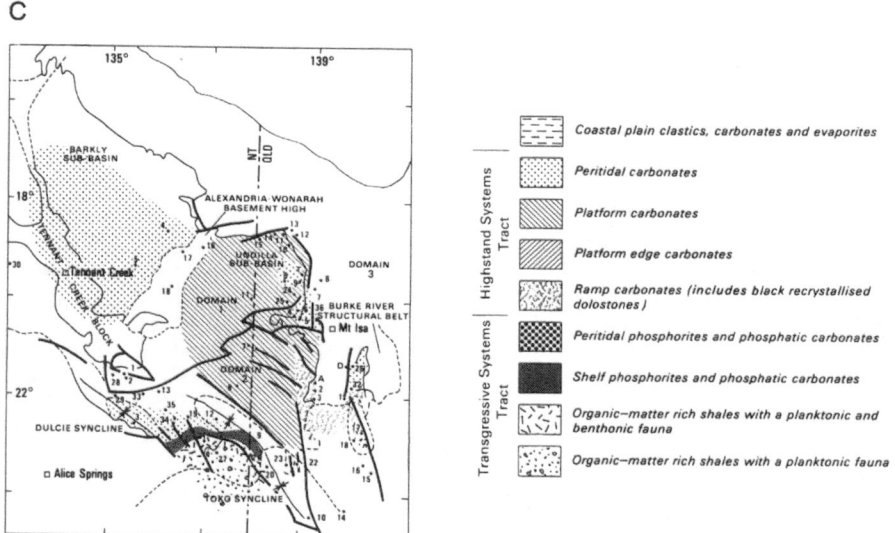

Fig. 6.15 (cont.) For explanations see previous page.

Outstanding is the carbonate suite of the Thorntonia Limestone and Gowers Formation, which occurs in spatial contrast to the siliciclastic suite. The Thorntonia Limestone represents a partly bioturbated carbonate mudstones sequence which includes a succession of phosphatized stacked discontinuity surfaces (Fig. 6.16, 6.17). The preservation of these surfaces with the associated erosional and microbial fabrics presents a precise picture of the sedimentological processes during phosphogenesis and particle generation in the carbonate deposystem. Along these surfaces, structureless or laminated carbonate mudstone is being phosphatized forming a several centimeter thick penetration zone of cryptocrystalline phosphate under preservation of primary sediment particles and structures. A phenomenon which is observed in most Phanerozoic phosphatic hardgrounds. The grade of phosphatization decreases with depth. The mineralization is subsequently or contemporary followed by further erosion forming cavities and generating round grain products from this sediment. Sediment structure, e.g. desiccation cracks, indicate at least partly a semidurated sediment consistence of the carbonate muds while being phosphatized. The spectrum of phosphatized sediments includes also fenestral mudstones.

The phosphatic hardground surfaces are covered to various degree by laminated, completely phosphatic crusts which prove an intense microbial activity. These phoscrets again show desiccation cracks which indicate partly subaerial exposure during rapid phosphatization (SOUTHGATE 1986a,b). The microbial mats find maximum growth conditions in protected depressions within the hardground surfaces. The presence of sediment structures, which indicate emergence, supports the interpretation of a depositional environment from the intertidal to peritidal zone with rapidly changing conditions and the formation of shoals (SOUTHGATE 1983, 1988, NORDLUND & SOUTHGATE 1988). Very distinct phosphate fabrics are columnar stromatolites within the suite (SOUTHGATE 1980, 1986). A widely consistent growth of these microbial structures characterizes the Bronco Stromatolite Bed.

Mechanical erosion of the mostly semi-durated sediment by turbulent water and possible erosion by bioturbation generate phosphate grains and larger flat pebbles. All stages of grain formation can be observed. The products are structureless, but have abundantly carbonate relicts incorporated. Commonly, grains are additionally coated by phosphatic envelopes. SOUTHGATE (1983) assumes a fecal pellet origin of some regular ellipsoid phosclasts. A peculiar grain type, which was not recognized in other deposits, are irregular phosphatic tubes of less than 1 mm size. These may represent partly phosphatized burrows or fossil fragments.

The phosphate grains derived from the erosional surface are winnowed and concentrated in thin horizons without a direct spatial association with their source beds. The concentrates are sporadically crossbedded. The association with the hardgrounds indicates grain formation and granular bed formation in association with sea-level fluctuations. The resulting sediments are phosclastic packstones and wackstones. Erosion-soled phosphorite concentrates which rest on top of the individual hardgrounds are mostly composite or multi-event con-

Fig. 6.16. (next page) The phosphorite facies in the Georgina basin in outcrop view. (A) Cross-bedded granular phosphorites of the siliciclastic dominated phosphorite lithofacies from the shallow subtidal realm. (Duchess deposit). (B) Outcrop view of the Duchess phosphate pit. The phosphorite shale succession, which is black in deeper drilling cores, is deeply weathered at the surface. (C) Polished slab of the peritidal phosphate facies. An erosional relief is overgrown by a phosphatic stromatolite crust. The remaining space is filled with several layers of granular phosphorite packstone. (D) Thin layers of phosphatized carbonate mud are ripped up and document very early phosphatization. (E) An erosional (mechanical and biological) surface is phosphatized to various degree. The phosphatization is indicated by the whitish color. The underlying sediment consists of phosclastic wackestone. The concentration of grains in discrete layers is thought to represent event concentrations. Moderate burrowing activity disturbed locally the bedding fabrics. Phosphate grain enrichment in depressions on the erosional surface documents grain trapping. (F) Outcrop appearance of the phosphate crust above an erosional surface. (G) Multiple phosphatization fabrics along an erosional surface. A succession as described under (E) documents higher degree of reworking and phosphate particle generation by erosion and is again truncated by a second surface, on which a thick phoscrust came to rest. (H) Outcrop view of a phosphate crust. Erosion generates preferential preservation in depressions on the underlying erosional surface.

Fig. 6.17. (2 pages further) Phosphorite facies of the carbonate, peritidal phosphogenic system. (A) Fe-phosphate crust with well developed stromatolite fabrics. (B) Laminated (?microbial binding) of event layers forming thin layers of phosclasts pavements. (C) Fe-stained, irregularly coated phosphate crust fragment in a silicified phosphatic packstone. The fabric was fragmented during early stage of deposition (?desiccation cracks). (D) Calcareous phosphate crusts. The upper surface shows desiccation cracks. (E) Amalgamated crust of bioclasts and phosphatic microbial mats, and a broad spectrum of phosphatic and non-phosphatic clasts. (F) An erosional surface on carbonate mudstone was slightly phosphatized and capped by a phosphatic microbial crust. (G) The phosphatic and non-phosphatic grain spectrum includes large (approx. 10 mm) marginally phosphatized carbonate lithoclasts.

Fig. 6.18. (3 pages further) Phosphorite facies of the granular subtidal system. (A-B) Bioclastic wackestone from redeposited, phosphate sediment bearing bioclasts (A) and phosphatized bioclasts (B). The grain spectrum further includes non-phosphatic bioclasts, and single, structureless phosclasts. The matrix is a carbonate mudstone. The phosphatic fabrics are indicated by their brown color. (C) Completely silicified phosbiowackestone to packstone. The intensive overprint by silicification of the common phosphorite type in the Georgina basin leaves the primary or secondary nature of the phosbioclasts open. Additionally, structureless phosclasts. Grain size approx. 1 mm. (D) Intensively amalgamated packstone from phosclasts and phosbioclasts. (E) Irregularly coated phosclasts are forming packstones. (F) Polymict packstone of various types of phosclasts and ooids and coated grains. The phosclasts are particle free or contain micro-bioclasts which indicates reworking from different phosphatized sediments. The various grain size and grain type document intensive mixing.

See also color copies of these plates in the appendix.

Fig. 6.16.

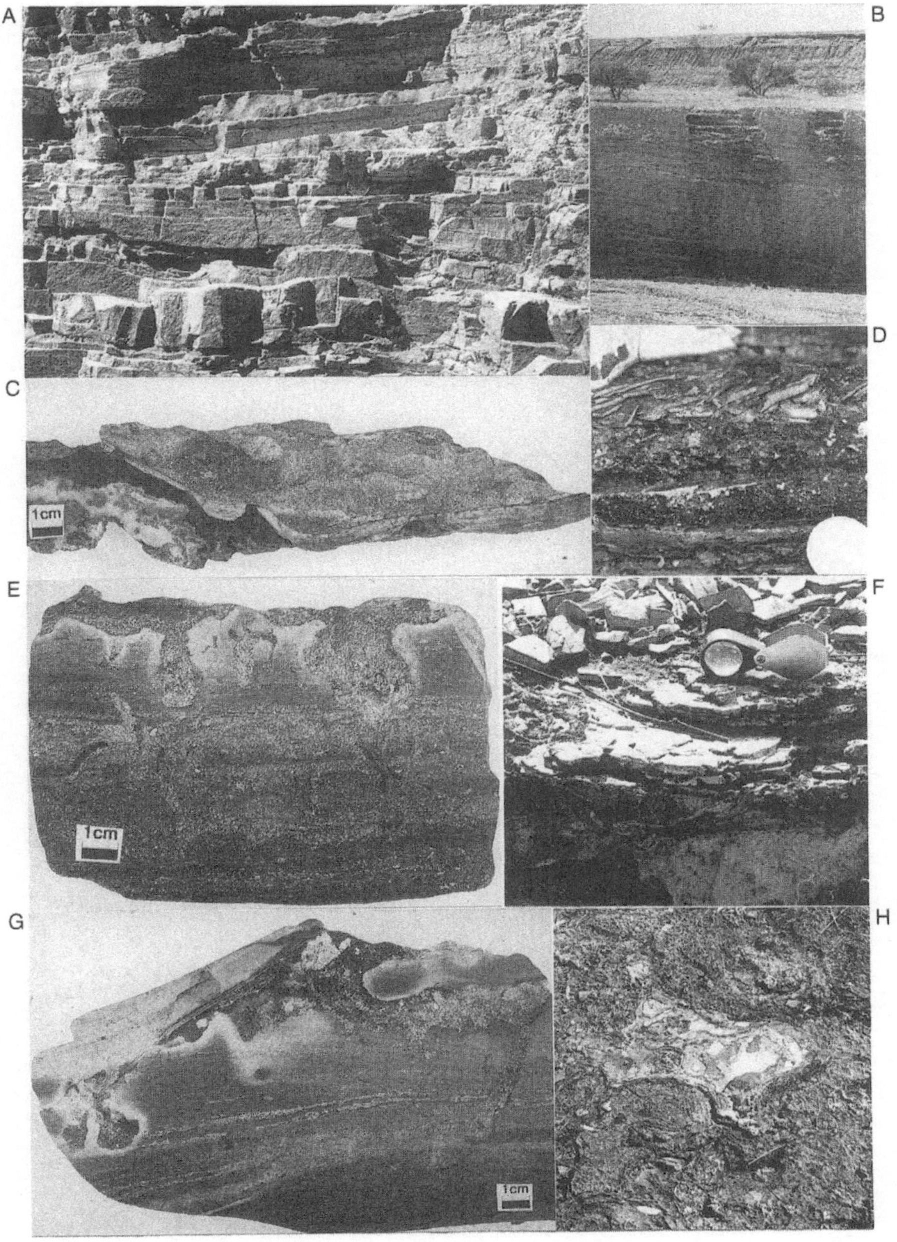

Fig. 6.17.

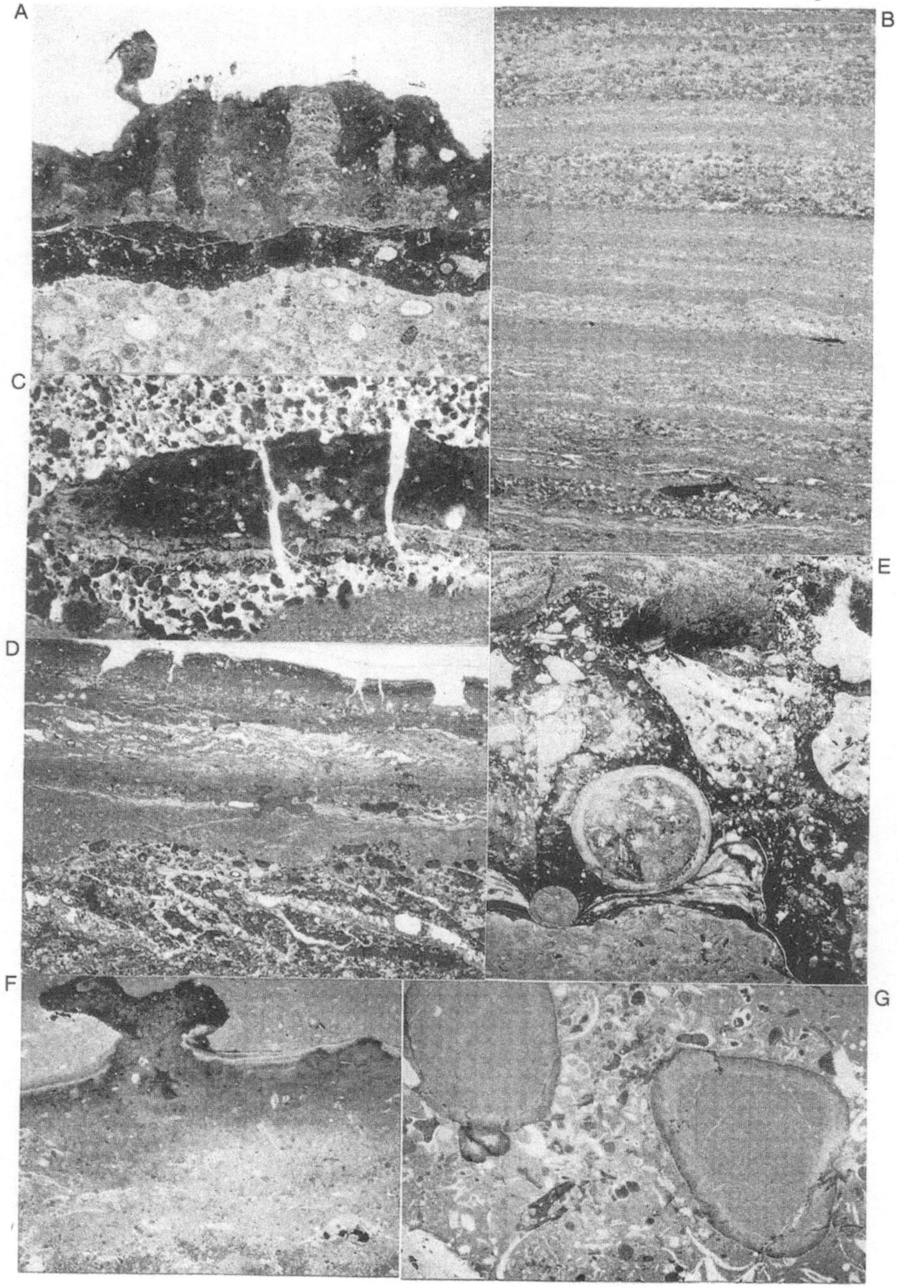

Fig. 6.18.

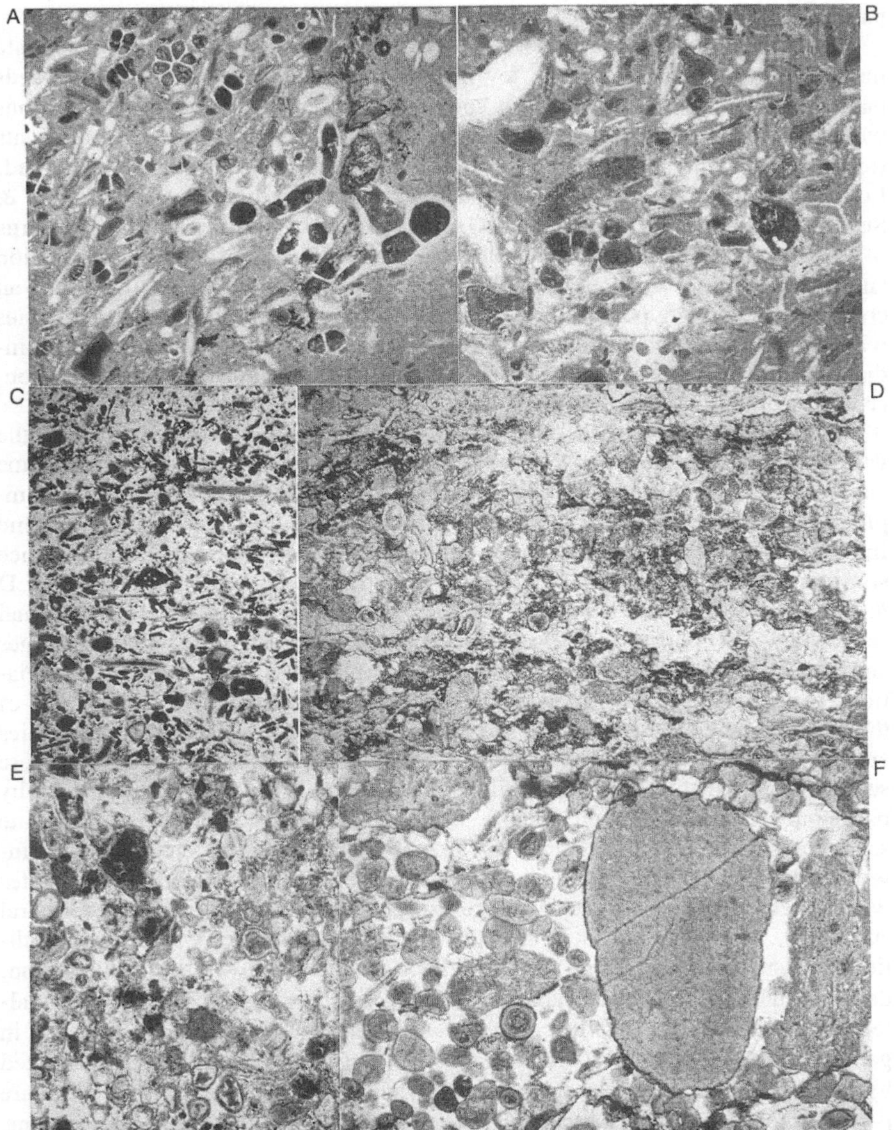

centrates (see chapt. 7). These are characterized by a poorly sorted phosphate particle spectrum, however, composed of mainly one grain type.

A distinct sediment type consists of a meter thick suite of carbonate mudstones with thin (one to a few grains) phospackstone layers. These beds have a relative homogeneous grain spectrum and represent event concentrations which result from reworking, winnowing, and transportation during storm events (tempestites) with an interparticle space filled with resettling carbonate mud. Low order cyclicity is also evident in the peritidal sediment. SHERGOLD & SOUTHGATE (1986) and SOUTHGATE (1986a,b) describe stacking patterns of various scales in the carbonate-phosphate suite. Phases of inundation, erosion and sedimentation are controlled by cyclicity of different order, affecting an complicated stacking pattern. The granular phosphorites in the siliciclastic facies realm are commonly characterized by a larger phosphate grain spectrum indicating more intense mixing. These sediments are mostly composite concentrates of the shallow subtidal.

The Middle Cambrian rocks of the organic-rich pelitic/siliciclastic facies, the various carbonates and phosphorites occur in isolated locations, leaving time and space relationship widely open. SOUTHGATE & SHERGOLD (1991) compiled the lithologic outcrop and core data as well as the biostratigraphic framework of the Middle Cambrian suite of the Georgina basin. The sequence stratigraphic concept revised the early stratigraphic concepts of SHERGOLD & DRUCE (1980), and identifies sedimentation during two distinct depositional sequences. The phosphorite occurrences reveal different biostratigraphic ages and appear in both sequences within different positions and lithofacies associations (Fig. 6.14). In this concept, the majority of the granular phosphorite of the pelitic/siliciclastic lithofacies association and the equivalent carbonate facies are interpreted to be "shelf phosphorites". It is important to note that also these sediments originate from the shallow water realm, which is documented by packstone texture and high energy sediment structures. But these phosphate sediment suites have in common to lack any indications of emergence or intertidal deposition. The "peritidal phosphorites" are all phosphate sediment suites which include phosphatic stromatolite crusts, phosphatic fabrics with fenestral structures, and other sediment structures indicating temporary intertidal conditions or subaerial exposure. As an effect of intensive sea-level fluctuation, depositional conditions are rapidly changing and generate a succession of alternating intertidal to shallow subtidal sediments. This facies distribution in general identifies a ramp setting. Transgressive systems tracts are characterized by phosphorite facies and organic matter-rich shales. Highstand sediments are peritidal to outer ramp carbonates. The stacking pattern within the systems tracts differs with topography and synsedimentary tectonics. Local tectonic activity during the early depositional sequences causes the formation of peritidal phosphorites on elevated blocks. But most of the basin occurs to be tectonically stable during most of the Middle Cambrian. The low inclination of the presumably homoclinal ramp causes the development of progradational lowstand wedges instead of a lowstand prism.

In comparison to other deposits, the facies architecture of the Georgina basin deposystem is extraordinary because of its broad phosphogenic peritidal belt. The widely stable tectonic conditions have further supported persistently repeated development of this phosphogenic environment. The ramp setting and lateral facies relationships as well as the general pattern of the sequences are typical for other occurrences, e.g. Phosphoria Rock Complex (see Case Study 4), but the broad peritidal carbonate-phosphate belt is widely suppressed and the shelf phosphorites more emphasized.

The extent of phosphogenesis in the peritidal belt, however, leaves the early steps of the box model of phosphogenesis (supply and trapping of phosphorus) enigmatic. All concepts for organic matter deposition in the peritidal zone require at least a semi-closed lagoonal setting. The application of both, preservation or bioproductivity models, are not able to explain satisfactory the elevated organic matter deposition which is necessary to trap sufficient amounts of phosphorus. Modern equivalents are unknown. Detrital organic matter deposition in a number of lagoonal puddles in a generally flat intertidal belt would be a conceivable scenario, but even more important would be the intensive growth of microbial mats. These mats are only exceptionally preserved in calcareous muds (own observation in Andros Island, Bahamas). But in both cases, the short lifetime contradicts the extent of phosphatization. Other sources of phosphorus than organic matter, e.g. volcanic or fluvial, are not supported by field evidence. The trapping process in this scenario is achieved by rapid burial under dense carbonate mud layers further locked by younger microbial mats. Decomposition under oxic to anoxic conditions would liberalize phosphorus and cause enrichment in the porewater, when sedimentation rates for carbonate are relatively low. A carbonate replacement pathway is completed with the precipitation of apatite.

The "shelfal phosphorites" are nearly unexceptional granular and deeply weathered in all near surface occurrences. The types of phosphate fabrics are substrate controlled or of primary biologic origin. Additionally, intense dispersed apatite precipitation leads to the formation of phosphatic mudstone. Most phosphorites are composite concentrates and originate from the multiple sediment mixing. The phosphogenic pathway for all granular and non-biological phosphorites remains widely undocumented due to the intensity of reworking, weathering and diagenetic overprinting including silicification to large extent. But the abundance of structureless phosclasts and the spatial association to black shales indicate microconcretionary growth of phosphate particles which are subsequently winnowed. Phosphate mudstones in these sediment suites document the second potential source of these grains. Additional processes are biologic precipitation of apatite skeletons and the formation of phosphate ooids.

The Georgina Basin phosphorites with their well preserved phosphate facies give distinctively evidence to the box model for phosphogenesis (chapt. 5). In a deposystem with generally favorable phosphogenic conditions, the specific substrate and depositional environment control the development of significantly different phosphate facies in the peritidal and permanently shallow subtidal zone of a ramp. The microfacies of the various phosphate fabrics demonstrate the

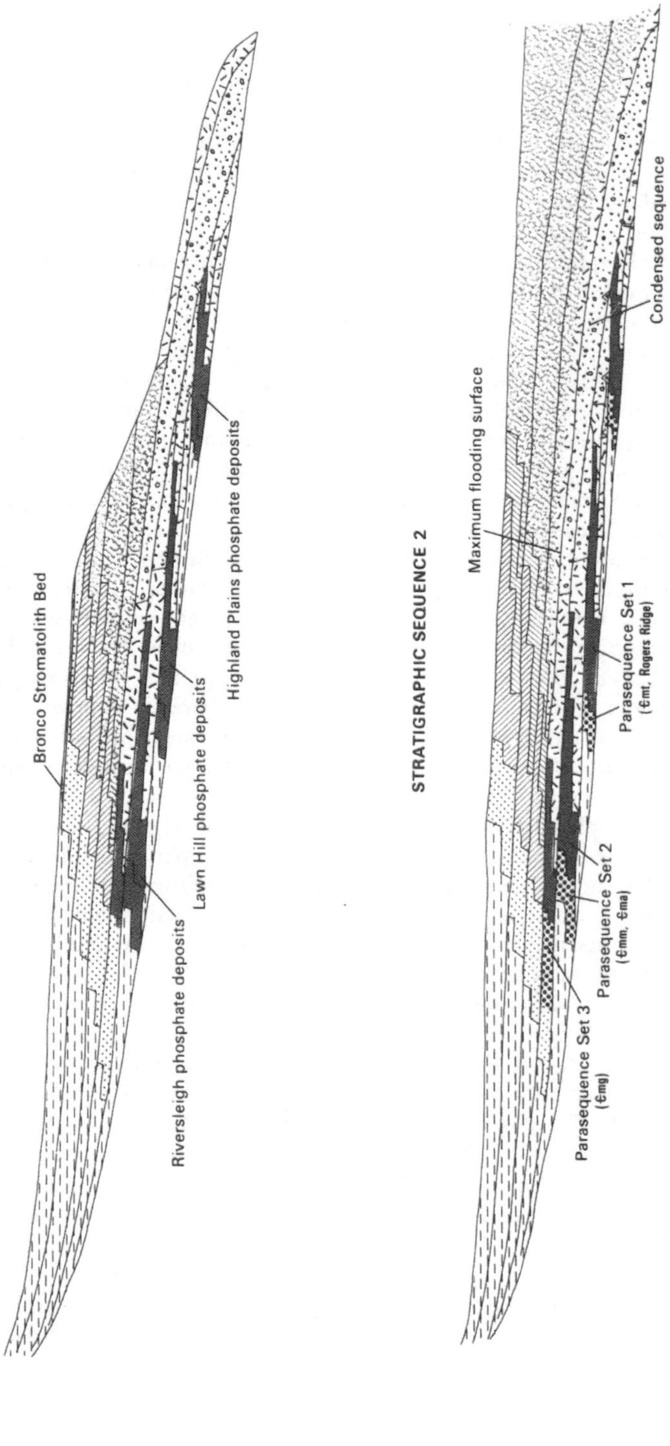

Fig. 6.19. Sequence stratigraphic context of the peritidal and subtidal phosphorite sedimentation in the Georgina basin during the two Middle Cambrian depositional sequences based on SOUTHGATE & SHERGOLD (1991). For legend see Fig. 6.15.

diversity of phosphogenic pathways which may exist within a single deposystem. It shows also that by the presence of a general high supply of phosphorus, phosphogenic environments are able to respond to various depositional conditions instead of being strictly limited to a single setting. But not only the sedimentary systems are responding, but also the biologic systems. Various microbial phosphate fabrics, preferentially in the pertidal zone, are supplemented by abundant skeletal phosphate formations in the more open marine environments.

In the peritidal realm, the larger effect of sea-level fluctuations documents furthermore the limitation of response potential of phosphogenic regimes and the dependence on very specific conditions. Phosphate mineralization here is strictly limited to low sedimentation rates and erosion, respectively low net sediment accumulation. The preservation of the peritidal and subtidal phosphate fabrics also allow reconstructions of the grain formation processes which are outlined in chapt. 7.

CASE STUDY 3
Starved basin concretionary phosphate formation, "Liegende Alaunschiefer" Fm. in Central Germany, Early Carboniferous

Anatomy of the deposit

Location: Several occurrences in the Harz Mtns. and the northern portion of the "Rheinisches Schiefergebirge"; 7°30' to 11°E and 51°to 52°N. Part of a larger system from Poland to the Pyrenees.
Stratigraphy, Age: Phosphate sediments occur within the black shale suite of the "Liegende Alaunschiefer" Formation and stratigraphic equivalents (PAPROTH & ZIMMERLE 1980, ZIMMERLE et al. 1980, STRUCKMEYER 1982), additional phosphorites in the overlying Erdbach Limestone, Age: Lower Carboniferous (middle Tournaisian, CdII); *Pericyclus* zone, respectively *sandbergi* to *anchoralis* zone (CLAUSEN et al. 1989)(Fig. 6.20).
General paleogeography: Large shelfal sea area of the Laurussian continent including the active Hercynian orogenic belt (ZIEGLER 1989, 1990, GOLONKA et al. 1994)(Fig. 6.21); Paleolatitude: 5°N, warm, humid, no coastal upwelling predicted (GOLONKA et al. 1994). Subsiding shelfal basin (Kulm facies) offshore from a fringing carbonate ramp complex (Kohlenkalk facies) along the south coast of the Old Red continent, subtidal, drowned outer shelf.
Paleotectonic setting, basin geometry: Rapidly subsiding subbasin within late syntectonic Variscan foreland basin (inner shelfal sag basin to back arc basin, distal flysch basin); Rhenish subbasin of the Rhenohercynian belt.
Depositional system, facies architecture: Siliclastic starved sedimentation filling a paleotopography, interfingering shorewards with subtidal and peritidal carbonates and offshore with a calciturbitide facies (Gladenbach Fm.), suppressed carbonate production, lacking benthos due to oxygen deficient bottom conditions.
Sea-level control, sequence stratigraphic interpretation: 3rd order transgressive system tract to maximum flooding, starved sedimentation in a drowning subbasin, lower order-cyclicity absent, but evident in calciturbidite facies (HERBIG & BENDER 1992).
Type of phosphorite particles: Pristine phosphate concretions.
Model for the phosphogenic processes: Apatite cement growth around aggregates of organic matter.
Phosphorite formation: Weak reworking on topographic highs, ?storm- or current-derived.
Reserves, resource quality: Non-economic.

This phosphate deposit was selected to investigate and explain the development of phosphogenic pathways in siliciclastic systems. The Early Carboniferous Alaunschiefer phosphorites are a typical example for concretionary phos-

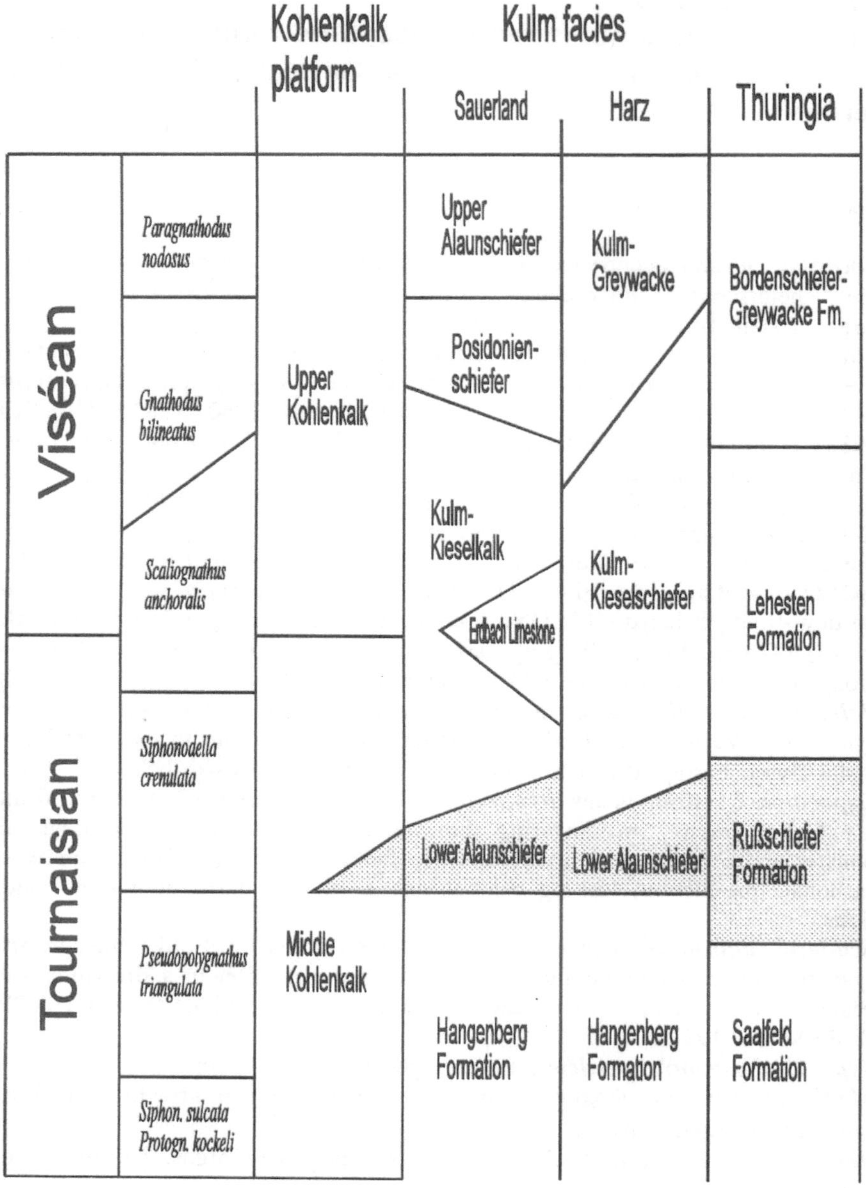

Fig. 6.20. Stratigraphy of the Early Carboniferous strata in central Germany and the stratigraphic range of equivalent oxygen deficient sediments in Laurussia based on CLAUSEN et al. (1989) for the Rhenish trough, other: various sources.

A

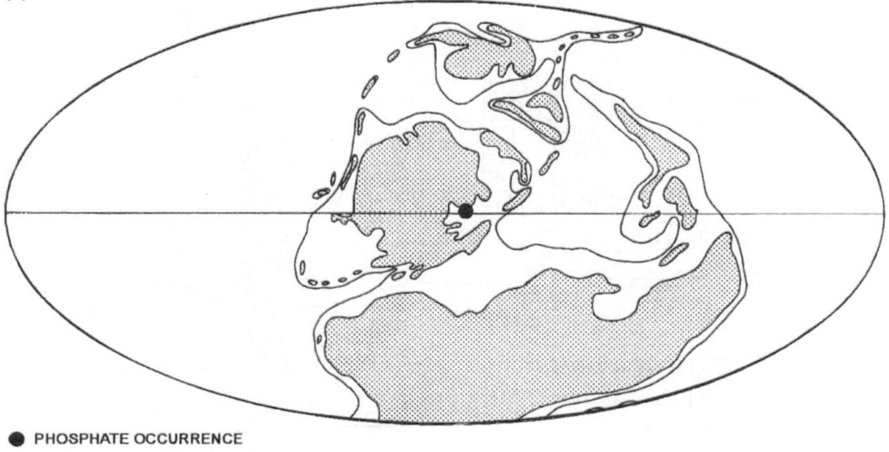

B

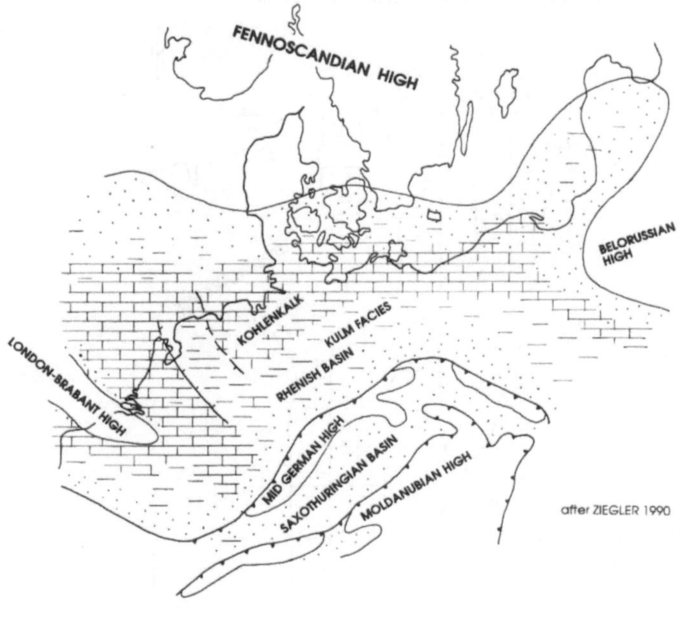

Fig. 6.21. A: Global paleogeographic location of the central German Alaunschiefer occurrences and equivalent deposits, based on paleogeographic reconstructions from GOLONKA et al. (1994). B: Paleotectonic configuration of the Rhenohercynian basin (after ZIEGLER 1990).

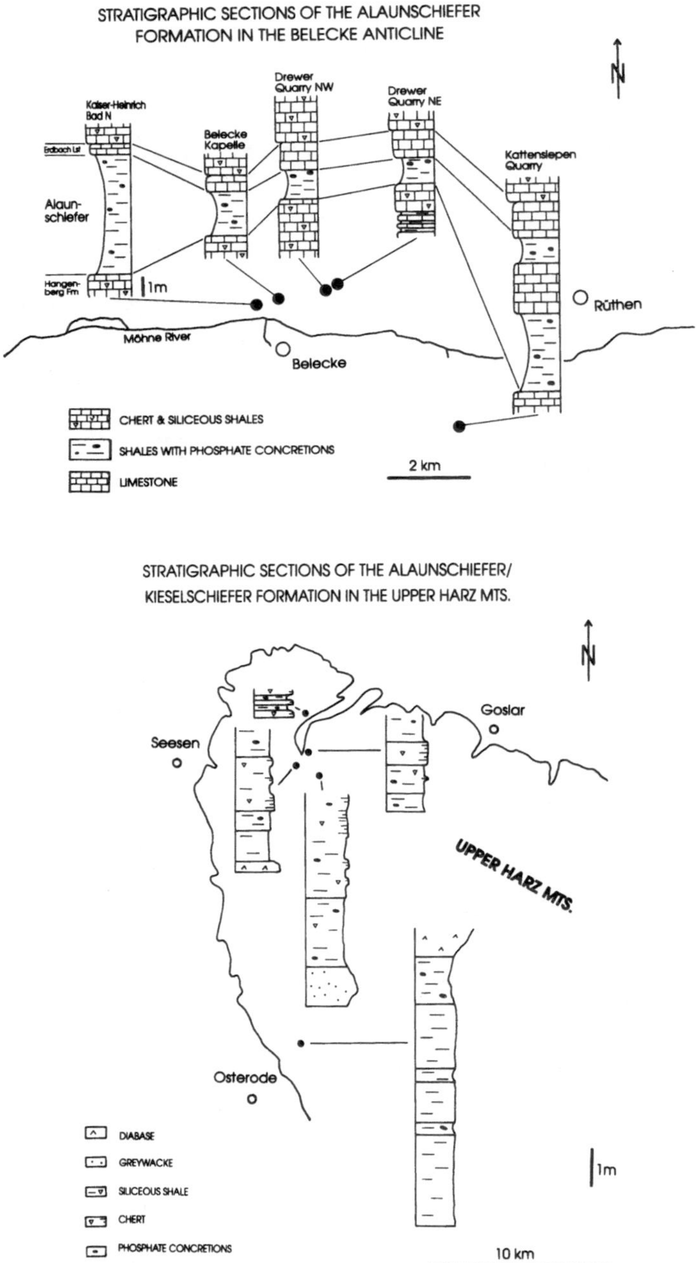

Fig. 6.22. Lithologic sections of the investigated "Liegende Alaunschiefer" occurrences in the Belecke Anticline and the Harz Mtns.

phogenesis in an organic matter-rich siliciclastic deposystem with pristine preservation. Their relative uniform occurrences have a wide distribution in Europe (BIDAUT 1953, BÖGER 1962, EICKHOFF 1962, SCHEERE & VAN TASSEL 1968, TIMMERMANN 1974, 1976, PAPROTH & ZIMMERLE 1980, STRUCKMEYER 1982, SLANSKY 1989, ZIMMERLE 1992) and in the Mid-continent/USA (CONKIN & CONKIN 1972, and own observations). These fabrics are also known from occurrences of other time intervals (USA: HECKEL 1977, KIDDER 1985, China: SIEGMUND 1995).

The petrology of these concretions was investigated to reconstruct the phosphogenic pathway and mineralization processes in organic rich sediments as a potential example for micro-concretion derived phosphate grains (see chapt. 4). The latter - commonly sand sized particles - are composed of densely packed non-oriented fibrous apatite crystals with various amounts of detrital and organic material. The preservation of an evidently pristine phosphate fabric to host sediment relationship furthermore offers the possibility to test geochemical techniques for paleoenvironmental reconstruction, which have been applied to granular phosphorites. It was investigated how far trace element geochemistry and REE-pattern within the phosphorites are representative for the general host rock geochemistry as thus for the geochemical environment.

The Early Carboniferous phosphate occurrences in Central Germany are only part of a large system of oxygen deficient deposition with abundant formation of phosphate concretion along the southern Laurussian continental margin. "Liegende Alaunschiefer" equivalents are known from North America, western Europe, northern Africa, and eastern Europe. Their origin seems to represent at least a major regional oceanographic phenomenon. The extent and origin are still unknown. Two study areas have been selected in Germany, the Belecke Anticline in the northern "Rheinisches Schiefergebirge", and the northern Harz Mountains in north-central Germany.

Fig. 6.23. (next page)(A) Outcrop wall of the Drewer section in the Belecke anticline. The phosphate-bearing "Liegende Alaunschiefer" beds are the upper black shale horizon which is condensed in this section. (B) Various shapes of phosphate concretions from the "Liegende Alaunschiefer" formation of the Belecke anticline and the Harz Mtns. (C) An upper phosphate bed in the Drewer outcrop contains large, silicified phosphate concretions of a patchy appearance.

Fig. 6.24. (2 pages further) Microfacies of the "Liegende Alaunschiefer" phosphorites.(A) Structureless phosphatic micro-concretions are found in the Drewer section, base of photograph, 2.5 mm. (B) Phosphatic mudstone layer in the Drewer section. Thickness of the bed approx. 7 mm. (C) Internal texture of the common phosphate concretions. Apatite palisade cements envelope radiolaria and other sediment constituents. Base of the photograph 10 mm. (D) In parts of the phosphate concretions, the cements intergrow to a homogeneous phosphate layer. Black areas are late precipitates of pyrite. Base of the photograph 15 mm. (E-F) SEM photographs of the palisade cements.

Fig. 6.23.

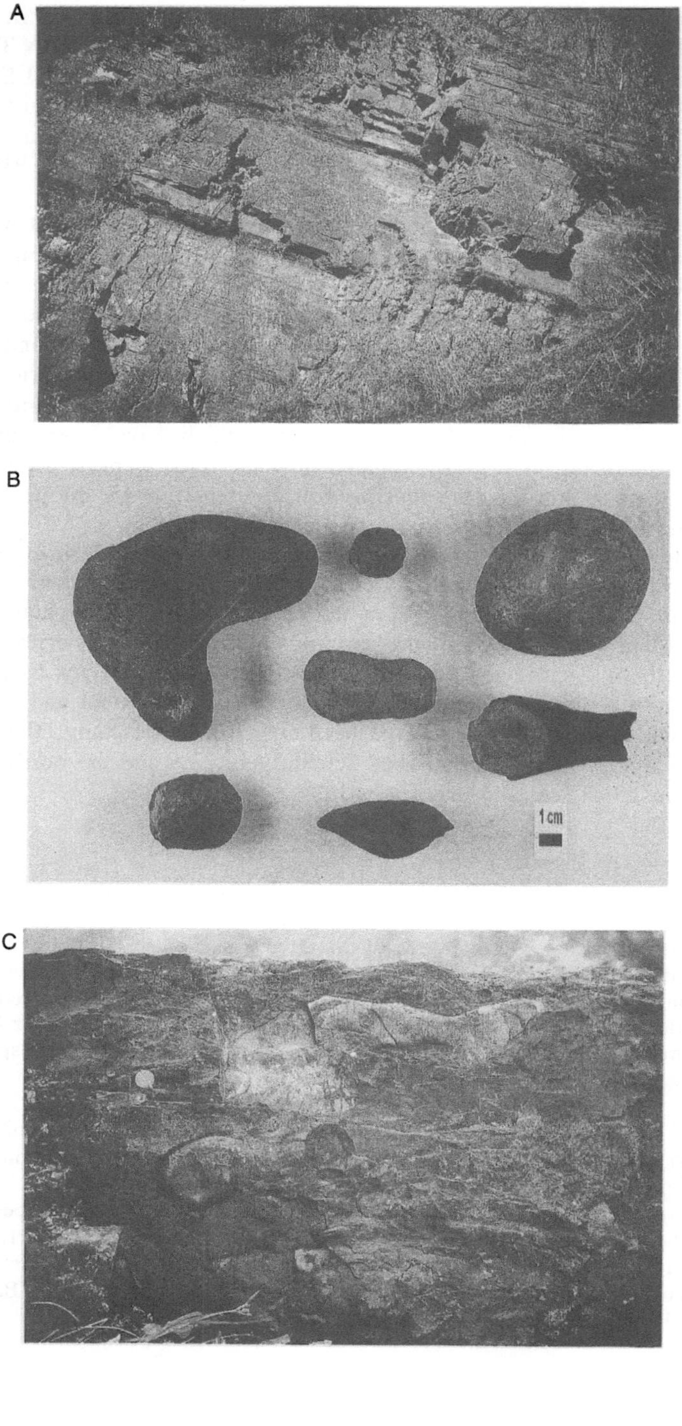

Fig. 6.24.

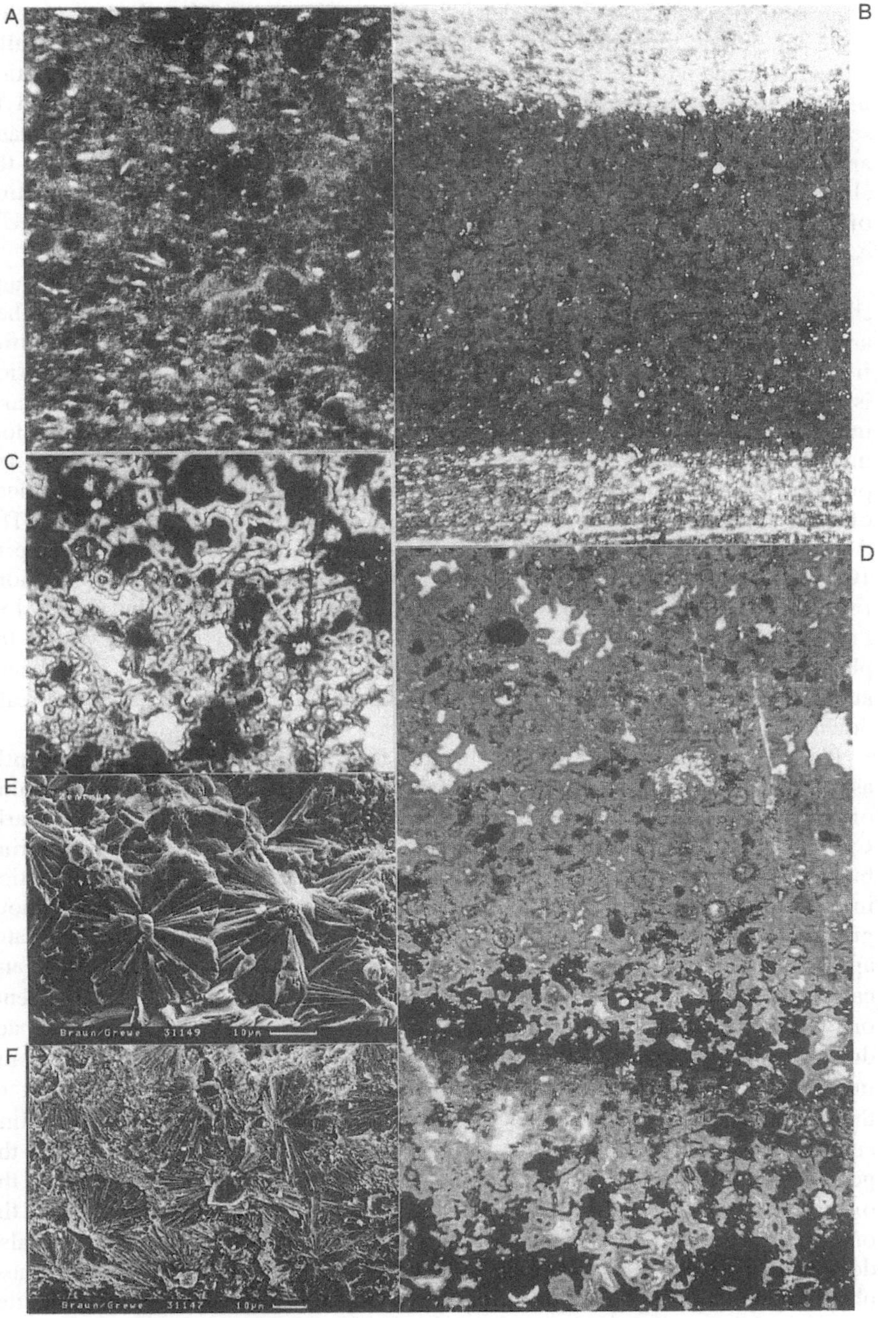

The "Liegende Alaunschiefer" Formation represents an organic matter dominated deposition with low detrital influx and suppressed carbonate production. The lateral transition in a fringing carbonate bank complex ("Kohlenkalk" facies) demonstrates the general favorable conditions for carbonate deposition ind the sea area. The absence of benthic carbonate production is related to sea-floor poisoning as the major cause. Sedimentation rates in central Germany are low in a scale of 0.04 to 0.2 mm/1000yr. A characteristic feature is the abundance of radiolaria, which also provide a source of silica for silicification of major parts of the sediment suite. ZIMMERLE et al. (1980), and BRAUN & GURSKY (1991) discussed volcanic ashes as the major source.

The phosphate concretions occur in various sizes (several mm to dm) and shapes (Fig. 6.23, 6.24). Intergrowing of several concretions is common. They are found scattered within the black shale suite as well as abundant in specific, from locality to locality changing horizons. Bioerosion or mechanical abrasion is absent. The nodules are silicified to various degrees and contain radiolaria in 3-dimensional preservation, whereas these are always flattened in the enclosing shales. This dates the growth of the concretions before compaction of the primary organic mud. The rapid growth of small concretions in organic muds off Peru was confirmed by precise dating (BURNETT et al. 1988). The Alaunschiefer sediments have C_{org} contents of usually less 1%, but with up to 10% in single horizons (STRUCKMEYER 1982). The phosphate concretions from the Harz Mountains always have C_{org} values less than 1%, but the enclosing shales up to 6%. The C_{org} contents within the concretions and within the phosphate bearing horizons are always lower than in the enclosing black shale and may indicate their development in areas of more intensive bacteria breakdown and P-liberalization of organic matter.

Phosphate mineralization, in the present form fluorapatite, occurs dominantly as palisade cements which have grown on radiolaria skeletons or remaining organic matter substrate. The untypical fluorapatite mineralogy of the Early Carboniferous phosphate fabrics is thought to result from carbonate loss during burial diagenesis (McARTHUR 1985). Zones of various degrees of pigmentation indicate precipitation generations, with the apatite crystals showing continuous growth. The formation of fibrous apatite cements around small nuclei generates apatite spherulites. Remaining pore space was later cemented with quartz and carbonate block cements. The mostly undisturbed growth of the apatite cements on various substrates requires the presence of a larger volume of pore space during growth. The formation of larger quantities of pore volumes in organic matter-rich muds is achieved by organic matter breakdown and dewatering of the sediment. The maintenance of the porous texture requires a coinciding cementation in the beginning stage. This developing framework stabilizes the generally settling sediment. The early cementation consequently preserves the original sedimentary texture, whereas the enclosing sediment is dewatered, the organic matter partially decomposed, and the sediment compacted. This is also documented in the different preservation of the radiolaria and dates the phosphate precipitation as a synsedimentary phenomenon. The presence of goniatites or plant remains in some concretions indicates fossil remains as a potential

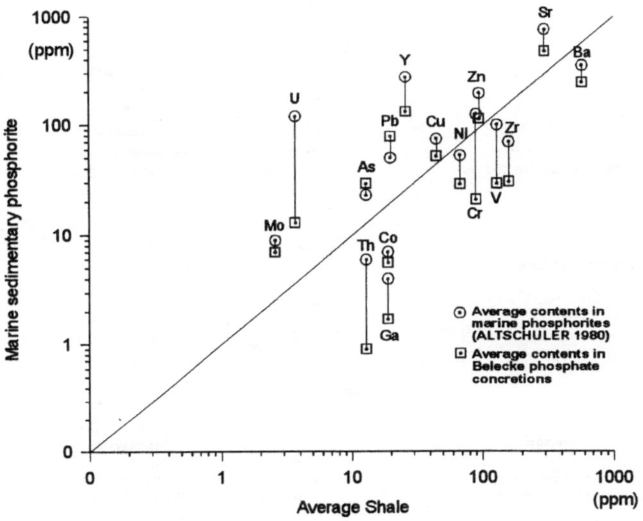

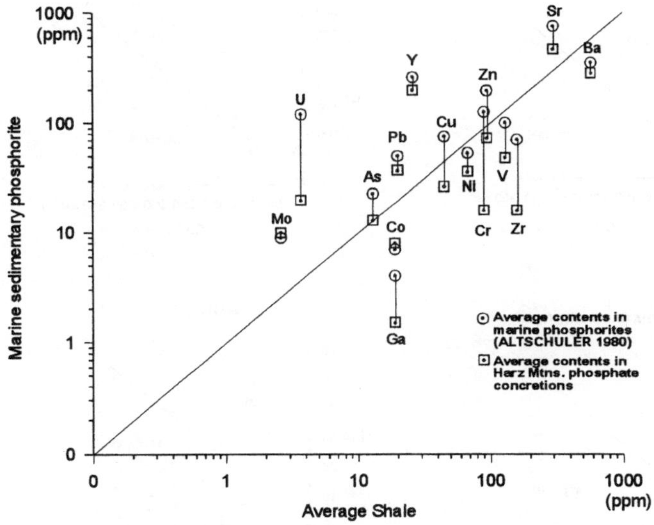

Fig. 6.25. Trace element characteristic of the Alaunschiefer phosphorites in comparision to the Average Marine Phosphorite (ALTSCHULER 1980) and Average Shale (TUREKIAN & WEDEPOHL 1961).

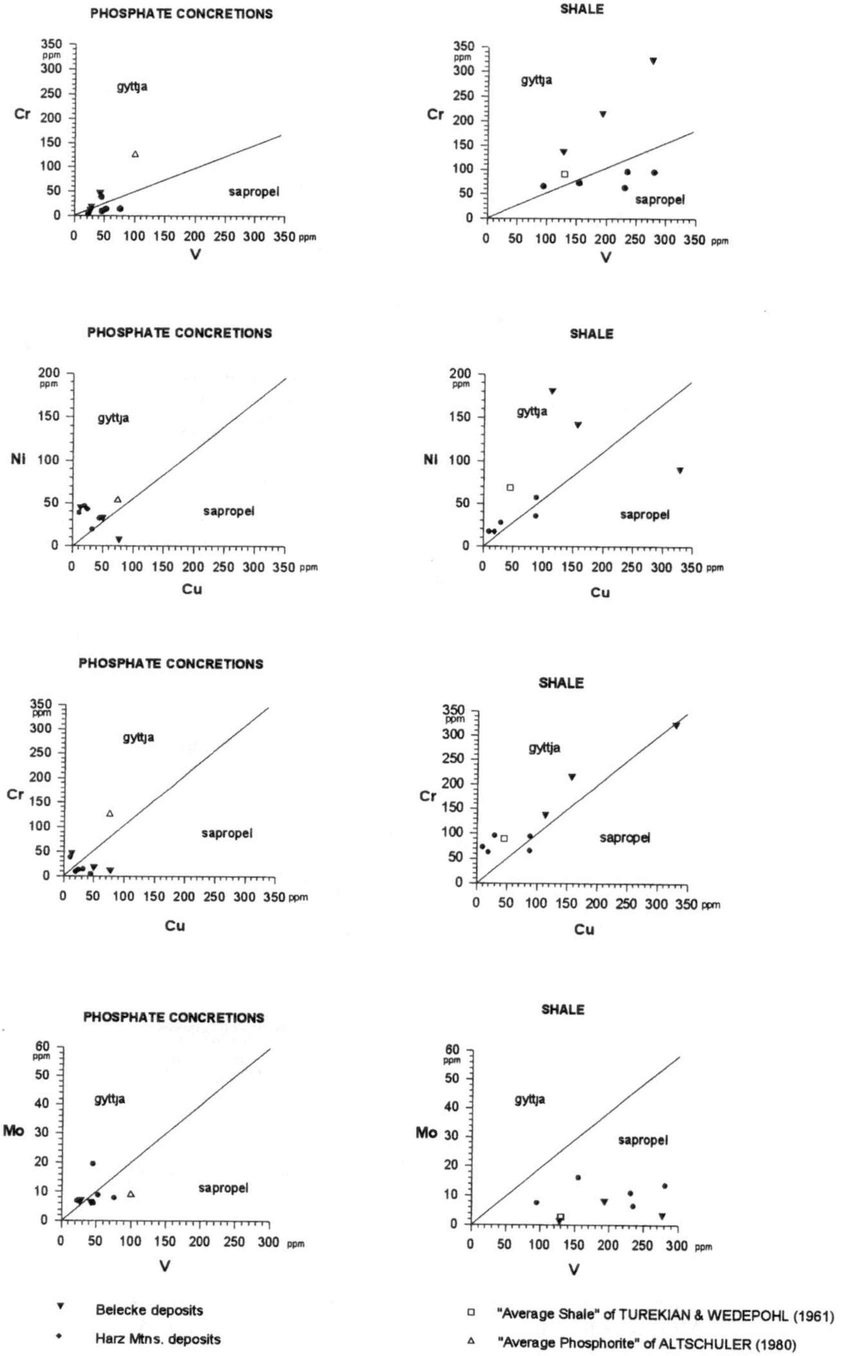

For explanations see next page

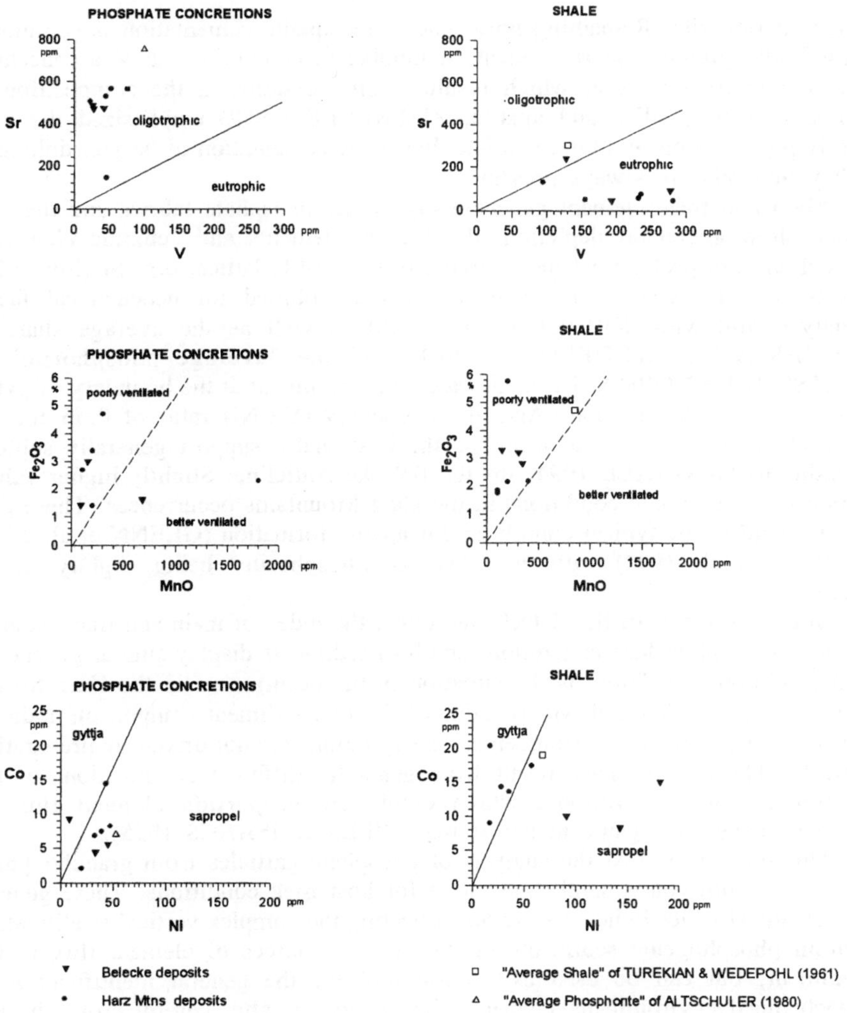

Fig. 6.26. Geochemical facies analysis after KRECJI-GRAF (1966) from the major and trace element composition of the investigated phosphorites and corresponding black shales from the Harz Mtns. and the Belecke Anticline. The various element ratios indicate partly different geochemical environments of the phosphogenic Alaunschiefer sediments, but all index ratios encirle a phosphogenic environment in the transition zone of sapropel to gyttja (suboxic to anoxic) sedimentation. It is important to note that also phosphate fabrics and corresponding shales may record different geochemical signals from a single index ratio. The V/Sr ratio in phosphate concretions is mainly driven by Sr substitution in the apatite lattice and cannot be used for geochemical environment analysis.

precipitation site. Remaining pore space after apatite cementation is commonly filled with granular quartz cement. A number of concretions show a concentric layered internal texture, which results from variations in the composition of organic matter, silica and apatite. ZIMMERLE (1992) emphasized the very early precipitation of albite, and late diagenetic cementation of baryte, dolomite. Clay minerals are always present.

Main and trace element geochemistry of the phosphate fabrics and the host rock show a general depletion of all trace elements and "euxinic elements" which are not preferential incorporated in the apatite lattice, e.g. Sr (Fig. 6.25, 6.26, see also chapt. 3). Element ratios are plotted for geochemical facies analysis following KREJCI-GRAF (1966) as well as the average shale of TUREKIAN & WEDEPOHL (1961) and the "average phosphorite" of ALTSCHULER (1980). These indicate an environment at the boundary of gyttja to sapropel sedimentation. Also the average V/(V+Ni) ratio of 0.56 for the phosphate concretions and 0.59 for the host shales support generally suboxic conditions (WIGNALL 1994) for the Belecke Anticline. Slightly higher values indicate more anoxic conditions in the Harz Mountains occurrences. These data correspond to the typical conditions for apatite formation (GLENN et al. 1994, JARVIS et al. 1994) with the "Average Phosphorite" being slightly anoxic (0.65).

In the sediments of the Belecke anticline, the index of main and trace element ratios of the phosphate concretions and host sediments display similar geochemical conditions. Different is the situation in the occurrences in the Harz Mountains. There, Cu/Cr and Ni/Co ratios of the two sediment components indicate different depositional environments. The variations are not driven by preparation effects. The discrepancies result from generally different compositions of the sediments and concretions, which emphasize proportion element flux by hydrogeneous or organic matter source (PIPER & ISAACS 1995).

This documents that the analysis of phosphate particles from granular phosphorites is not consequently indicative for host rock conditions. These general geochemical facies indicators are not reflecting the complex vertical stratification within phosphogenic sediments or the various sources of element flux in the sediment, but can be used as a basic tool for the general identification of geochemical environments favoring phosphogenesis after careful cross checks.

Different is the situation of the REE pattern of phosphate fabrics and host rocks (Fig. 6.28). The host rocks are showing generally a shale flat pattern and are depleted in comparison to the phosphate fabrics. The phosphate fabrics always have a distinct MREE (middle REE) enrichment and a relative HREE (heavy REE) depletion. This pattern was also recently observed in petrologically similar Pennsylvanian phosphate concretions of the Midcontinent/USA (KIDDER & EDDY-DILEK 1994). Both occurrences are characterized by the absence of a Ce-anomaly. The pattern is enigmatic and is not corresponding to any seawater composition. KIDDER & EDDY-DILEK (1994) discuss a fecal nature of the sediments, which emphasizes MREE enrichment. Fecal pellets have not been found so far in these sediments. MREE enriched and HREE-depleted patterns are known from all biologic phosphate precipitates (see chapt.

TRACE ELEMENT CHARACTERISTICS
OF DIFFERENT PHOSPHORITES

	Egypt (L. Cret.)		**Average phosphorite**	**Germany** (E. Carb.)	
	Restricted marine to estuarine type	Upwelling type	ALTSCHULER 1980	Belecke Anticline	Upper Harz Mts.
Neg. Cer-anomaly	weak	distinct	distinct	absent	absent
Ce/La	> 1.5	< 1.5	0.8	1.9 (1.9)	2.4 (3.0)
Y/ La	<1.5	>1.5	1.7	1.2 (1.0)	1.6 (1.0)
Th/U	> 0.1	< 0.1	0.07	0.06 (1.2)	0.04 (0.8)
V/Cr	1.5	1.0	0.8	1.6 (2.2)	3.7 (2.3)
	Egypt: GERMANN et al. 1987			(shale)	(shale)

Fig. 6.27. Trace element and REE ratios for the Belecke and Harz phosphorites in comparison to the ratios for esturarine and upwelling phosphorites in Egypt (GERMANN et al. 1987) and the "Average Phosphorite" of ALTSCHULER (1980). The latter indicates an upwelling nature for the marine phosphorite, but not the Alaunschiefer phosphorites. The variations are thought to be an effect of detritus content in the concretion.

3) and also from microbial phosphate ooids, both indicating a biologic mediation. The non-seawater pattern in the Alaunschiefer phosphorites, which evidently developed during a very early stage of organic matter breakdown in the sediment, must reflect a porewater composition. The correspondence to biologic REE patterns implies a mediation by microbial activity in the more and more enclosed microenvironment of the rapidly forming concretion. With the incor-

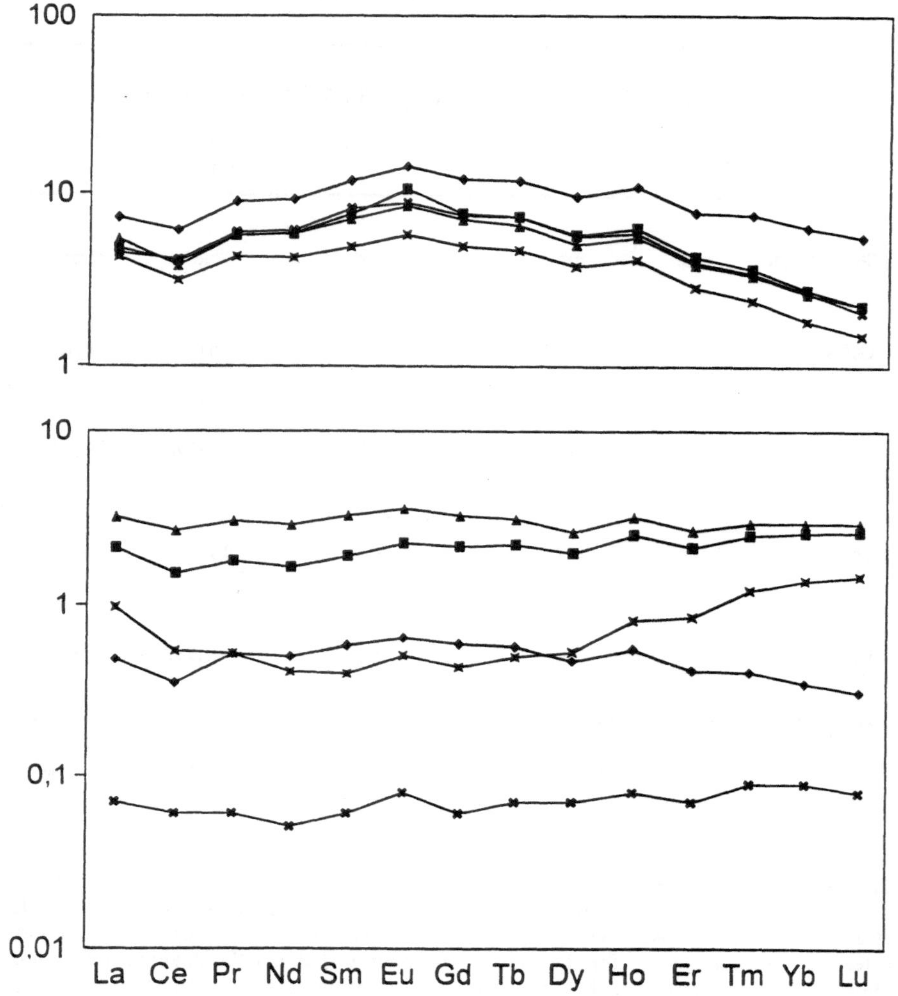

Fig. 6.28. NASC-shale normalized REE distribution of the phosphate fabrics (upper diagram) and the corresponding host rock (lower diagram) of the "Liegende Alaunschiefer" Formation in the Harz Mtns. and the Belecke Anticline. Sample pairs are indicated by the same icon. The fairly consistent MREE-enriched and HREE-depleted pattern of the concretions contrast to the inconsistent shale-like pattern of the host rocks.

poration of shale components, the REE pattern approaches progressively a shale distribution. It is important to note that REE patterns of phosphate fabrics not consequently correspond to a host rock distribution.

The absence of Eu- and Ce-anomalies in phosphate fabrics as well as in the host rock indicates an offshore environment, which lacks deeper oceanic water properties or influence. The element proportions of GERMANN et al. (1987) for the differentiation of fluvial and oceanic supply of phosphorus (see chapt. 3) give contradictory results.

The nature of the fabrics differ from micro-concretions found in other occurrences by their well developed cement (chapt. 4, Case Studies 4,5). But these fabrics show also some similarities. Both fabrics have a similar host rock and incorporate sediment components with micro-concretions usually having high sediment contents and non-oriented fibrous apatite crystals of much smaller size. The micro-concretions seem to reflect phosphate mineralization in sediments in which lacking pore space inhibits cement formation and triggers a more dispersed apatite formation.

7 Phosphorite genesis

In the majority of all phosphate sediments, phosphogenesis - the processes of authigenic and biological phosphate mineral precipitation - is followed by synsedimentary to postsedimentary mechanical alteration. These processes which result in the transformation of in-situ phosphate fabrics into clastic grains are summarized under the term phosphorite genesis. The alteration of the phosphate sediments in agitated water coincides with submarine weathering (e.g. organic matter oxidation), further abrasion of grains, bioerosion, additional coating of grains, as well as diagenetic overprint, e.g. silicification.

Grain formation by mechanical reworking and winnowing as a result of temporarily or permanently agitated water conditions is commonly accompanied by significant sorting and transport processes. These events affect the total reorganization of the primary sediment texture and bulk composition. The latter is caused by the removal or supply of external carbonate and siliciclastic sediment and furthermore by an accumulation of phosphate sediments in non-phosphogenic environments. As a result of sediment transport, autochthonous and allochthonous granular phosphorite deposits are forming. Again, paleogeography and eustasy define fundamentally different genetic pathways of phosphorite genesis. These pathways determine the grade of relative concentration of phosphate particles in granular phosphorites, the magnitude and direction of transport, as well as their final accommodation within the stratigraphic context. Considerable factors affecting intensity and character of the processes and the resulting facies architecture are scale of sea-level change, paleotopography of the sea areas, time, and frequency of events.

The relative concentration of phosphate grains is a substantial process for the formation and quality of valuable economic resources. The transport components significantly affect exploration strategies.

7.1 Grain formation

Two distinctively different mechanisms cause grain formation which are reworking and winnowing (SHELDON 1981). Reworking is defined as the physical destruction of lithified or semidurated sediments and is commonly associated with grain displacement. Winnowing describes a sorting process, which removes a non-lithified fine-clastic matrix and exposes the enclosed pristine phosphate fabrics.

Most in-situ phosphate deposits yield only low grade ores with pristine phosphate fabrics widely scattered in a host sediment. High-grade stratiform fabrics

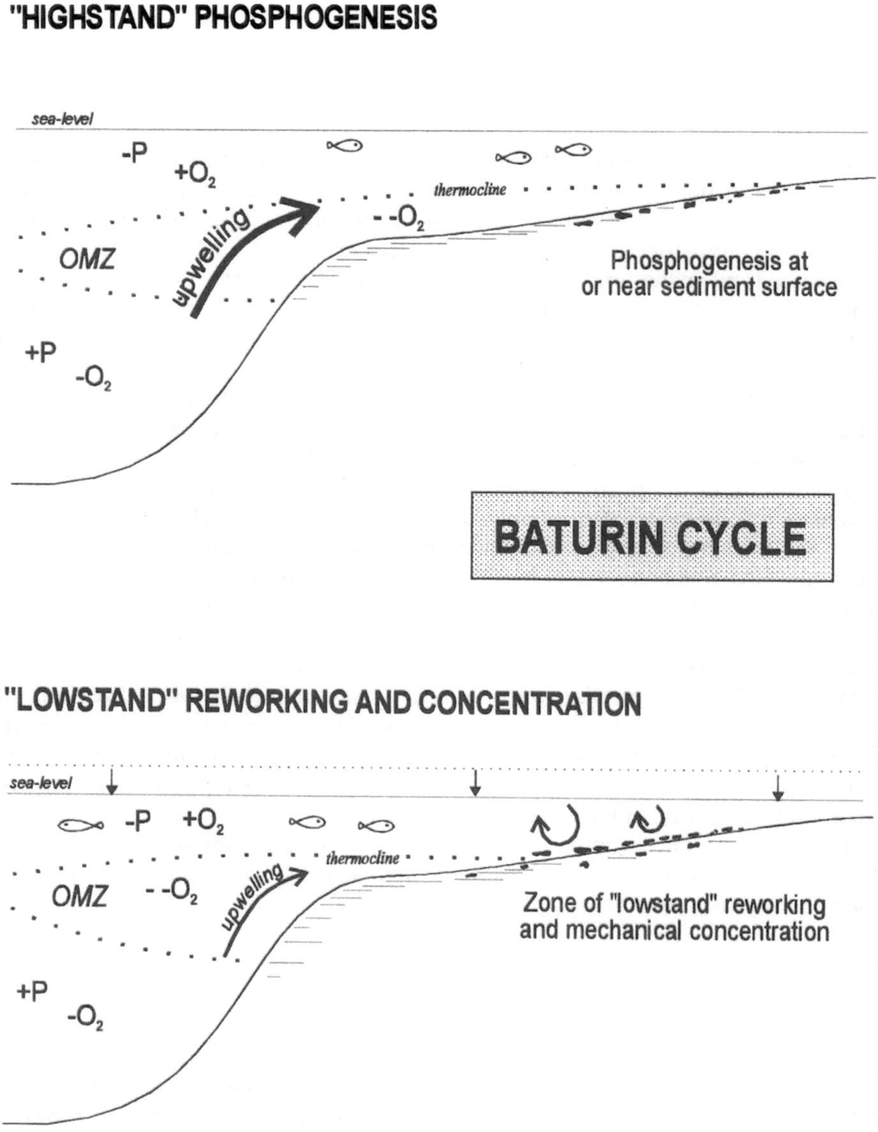

Fig. 7.1. BATURIN cycling. The mechanism describes the reworking or winnowing of "highstand" phosphorites during "lowstand". The mechanism is very much modified in most modern models of phosphorite formation.

such as phoscrusts, phosphatized hardgrounds or phosphatic stromatolites have only little thickness and are consequently also of only little economic importance. Most mineable deposits have been described as "pelletal" and microfacies investigations identify their granular textural compositions (RIGGS 1979a,b, TRAPPE 1992a,b, GLENN et al. 1994). The textural and compositional properties of phosphate particles clearly show evidence of being part of the allochem spectrum (see chapt. 4). The discussion of the formation of phosphate fabrics in sediments has identified the authigenic, microbial, and biological origin. All phosphatic allochem grains must have been derived from

(1) the reworking and winnowing of the initial "pristine" fabrics, or
(2) the phosphatization of grains being already allochems, or
(3) are bioclasts which are directly supplied by phosphatic shells, e.g. Ordovician deposits in Estonia (see Case Study 1) or basal Meade Peak sediments of the Phosphoria Rock Complex (see Case Study 4, SHELDON 1981).

The transformation of the pristine or in-situ fabrics into allochemical grains, and all subsequent clastic processes of mechanical alteration of phosphate grains, the mechanical concentration, grain transport, and redeposition are included in the process of phosphorite genesis. As the first, BATURIN (1971a,b) postulated a depositional cycle of reworking of primary phosphate fabrics, a process named later BATURIN cycling (Fig. 7.1). The mechanisms of phosphate mineralization, grain formation, and the subsequent depositional processes may be strictly successive or significantly overlapping. The resource quality is mainly determined by these depositional processes. The first stage, the grain formation is discussed in this chapter, the subsequent depositional processes in the following chapter 7.2.

Similar to the process of phosphogenesis, depositional environment and substrate determine various pathways of phosphorite genesis, which affect facies as well as paleogeographic and stratigraphic position of the resulting sediments. During grain formation, the most important controlling parameters are the different physical sediment properties, the types of phosphate fabrics in carbonate and siliciclastic sediments, and further the sediment dynamics in the different environments.

7.1.1 Grain formation in carbonate deposystems

Phosphogenic processes in carbonate deposystems develop mostly stratiform fabrics in form of phosphatic crusts and phosphatized erosional surfaces. The timing between phosphogenesis and grain formation is evidently synsedimentary in some cases, or follows after a significant time span up to several million years on omission surfaces (hiatus deposits) and during erosion of older strata. Three major physical mechanisms are responsible for phosphate grain formation in carbonate systems. Potential bioerosion was observed to be always important, but not burrowing.

Synsedimentary reworking of phosphatized carbonate mud and microbial crusts

In peritidal and shallow subtidal marine environments rapidly phosphatizing, non-lithified or semi-durated sediments, or fixating microbial colonies are ripped up by wave action during storm events. The mechanism is commonly known from carbonate sediments in peritidal settings and results in intraclast formation (RIGGS 1979, SOUTHGATE 1986, 1988). Recent carbonate examples are found along the shoreline in northwestern Andros Island Bahamas (HARDIE 1977, own observations). Due to the rapid hardening or lithification of carbonate muds, abundant structureless grain formation is dominantly generated during synsedimentary reworking.

The process of synsedimentary grain formation from phosphatized carbonate muds is best observed from the Middle Cambrian deposits in northern Australia (s.a. Case Study 2). Carbonate mud deposition on an intertidal to peritidal platform is frequently interrupted by oscillating sea-level changes resulting in formation of stacked hardground sequences. The hardground surfaces are rapidly phosphatized during continuous erosion. Undermining and cavity formation on these hardgrounds generate structureless and rounded phosphate grains which continuously are becoming spherical during winnowing. The grade of bioerosion by microorganism is uncertain in these middle Cambrian occurrences. The extraordinary preservation in the Georgina Basin occurrence documents all stages of the process.

Exhumation of older phosphatic grains through reworking or reworking of older completely phosphatized strata

Phosphatization of hardground surfaces which develop on completely lithified rock was commonly recognized. These disconformity surfaces in carbonate deposystems generally result from a major relative sea-level drop and the removal of the soft sediment column (RIGGS 1979a,b, 1980, 1984, JARVIS 1980a, 1992, PEDLEY & BENNETT 1985, POMONI-PAPAIOANNOU & SOLAKIUS 1991, CARSON & CROWLEY 1993, POMONI-PAPAIOANNOU 1994, REHFELD & JANSSEN 1995). The surfaces record various amounts of time according to the intensity and duration of erosion. The exposure of the lithified surface over longer time spans give way to physical and biological erosion. Besides reworking strictly by wave or current energy, bioerosion plays a significant role in most hardgrounds related particle forming processes. The significance of the latter increases with the development of hardground dwelling faunal assemblages during the Phanerozoic. Formation and erosion on hardground surfaces in carbonate systems was sumarizing described in BROMLEY (1975) and FÜRSICH (1979). Phosphatization of the surfaces under ongoing erosion results in the formation of marginally phosphatized carbonate intraclasts. The process is also of large scale importance for the formation of the Neogene Florida phosphorite deposystem (see Case Study 5).

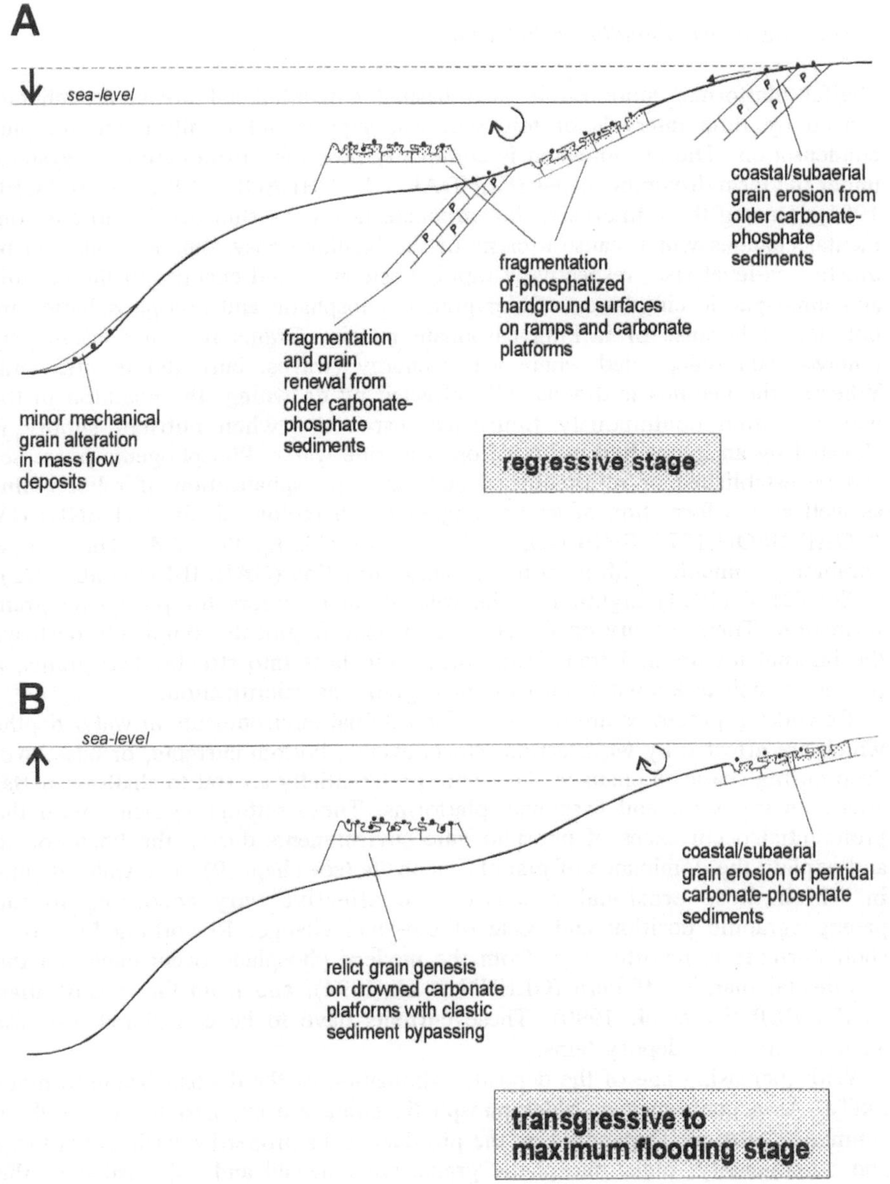

Fig. 7.2. Processes and mechanisms of phosphate grain formation in a carbonate deposystem. A: Processes during regressive stages, B: Processes during transgressive to maximum flooding stages.

Recycling of vagabonding relict grains

Shelfal platforms, more rarely also basinal epicontinental areas, are characterized by time intervals of low sediment supply and resulting stratigraphic condensation. The phenomenon is commonly described from carbonate systems under the term drowning stages (KENDALL & SCHLAGER 1981, SCHLAGER 1981). During these intervals, the carbonate factory is shut off due to environmental changes which cause a crisis of the benthic ecosystem. Reasons can be drastic sea-level rise, respectively rapid subsidence, and changes in the oceanic and atmospheric circulation. Older grains, phosphatic and non-phosphatic, are not buried because of lacking carbonate matrix. Grains remain continuously renewed and redeposited where water energy (waves, currents) is sufficient. Whereas the benthos is dramatically affected by drowning, the plankton in the water column continuously flourishes, especially when nutrient supply is elevated by an intensified influx of open marine water. Phosphogenic processes can be established or re-established and cause phosphatization of relict grains as well as the formation of apatite aggregates in moldic bioclasts (KENNEDY & GARRISON 1975, SHELDON 1981, see also CASE STUDY 4). The process commonly coincides with local hardground formation (GARFIELD et al. 1992).

SOUDRY (1994) highlighted the role of micro-borers for phosphate grain formation. Their activity on the surface of bone fragments completely destroys the internal texture and transforms former bioclasts into structureless grains, a process which is known from carbonate grains as micritization.

Reworking processes are bound to depositional environments in water depths which are affected by wave action, storm events, bottom currents, or base-level drop during sea-level changes. These are preferentially coastal to shallow shelfal areas, epeiric seas, and carbonate platforms. These settings coincide with the preferential occurrences of phosphogenic environments during the Phanerozoic and explain the dominance of granular deposits (see chapt. 9). The water depths in which these erosional processes are affective vary according to the paleogeographic position and scale of sea-level change. Reworking by cross-shelf currents is reported e.g. from the modern phosphate occurrences on the continental margin off Peru (GLENN et al. 1994), and from the E-Australian shelf (O'BRIEN et al. 1990). These currents have to be considered also for ancient carbonate deposystems.

With increasing age of the deposit, exhumation of the deposit becomes more likely. As a consequence, older phosphatic strata are exposed to subaerial or submarine erosion. Depending on the physical rocks properties such as hardness and homogeneity, older phosphatic grains are renewed and redeposited or the entire phosphatic rock is reworked and grains are derived from the complete rock. The processes are commonly known from mixed carbonate siliciclastic systems where soft marls or argillaceous carbonates are exposed to erosion. Examples are the Pliocene phosphate beds in Florida and the Carolinas which originate from Miocene rocks (SCOTT 1988), or common phosphate grains on the shelf off Morocco (SUMMERHAYES et al. 1972).

7.1.2 Grain formation in siliciclastic deposystems

Grain formation in siliciclastic systems follows different mechanisms than in carbonate systems. This contrast results from the different physical properties of the two sediments, from the different phosphogenic pathways, and the different primary phosphate fabrics. Most phosphogenic environments in non-carbonate deposystems develop in organic matter-rich mudstones, e. g. the Phosphoria facies of the Permian Phosphoria Rock Complex (McKELVEY et al. 1959, SHELDON 1981, see also Case Study 4), the Neogene Monterey deposystem in California (MULLINS & RASCH 1985, GARRISON et al. 1990, FÖLLMI & GARRISON 1991, GRIMM 1992, SCWENNICKE 1992) or the modern sediments off Peru (GLENN et al. 1994). Pristine phosphogenesis in sand or sandstones is more exceptional (SWETT & CROWDER 1982, FÖLLMI 1989a,b, 1990).

The dominating in-situ phosphate fabrics in fine clastic sediments are micro-concretions, microsphorite horizons (phosphatic mudstone and phosphatic crusts), microbial mats, and phosphatized coprolites to uncertain degree. Micro-concretion growth and phosphatic mudstone formation is a phenomenon of the entire subtidal shelf and of central portions of epicontinental basins. In coarse clastic sediments, local phosphate cementations along surfaces or patchy within the sediment are the most important fabrics (FÖLLMI 1989a, see also Case Study 6). The formation of phosphatic ooids and oncoids is also known from these lithologies (MABIE & HESS 1964, HORTON et al. 1980, SWETT & CROWDER 1982, SCHWENNICKE 1992, TRAPPE & ELLENBERG 1992). Four processes of grain formation characterize this depositional system.

Winnowing of soft pelitic muds

Slow lithification behavior of siliciclastic muds serves for long time spans in which these sediments can be winnowed. The limiting factor is mostly the cohesive behavior of clays. In areas of low subsidence, which means low burial and low tectonic stress, shales can remain soft for more than 500 Ma. A good example are soft Lower Cambrian clays in Estonia which contrast to overlying completely lithified carbonates (see KALJO 1990). The most common phosphate fabrics in pelitic, organic matter-rich muds are micro-concretions. BURNETT et al. (1988) could show that in modern sediments of similar lithofacies on the shelf off Peru, these micro-concretions grow and become hard within months or years. This behavior is also supported from observations in ancient deposits (see also Case Study 3). In comparison to the phosphate fabrics, the enclosing shales remain soft much longer, especially if these are water saturated. In the association of harder phosphate fabrics in soft siliciclastic muds, an increase in water energy due to currents, relative sea-level drop, or storms affect selective sediment transport. The mud is "winnowed" and leaves the micro-concretions as larger and heavier particles behind. As a consequence, the phosphate

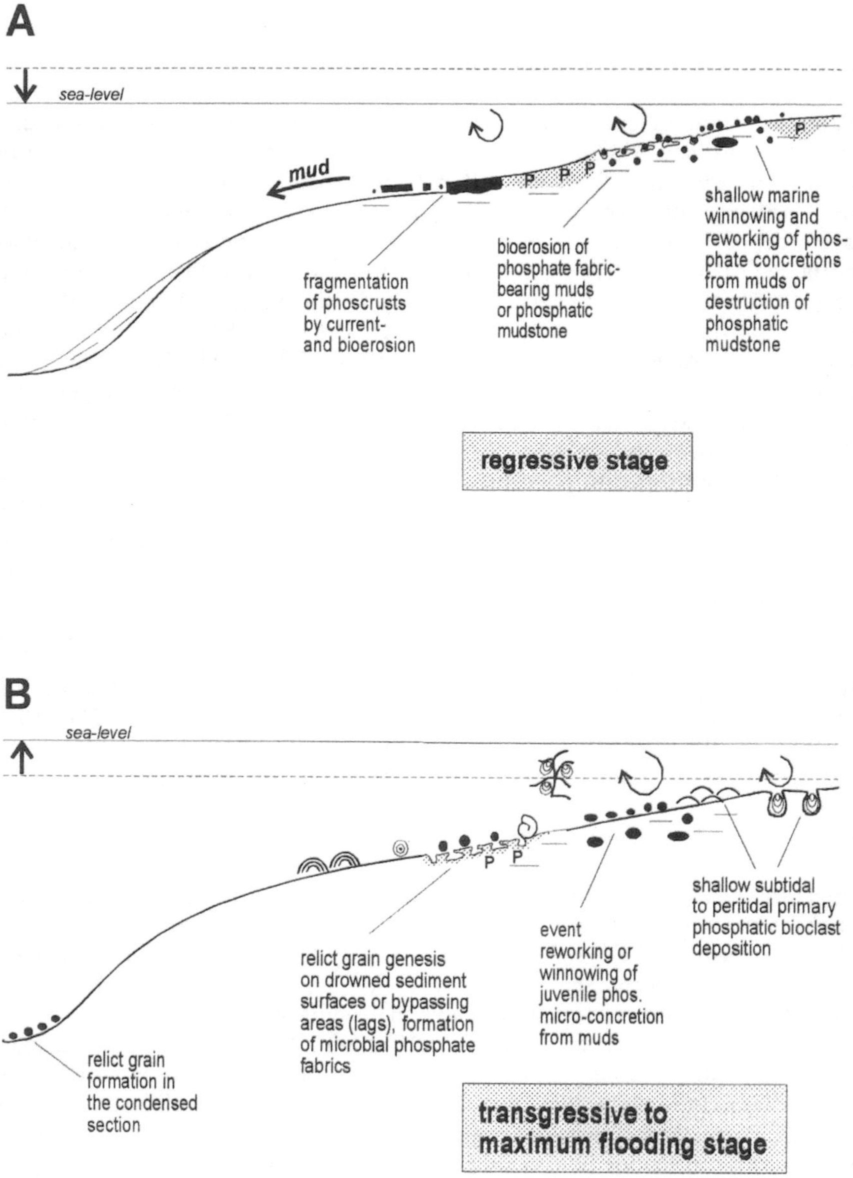

Fig. 7.3. Processes and mechanisms of phosphate grain formation in a siliciclastic deposystem. A: Processes during regressive stages, B: Processes during transgressive to maximum flooding stages.

particles become gradually concentrated. This sedimentary process is also evident in most ancient black shale related granular deposits and is a phenomenon of most shallow marine and peritidal environments. The Phosphoria Formation in the Western Phosphate Field/USA displays nicely the sedimentary fabrics in all stages (CRESSMAN & SWANSON 1964, SHELDON 1981, see also Case Study 4). Winnowing is the most significant process of the mechanical concentration of phosphate fabrics and the formation of valuable economic phosphate deposits. In many deposits, the process coincides with substantial transport of the grains, which is being discussed in the following chapter 7.2.

Reworking of phosphate fabrics from siliciclastic rocks

In coarse grained siliciclastic sediments, phosphate mineralization occurs as patchy cementations of the pore space and causes varying resistances to erosion (see also chapt. 4.3.4). Exposure and reworking of these rocks selectively preserve the harder phosphate cemented portions of the sediments. Further abrasion produces gravel. The mechanism is documented in the mid-Cretaceous Graschella/Brisi Fms. of the Helvetian Alps (FÖLLMI 1989a, 1990), but was also recognized in the Neogene Florida system (see Case Study 5).

Renewal of relict grains and fossils on omission surfaces

Another process is the phosphatization, renewal and redeposition of relict grains in condensed sections or on omission surfaces. Low net sedimentation rates of siliciclastics or pelites, but a continued rain of organic matter generates in some cases oxygen deficient conditions and an environment favorable for phosphogenesis (see chapt. 6.3). In most occurrences phosphatization of relict grains and formation of microbial phosphate fabrics, phosphate impregnation of cementation of fossils on the sediment surface is commonly associated with authigenesis of glauconite and corrosion (FÖLLMI 1990). The mechanism, which is also known from carbonate platforms (see above), is associated with the drowning of shelves (FÖLLMI 1989) or is related to sediment starvation in distal parts of basins, respectively outer shelf to slope. The sediment supply in siliciclastic systems is allogenic controlled. The main source are the continental areas. During rising and relative high sea level stages, material from the continents is deposited in the marginal part of the basin or shelves, where in the ideal text book situation, meaning accommodation space is available, the maximum deposition rates are achieved. In the basin center or on the outer shelf, sediment starvation develops during the maximum transgression when the source area is shifted in maximum distance. Omission surfaces or condensed sections are formed (LOUTIT et al. 1988). Bottom current activity may emphasize the granular appearence of these sediments. Regressions remobilize the newly deposited sediments in the marginal sea areas and initiate downslope sediment transport which terminates the starvation phase.

Destruction of phoscrusts and microsphorite horizons

GARRISON & KASTNER (1990) and GLENN et al. (1994) report irregular phosphate crusts from the modern occurrences off Peru which are formed at or near the surface of a soft organic matter rich, fine clastic sediment. These crusts are undermined and fragmented by bottom currents and bioerosion generating gravel and sand-sized grains. In the Lower Carboniferous occurrences in Germany (see Case Study 3), but also reported from some other deposits (Australia: DE KEYSER & COOK 1972, Israel: SOUDRY 1992), microsphorite or phosphatic mudstone layers are developed within the suite of phosphorite-bearing sediments. In the German occurrences (see Case Study 3), the microsphorite beds are partly ripped-up and contribute to structureless grain spectrum.

In all four types of grain formation in siliciclastic systems, strong currents during frequent winnowing intervals finally create relict deposits of phosphate grains. The process is best developed when sea-level drop is intense and rapid. Grain formation during the lowstand is followed by the accumulation of relict phosphate sands during the earliest transgressive stage. These beds are known as transgressive lag concentrates (VAN WAGONER et al. 1990, LOUTIT et al. 1988, GLENN et al. 1994). Sea-level changes in shelfal and epicontinental setting show greatest impact in the more marginal and shallow portion of the sea. Consequently, these beds are best developed in the marginal parts of a basin.

A special role for grain formation in siliciclastic systems play burrowing macroorganisms. The significance of etcobentos for the development of carbonate hardground surfaces was pointed out earlier. In siliciclastic systems, with mostly softground environments, the dwelling organisms are mostly endobenthos. Alteration of the geochemical micro-environment in burrows leads not only to phosphogenesis (ZIMMERLE 1994), but also to phosphate grain mobilization. The activity of sediment feeders in organic matter rich, phosphate bearing sediments commonly concentrates the harder phosphate grains in nests or traces during their burrowing activity (Fig. 4.3).

7.2 Concentration - the formation of autochthonous granular phosphorite beds

The total P_2O_5 values of pristine phosphate rock are determined by the relative apatite contents of the phosphate fabrics, but even more by the ratio of phosphate fabric to host rock volume. The first condition is a consequence of the phosphogenic process, respectively of the substrate incorporation. Slight chemical alteration and enrichment is only affected by weathering, in particular by organic matter decomposition (see also chapt. 3.8). More significant for the relative enrichment of the phosphate constituent of the sediment is the reduction of matrix as a coinciding affect during grain formation.

Larger pristine phosphate fabrics, e.g. nodules of the Alaunschiefer Fm. in Germany, or microsphorite beds with P_2O_5 contents of 30% occur usually in very thin beds or are found widely scattered in a non-phosphatic host sediment. As a consequence, these deposits are not economically mineable. Economic P_2O_5 values of more than 20% for total phosphate rock of larger sediment bodies are only accomplished in granular deposits in which the transformation from in-situ fabrics to clastic grain coincides with a relative concentration of the phosphate particles in the resulting sediments. Nearly all economic phosphate deposits belong to the latter group. Concentration processes were recognized in many, but not in all deposits (BRITISH SULFUR CORPORATION 1987, GLENN et al. 1994).

The relative concentration of phosphate grains by matrix reduction is achieved by two fundamental processes which are (1) the mechanical concentration of apatite particles or (2) the synsedimentary destruction of parts of the matrix sediment via chemical processes. Mechanical concentration of phosphate particles results from the various forms of sorting in aquatic environments. By the predominant occurrence of phosphogenic systems in the marine realm, these are processes in marine high energy environments. Fluvial concentrates are exceptional and only important for bone bed formation. Processes differ in response to the substrate.

The partial or complete removal of the fines by winnowing is important in sediment compositions of hard or semidurated phosphate particles within a non-lithified mud. The process results in relative concentration of the larger, harder and heavier apatite particles. The process was commonly observed in pelitic or mixed carbonate-siliciclastic deposystems (SHELDON 1957, 1963, 1981,

Table 7.1. Important concentration mechanisms within specific environment

supratidal:	karstification, weathering, fluvial sorting (bone beds)
peritidal:	offshore shift of wave base during sea-level drop, storm events, landward shift of shoreline above lagoons during sea-level rise
shallow shelfal/ basinal subtidal:	enhanced bottom current activity by climatic change, tropical storms, sea-level drop of various scale of cyclicity, event concentrations
basinal:	enhanced turbidite activity, carbonate dissolution

McKELVEY et al. 1959, CRESSMAN & SWANSON 1964, RIGGS 1979a,b, BATURIN 1971a,b, 1982, COOK 1976b, BATURIN & BEZRUKOV 1979, SCRÖTER 1986, GLENN 1990, GLENN & ARTHUR 1990, TRAPPE 1991, 1992a, GLENN et al. 1994, SIEGMUND 1995).

Sorting by specific weight within the sand-size fraction during a winnowing process occurs in coarse siliciclastic deposystems (COOK 1967). The concentration process has been recognized in Ordovician strata in Australia (COOK 1967, 1972) and has also been considered for granular phosphorites in the Shedhorn Sandstone of the Permian Phosphoria Rock complex (SHELDON 1971, THORNBURG 1990). Winnowing and sorting were also described from the mid-Cretaceous strata in the Helvetian Alps (FÖLLMI 1989a,b, 1990).

Driving forces are periods of elevated water energy which are provided by enhancement of bottom current activity (FÖLLMI 1990, GLENN et al. 1994) or wave action (SHELDON 1981, GLENN & ARTHUR 1990, TRAPPE 1991). The resedimentation in the high energy regime generates typical sediment structures such as cross-stratification, submarine dunes, graded bedding etc. (see chapt. 4.1). The grade of mechanical concentration depends on the intensity, duration and frequency of the winnowing intervals. The significance of single parameters are usually difficult to determine in ancient deposits. The processes of grain formation and concentration may overlap with sediment transport and repeated phosphogenesis, e.g. during phosphogenesis in relict grains.

Non-mechanical phosphate particle concentration results from chemical dissolution or microbial degradation of non-phosphatic rock constituents (SHELDON 1981). These processes are related to subaerial or submarine weathering, and their significance and existence remains mostly uncertain. In carbonate deposystems, exposure of phosphate-bearing rocks leads to preferential dissolution of the more soluble carbonate constituents and the relative concentration of the less soluble apatite along erosional surfaces. An example is the so-called residual "brown rock" of Tennessee and Alabama which is thought to be formed by Holocene leaching of Middle Ordovician phosphatic limestones of the Nashville Group (CATHCART 1989, 1991). Concentration of phosphate fabrics by karstification has also importance in insular phosphate deposits. In organic matter-rich sediments, degradation and leaching of the organic matter portions of the sediments by synsedimentary to recent surface weathering decrease the matrix to an uncertain extent. The process becomes evident by the contrast of white or light-gray phosphorite sediments in surface outcrops and in a dark organic matter-richer subsurface composition, e.g. in the Georgina Basin/Australia (DeKEYSER & COOK 1972, SHERGOLD & SOUTHGATE 1986) or central Morocco (pers.com. J. LUCAS & L. LUCAS-PRÉVôT 1994).

7.3
Sediment transport and redeposition - the formation of allochthonous phosphorites

Most mechanical concentration processes involve a relative displacement of grains by agitated water. The grain transport is understood herein as an observable redeposition of phosphate material in depositional environments or textural context, which are evidently different from the primary phosphogenic environment. The phenomenon summarizes various mass transport processes, and has significant consequences for the development of exploration strategies. Direction, intensity, extent, and stratigraphic context are controlled by the paleogeography and by sea-level changes.

7.3.1
Downslope sediment transport

Mass sediment transport in classical paleogeographic concepts is downslope directed. This gravity margin-to-basin transport of phosphate particles is best known from the Neogene deposits in California and Baja California (GARRISON et al. 1987, GALLI-OLIVER et al. 1990, FÖLLMI & GRIMM 1990, FÖLLMI & GARRISON 1991), where the paleogeographic setting widely corresponds to the modern occurrences off Peru. The sedimentology and ecology of the Neogene mass transport sediments in California were investigated in detail by SCHWENNICKE (1992, 1994), FÖLLMI & GRIMM (1990) and GRIMM (1992). Phosphatic turbidites and gravity flows have been also recognized in some other Cretaceous deposits in Columbia (FÖLLMI et al. 1992), Israel (KOLODNY & GARRISON 1994), Egypt (GLENN et al. 1994), and Jordan (ABED 1989, 1994). Trapping of gravity-transported granular phosphorites in structural lows was described from the Israelian and Jordanian deposits (BENTOR 1953, BEERBAUM 1977).

These deposits are characterized by isolated phosphate beds intercalated in a non-phosphoritic sediment suite showing graded bedding, allochthonous bioclasts, and a trace fossil assemblage typical for turibiditic beds (FÖLLMI & GRIMM 1991).

7.3.2
Vagabonding upslope transport

Many ancient phosphogenic deposystems developed in extremely shallow epeiric basins or broad platform-like shelfal environments with no significant slope or dipping margin-to-basin profile (see chapt. 9). The extent of broad and shallow epicontinental sea areas during Phanerozoic earth history emphasizes these paleogeographic and sedimentological constellations which are poorly developed in the modern world. Significant gravity transport abates with decreasing dip of the basins or platform profile. Other transport mechanisms than gravity transport are becoming more accentuated. Due to their shallowness, entire epeiric basins are intensively affected by eustatic sea-level fluctuations and storm events (HECKEL 1977, WIGNALL 1994). The granular phosphorites from these settings show evidence by sediment structures and textural composition of intense temporary or permanent remobilization and resedimentation in agitated water (e.g. SCHRÖTER 1986, TRAPPE 1991, 1992a,b). With increasing preciseness of paleogeographic reconstructions, granular phosphorites have been identified which have been accumulated in marginal non-phosphogenic environments of the basins. Source of these granular phosphorites are still shallow, but distinctively more offshore phosphogenic sediments (TRAPPE 1991, 1992a,b,c). These sediments document commonly underestimated vagabonding sediment transport or in the context of transgressive onlap also shoreward-directed sediment shift. Several sedimentary processes lead to this style of sediment transport:

(1) In shallow subtidal environments, granular phosphorites are accumulated in sand waves (GLENN & ARTHUR 1990). The formation of these sediments is commonly known from epicontinental siliciclastic deposystems (BRENNER & DAVIS 1974, SPEARING 1976, BRENNER 1978, SWIFT 1974). The sediment bodies migrate or decay with changing sea-level. Driving forces are tidal- and storm-induced currents. The accumulation of shoals from granular phosphorite sediments is also apparent in the peritidal/intertidal zone (see also Case Study 1).

(2) On broad, essentially flat platforms or extreme shallow ramps with low net-sediment supply, relict granular phosphorite blankets are formed during low sea-level stages. Common remobilization of pavements of phosphate grains from various sources leads to the formation of vagabonding phosphorite blanket sands. During transgressive onlap, these sands blanket sands are shifted coastwards by vagabonding sediment transport. These high-energy sediments finally come to rest on a distinct erosional surface as a transgressive lag concentrate. The process is commonly known from siliciclastic sediments in North American cratonic sedimentation (DAPPLES 1955, SLOSS 1963, DOTT & ROSHARDT 1972). The formation of transgressive lags of this type is also evident in the Phosphoria Rock Complex (see Case Study 4). Transgressive phosphorites were also recognized in transgressive black shale deposystems (WIGNALL 1991).

(3) Tempestites on carbonate platforms which cause winnowing and grain displacements with no defined transport direction may also be present in the Thorntonia Fm. in Australia, though sediment transport on this carbonate platform is most probably minor (see Case Study 2).

(4) Coastal parallel transport was noticed from phosphate-bearing sandstones of the Shedhorn Formation by SHELDON (1972), but is usually not bound to sediment shift into different depositional environments and consequently difficult to recognize.

In many basins, granular sediment concentration, remobilization and redeposition was observed. Although, a transport component was assumed, poor knowledge of paleogeography does not allow to determine the direction and degree of transport.

7.4 Pathways of concentration and sediment transport

All mechanical concentration and transport processes in granular phosphorites result from a shift of low-energy phosphogenic environments toward high-energy conditions. These environmental changes can be induced by short-term events, e.g. storms, or by long-term episodes of various temporal scale, e.g. sea-level change. Eustatic cyclicity affects fundamentally different pathways of mechanical reorganization during rising or falling sea-level.

The drop of base-level during regressive stages affects an increase of water energy over most parts of shallow basins, platforms of shelves. This causes permanent elevation of water energy as well as increased frequency of storm sediment mixing. In deeper environments, granular phosphorites are mobilized and concentrated by preferential removal of fines in siliciclastic dominated deposystems when bottom current activity in enhanced during the initial phase of sea-level fall. These highstand phosphorites have been recognized by GLENN (1990a). Continuing coastal offlap and the lowering of the erosional base on the continent enhances siliciclastic sediment supply and progradation of deltaic complexes. The increased clastic sediment flux begins to override the concentration of phosphate grains in shallow basins. In environments with a significant slope, increased sediment mobilization coincides with gravity sediment transport. But extensive sediment transport overwhelms the concentration processes. Isolated sheets of granular phosphorites are formed in basinal positions (low stand prism). In carbonate deposystems, base-level drop initiates hardground formation by removing the non-lithified sediment body. This leads to the formation of relict grain pavements. This is further enhanced when the carbonate production is reduced.

The following transgressive stage is characterized by low clastic sediment supply due to coastal onlap and a rising erosional base level on the continent, which supports concentration processes of granular phosphorites. In margin-to-basin profiles with significant dip or on isolated platforms, subtidal and peritidal phosphorite sand bodies are successively becoming drowned and buried. On shallow, broad ramps, these sediment bodies are transformed within the landward migrating zone of agitated water in relict blanket sands or basal conglomerates (e.g. TOURTELOT & COBBAN 1968, MILTON & BENNISON 1969, BRAID 1978, see CASE STUDIES 1 & 4). This reworking and concentration coincide in places with sediment shift in landward direction. In coastal positions, grain formation and concentration are affected by peritidal erosion as well as formation of tempestites (see Case Study 2). The redeposition of lowstand products initiates an episode which was identified being favorable for the formation of phosphogenic systems (see chapt. 6). The coexistence of phosphogenic sediments which are partially transformed into granular phosphorites

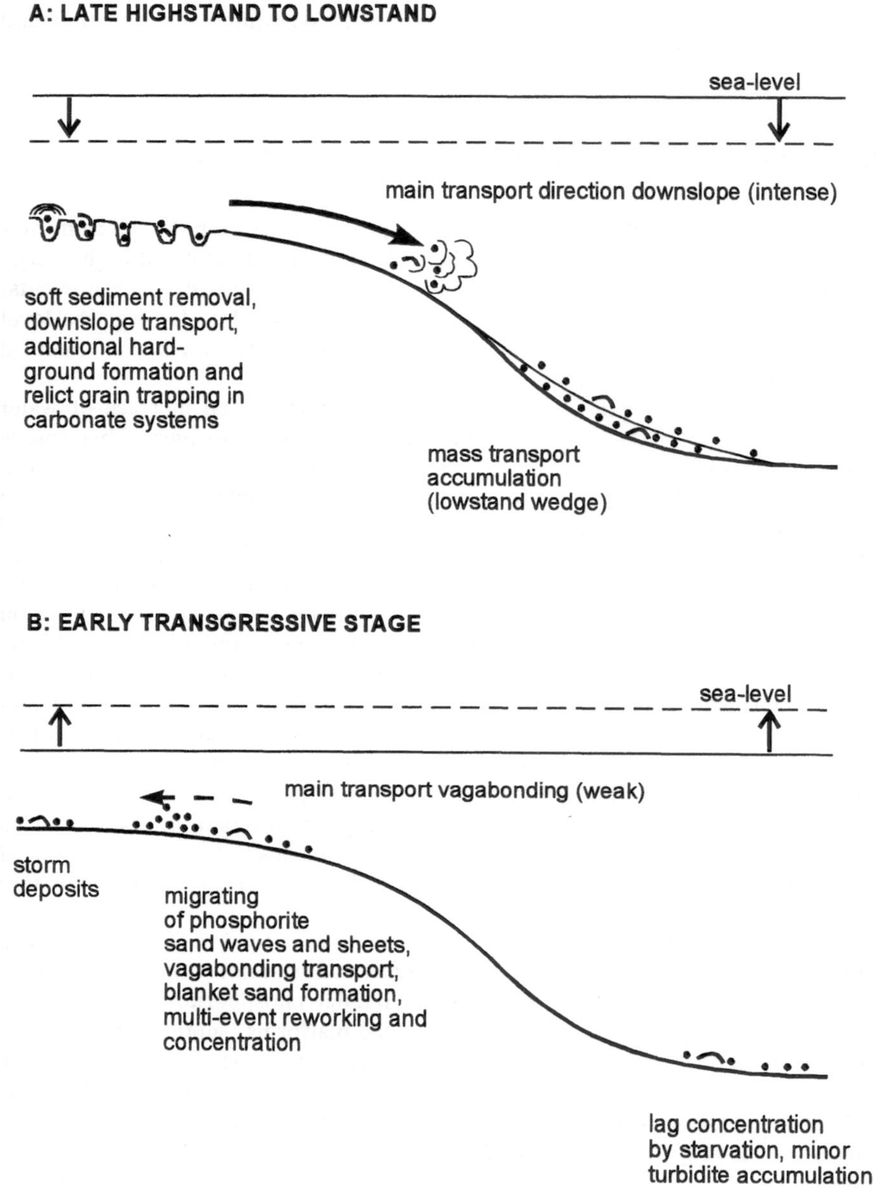

Fig. 7.4. Pathways of phosphate sediment concentration and transport in the various environments along a margin-to-basin profile in relation to sea-level change as the most important force. A: late highstand to lowstand interval. B: transgressive interval.

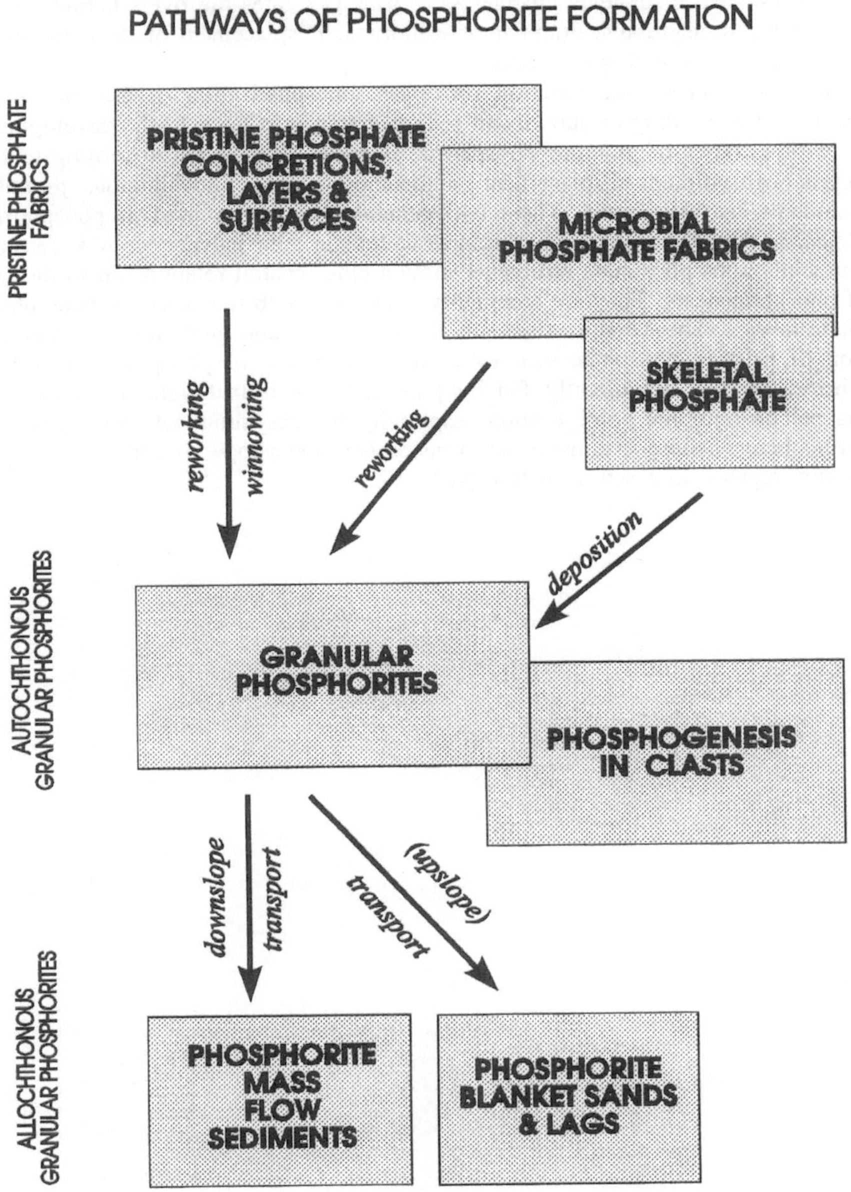

Fig. 7.5. Schematic diagram explaining the genetic relationships of the various types of granular phosphorites.

during low-order sea-level changes or storms generates the overwhelming portion of the transgressive granular phosphorites. Displacement of these grains is commonly vagabonding transport.

The mechanical concentration processes, phosphatization of granular sediments and coinciding or subsequent sediment transport leads to the development of various types of granular phosphorite deposits (Fig. 7.5). According to the degree of sediment displacement, autochthonous and allochthonous granular sediments are recognized. These two general types differ in their phosphorite facies, in their lithofacies association as well as in their paleogeographic setting. Autochthonous phosphate sediments show a close spatial relationship to the site of phosphogenesis. They are commonly associated with remaining pristine phosphate fabrics. These sediments result from chemical and mechanical reorganization of pristine phosphate sediments, the destruction of phosphatic microbial fabrics, or they are directly fed by phosphatic shell material. Depending on the paleogeographic configuration, especially the magnitude of slope and sea-level change, these sediments are transported downslope or are accumulated during vagabonding sediment transport.

CASE STUDY 4
Multi-lithological pericratonal sag-basin deposystem, Phosphoria Rock Complex (PRC), USA, mid-Permian

Anatomy of the deposit

Location: Wide distribution in the middle and northern Rocky Mountains and northern portions of the Great Basin in Idaho, Montana, Wyoming, Utah, westernmost Nevada; 114°30'to 107°W and 40° to 47°N (Fig. 7.6).
Stratigraphy, age: Lithostratigraphic system established by McKELVEY et al. (1959) in which the black shale-chert facies was restricted to the Phosphoria Formation while its calcareous equivalents and the sandstone facies were assigned to the Park City and Shedhorn Formations, respectively. A basal carbonate suite is named Grandeur Member of Park City Formation. The phosphorites, which occur in two distinct stratigraphic levels were named Meade Peak Member and Retort Phosphate Member. The massive chert succession within the marine suite was included in the stratigraphic system (Rex Chert Member, and Tosi Chert Member). McKELVEY's working group noted the complicated interfingering of the various lithologies which convinced YOCHELSON (1968) to summarize the Permian marine rocks informally under the term "Phosphoria Rock Complex" (PRC). Age: mid-Permian (Artinskian to Kazanian); Grandeur Member: middle to late Leonardian; Meade Peak Member: (late Leonardian) Rodian, all overlying rocks including Retort Phosphate Member: Wordian, ?Capitanian (Fig. 7.7).
General paleogeography: Westcoast position of the North American Continent (Fig. 7.6), Paleolatitude 13° to 25°N (GOLONKA et al. 1994), shallow marine large gulf between uplifts of the Ancestral Rocky Mountains (N: Milk River Uplift, SE: Uncomphagre Uplift), low relief coast to the east with transition into red beds (Goose Egg Fm., MAUGHAN 1965), MAUGHAN (1994), WARDLAW et al. (1995); 25° C average temperature and probable humid conditions in the southerly portion; probable upwelling sites at the continental margin and at the northern edge of the gulf (GOLONKA et al. 1994).
Paleotectonic setting, basin geometry: Pericratonic sag basin above cratonic basin, probably shallow bowl like, silled to the west; some inherited structural elements of the Ancestral Rocky Mountains (MAUGHAN 1990, 1994).
Depositional system, facies architecture: Mixed siliciclastic-carbonate-black shale-phosphorite, diagenetic overprint, intense silicification. Distinct lateral facies zonation, central portion (Sublett Basin), Phosphoria facies: black shale, phosphorite, spiculitic and diagenetic cherts; marginal portion, Park City facies: fringing carbonate ramp along the southern margin (Confusion Shelf) and eastern margin (Park City Shelf); Shedhorn facies: siliciclastic barrier island complexes along the Milk River Uplift. Maximum phosphorite occurrence: transitional Phosphoria to Park City facies. Cyclic vertical facies repetition.
Sea-level control, sequence stratigraphic interpretation:Transgressive/regressive cycle of 2-order scale resembling three 3-order cycles (Grandeur, Astoria

Hot Springs with Meade Peak phosphorites, Minnesela with Retort phosphorites), 4-5-order cycles evident (PETERSON 1984, ROSS & ROSS 1987).
Type of phosphorite particles: Wide range of granular, replacement and in-situ fabrics. Granular: phoslithoclasts, phosintraclasts (including rip-up clasts), phosooids, phosoncoids, phosbioclasts, nucleated grains; replacement: phosphatized intraclasts, phosphatized bioclasts; in-situ phosphate fabrics: phosphatized carbonate hardgrounds, phosphatic microbial mats, ?microsphorite beds, phosphate concretions.
Model for the phosphogenic processes: Replacement, biogenesis, authigenic-concretionary; bound to transgressive flooding.
Phosphorite formation: Sea-level derived during 3 to 5-order cycles, tempestites, nearshore: wave action.
Reserves, resource quality: The Phosphoria Rock Complex (PRC) represents the largest phosphorite resource of the world with resources of roughly 500 billion tons, but only 12 billions mineable in open pit mines and 13 billions in underground mines (less than 300 m below entry level)(CATHCART 1991).

The unique diversity of phosphate fabrics within a multi-lithologic deposystem, the diversity of phosphorite concentrates, and the clear facies architecture make the Phosphoria Rock Complex (PRC) to a key deposit for genetic studies on particle generation and concentration processes as well as for the general stratigraphic and paleogeographic context of large scale phosphorite formation. Although intensive research on the deposits produced a large database on stratigraphic, petrographic and geochemical results, the enormous extent, the complex lithologic facies relationships, and the diagenetic overprint including extensive silicification require an elaborated investigation to uncover the genetic processes. The diversity of phosphate particles againreveals the multifarious genetic nature expressed by the box model of phosphogenesis, which was broadly discussed in the Case Studies 1 to 3. In the course of this chapter, the focus will be here on granular phosphorite formation and its paleogeographic context.

The Phosphoria Rock Complex (PRC) was intensively investigated during the 50ies and 60ies by a USGS working group with McKELVEY et al. (1956, 1959), SHELDON (1963), SWANSON (1970, 1973), CRESSMAN & SWANSON (1964), MAUGHAN (1976, 1980, 1983, 1984, 1994), and numerous other researchers. Later, the paleogeography and depositional history were reinterpreted by PETERSON (1980a,b, 1984). In the following, TRAPPE (1992b,c,e, 1993) developed a new dynamic model for the phosphorite deposition in the PRC on the basis of intensive microfacies investigations.

The phosphorite facies within the Phosphoria Rock Complex is dominantly bound to two distinct stratigraphic intervals, the Meade Peak Phosphate Member and the Retort Phosphate Member (Fig. 7.7). In all other stratigraphic intervals, phosphorites are very exceptional. Phosphate mineralizations occur in three major forms:
(1) phosphate rock of granular appearance,
(2) phosphatized surfaces, and
(3) phosphatic shale.

A

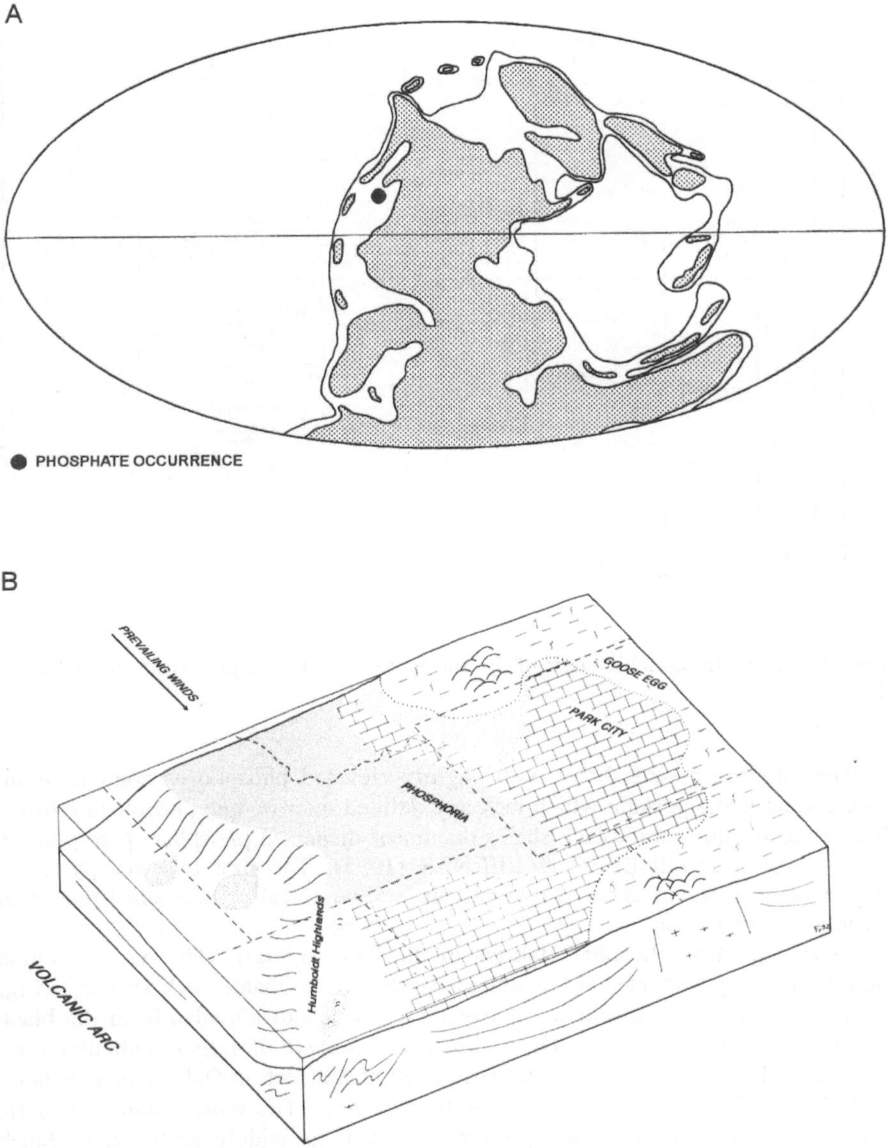

Fig. 7.6. A: Global paleogeographic location of the Phosphoria sea (Sublett basin) based on paleogeographic reconstructions from GOLONKA et al. (1994). B: Block diagram showing the paleogeography and general facies distribution of the mid-Permian Phosphoria sea.

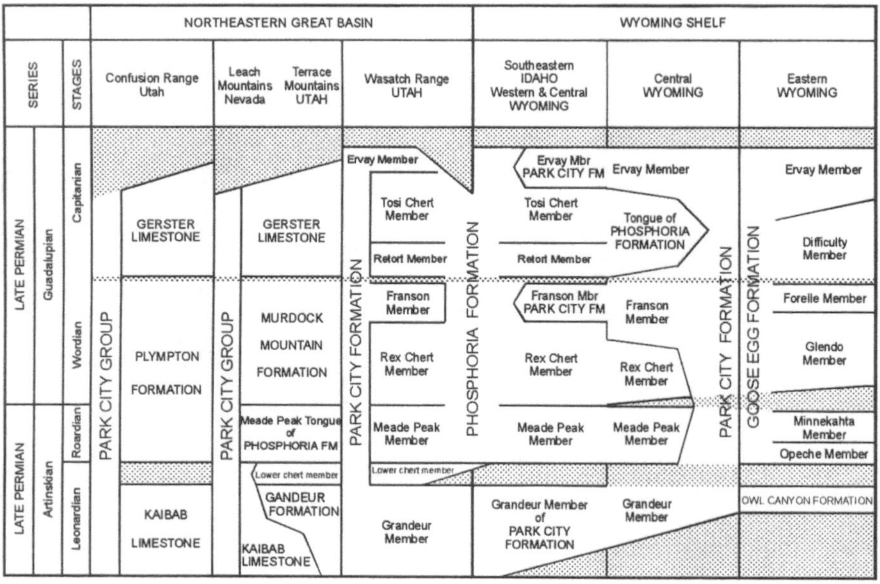

Fig. 7.7. Lithostratigraphic chart of the Phosphoria Rock Complex (from MAUGHAN 1994)

The latter represent shales with slightly elevated phosphorus contents compared to an average shale which lack any defined microscopic phosphate fabrics. These mostly dolomitic black shales document dispersed phosphate precipitation in sense of RUTTENBERG & BERNER (1993). The first two groups reveal phosphorite rocks of economic significance. Sporadically, biological/microbial structures are present.

Clearly pristine phosphate fabrics in the PRC are rare. The most common non-granular phosphate fabrics are phosphatized carbonate rock surfaces (Fig. 7.9). Structureless, phosphatic micro-concretions occur abundantly in the black shale host rock (Fig. 7.8). The sedimentary context of larger lenticular concretions, which are found in southwestern Montana (section Daly Spur/Montana, SWANSON 1970), indicate a pristine preservation. The more common spheric micro-concretions occur densely packed as well as widely scattered in black shale matrix. Densely packed thin horizons with the apparent absence of any intergrowth of concretions indicate concentration by winnowing of the pelitic mud andstratigraphic condensation (event concentration). The degree of matrix removal in scattered occurrences is beyond resolution of the sedimentary record and has to remain uncertain. Exceptionally, phosphate mudstone is found which shows a microbial fixation (Fig. 7.8). Stratiform mats are extremely rare in the PRC, but occur abundantly during earlier phosphatic events in the Rocky

Mountains and Great Basin (OBERLINDACHER & ROBERTS TOBEY 1986, NICHOLS & SILBERLING 1990, SILBERLING & NICHOLS 1991, SILBERLING et al. 1995, and own observations).

The phosphate facies in the PRC is dominated by phosphorite types of granular appearance. The formation of the various fabrics clearly responds to the offered substrate and evidently supports the box model. In adaptation of the depositional environment, structureless micro-concretions are formed which are definitively linked to the presence of an organic matter-rich, pelitic host rock. Phosooids are very abundant (Fig. 7.8). To a major portion, these are derived from replacement of carbonate ooids (see discussion in chapt. 4). Their origin as well as that of phosoncoids (Fig. 7.8) in other occurrences within the PRC indicates primary microbial growth of laminated coatings around structureless phosclasts. Evidently resulting from primary apatite precipitation is the appearance of phosphate shell fragments, which are very abundant in some beds (Fig. 7.8). Shells of foraminifera also act as a precipitation site for the growth of nucleated grains as well as quartz grains and other bioclasts (Fig. 7.8). Where carbonate allochems are present, these grains are being renewed and phosphatized according to their texture. Phosphate impregnations of the micropore space characterize crinoid fragments (Fig. 7.8). In bryozoans, clear and structureless micro-concretions are precipitated (Fig. 7.8). Both types are commonly associated with glauconitization. Mollusk shells and also bryzoan fragments are recycled with their internal molds being filled with phosphatic mudstone (Fig. 7.8). Carbonate intraclasts are marginally phosphatized, similar to carbonate erosional surfaces (Fig. 7.9). Most granular phosphate fabrics are less than 2 mm in diameter, but in coarse clastic beds phosphate fabrics reach up to several centimeters. The granular phosphorites are forming packstones,

---

Fig. 7.8. Microfacies of biological and microbial phosphate fabrics of the Phosphoria Rock Complex. (A) Silicified packstone composed of structureless phosclasts and large primary phosphatic shells with mircoboring along their surfaces. (B - C) Phosphatic mudstone was fixated by microbial mats. (D) Partly washed phosooidic grainstone with the remaining mud matrix being phosphatized. (E) Silicified phosooidic packstone. Ooids have large nuclei of sediment clusters and are charactized by irregular coatings. (F - G) Phosbioclastic packstone with particle sizes up to 2 mm. Bryozoan fragments have phosphatic sediment preserved in the zooecia. In crinoid fragments, easily recognized by their lattice structure, the interlattice space is filled with apatite precipitates.

Fig. 7.9. Microfacies of granular phosphorites from the Phosphoria Rock Complex. (A) Silicified, bimodal phosclastpackstone. The phosclasts of approx. 1 mm grain size show only slight pigmentation by organic matter and sediment incorporation. Some clasts have also been coated. The grain spectrum is complemented by well rounded phosclasts of similar texture by significant larger grain size. This textural composition contrasts the poorly sorted packstone of phosclasts with various stages of pigmentation and internal grain texture. (B) Poorly sorted packstone to rudstone with dominantly phosphatized carbonate intraclasts. (C) Phosphatic packstone with reworked bioclasts with internal phosphatized mud fillings (base of the photograph = 20 mm.)

See also color copies of these plates in the appendix.

Fig. 7.8.

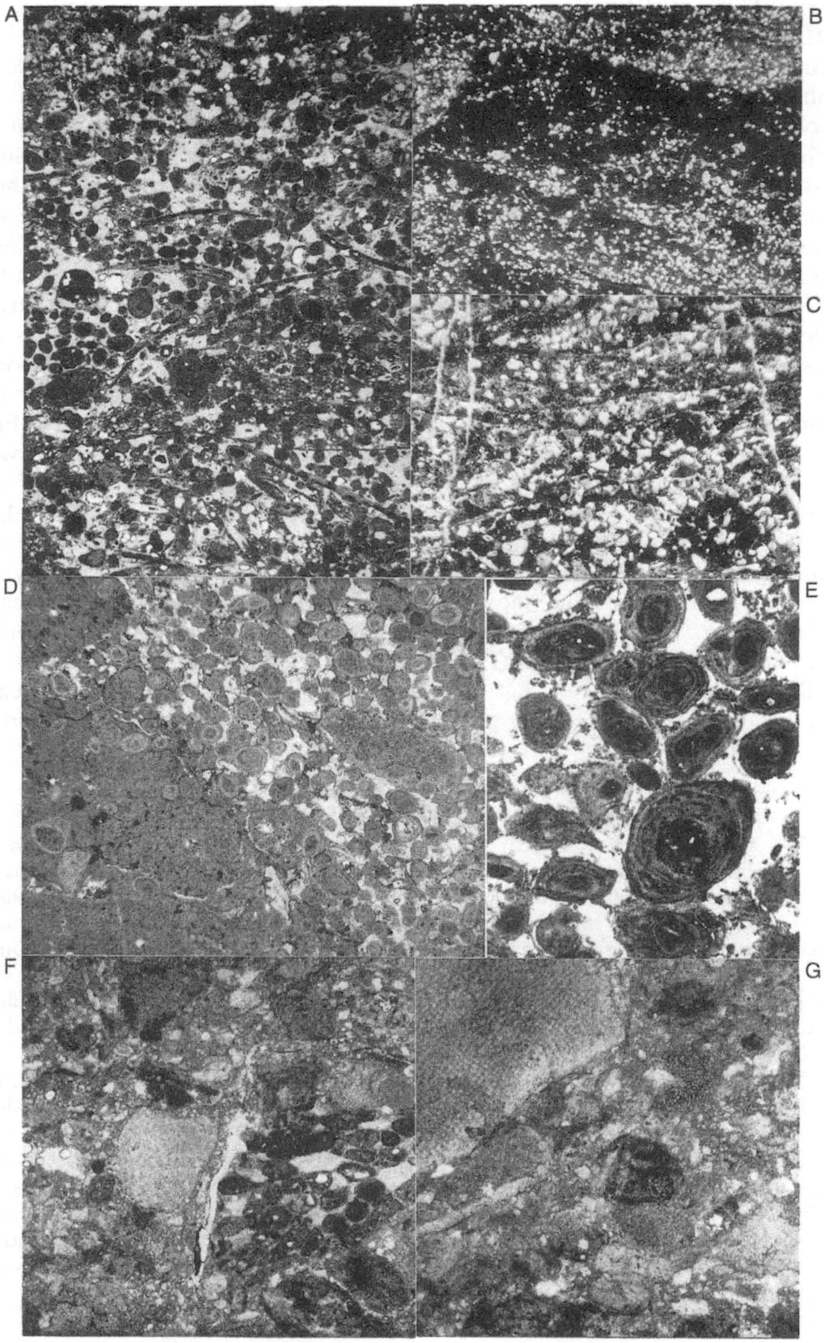

Fig. 7.9.

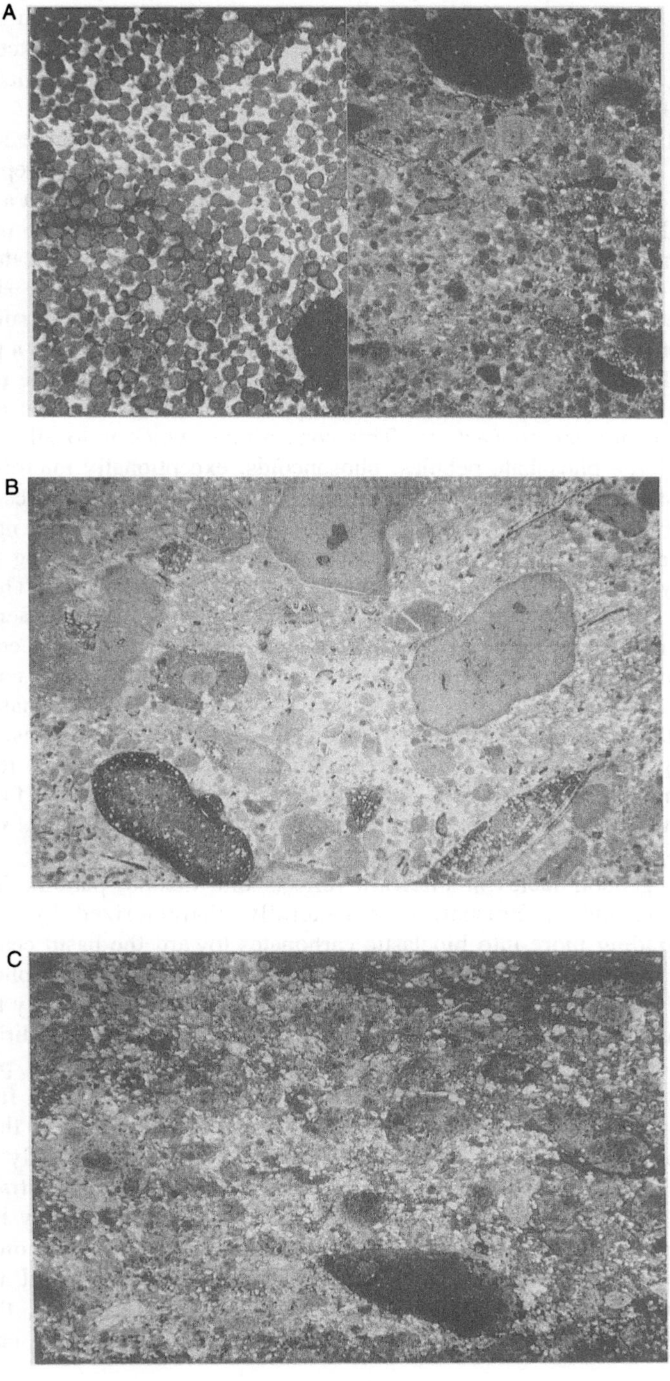

wackestones, rudstones, and rarely grainstones. The non-phosphatic grains composition in phosphorites comprises the complete carbonate allochem spectrum, glauconite, and detrital quartz. The matrix is calcareous or pelitic, exceptionally phosphatic, commonly diagenetically silicified, dolomitized, or calcified (COOK 1969, 1970).

The vertical and lateral facies relationships within the phosphorite suites and within the complete PRC are characteristic for many phosphate deposits and indicative for the stratigraphic emplacement of the phosphate facies in a deposystems. The Meade Peak and Retort Phosphate Members appear to be markerbed suites in most of the PRC, while the other lithologies are complicately interfingering. This persistence of the phosphate suites does not exclude significant internal facies variations. In most parts of the basin, the phosphate suites show distinct vertical successions (Fig. 7.10). This general pattern remains a persistent feature, independent from local and basinwide facies variations. The phosphate facies always rests on a more or less conspicuous erosional surface and starts with a phosphatic lag concentrate. This concentrate, which is locally silicified, consists of large phosphate pebbles, phosoncoids, exceptionally microbial mats, phosbioclasts, and subordinatly of all other granular phosphate fabrics. The lag at the base of the Meade Peak Member is well known under the name Fish Scale Bed and contains abundant vertebrate remains. This basal lag is superimposed by a phosphorite succession of composite concentrates. These beds form large lenticular sediment bodies which show intense lateral facies variations. The rapidly changing and lenticular nature of this succession contradicts any correlation effort. The upper portion of the phosphate suite consists of phosphatic shales with single, thin event concentrations of phosphate grains. The importance of this uppermost portion decreases toward the basin margin emphasizing the composite concentrate. In the central portion of the basin, phosphogenesis generally ceases. Lateral changes in the phosphate facies concerning particle type and degree of mixing coincide with lateral facies variations of the non-phosphatic facies from subtidal to peritidal conditions.

Also the general facies architecture reflects this distinct pattern. The non-phosphatic Grandeur Formation is generally characterized by carbonate mudstone grading more into bioclastic carbonates toward the basin center (Fig. 7.12). The two non-phosphatic successions superimposing the two phosphorite suites show a distinct lateral facies architecture, the upper one is partly truncated by post-Midpermian erosion. The basin facies of black shales, spiculitic cherts, as well as silicified carbonates to minor degree correlates with a phosphate facies, which is dominated by structureless phosphate grains derived from winnowing of micro-concretions. Advancing toward the basin margin, the uppermost portion of the non-phosphatic sediments is becoming gradually replaced by bioclastic carbonate rock (Fig. 7.11). The sediments in this transitional position are widely silicified (Rex Chert, Tosi Chert) or are slightly bioclastic carbonates with abundant chert nodules. The phosphate facies is becoming rich in phosphate ooids. The lag concentrate is rich in clasts derived from the reworking of phosphatized carbonates. In more marginal positions, the entire pelitic non-phosphatic suite is becoming progressively replaced by carbonate,

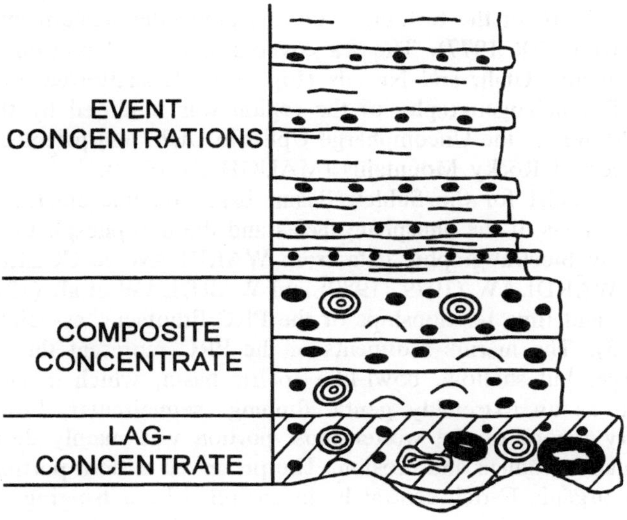

Fig. 7.10. Typical succession of a granular phosphorite concentrate in the Meade Peak or Retort Phosphate Member of the PRC. The basal lag-concentrate is commonly silicified and is composed of a wide range of phosphate grains and microbial fabrics. The grains spectrum in the composite concentrate is much lower, but packstone beds of phosphorite are still forming. Upsection, event concentrations form thin layers of uniform grain type in shales.

including calcareous mudstones with abundant nodular chert superimposing the phosphorites. These phosphorites are composite concentrates from structureless phosphate grains, renewed and phosphatized bioclasts, and phosooids. Glauconite occurs sporadically. The sediments in the most marginal positions consist of bioclastic carbonates, fenestral limestones, neomorphic dolostones with evaporite replacement nodules locally. Phosphorites are dominated by phosphatized and glauconitized relict grains and carbonate replacement grains. Structureless grains are locally also common, but are very poorly sorted. In this position, any phosphate host rock is absent and a cyclicity apparently occurs in the carbonate suite (GARFIELD et al. 1992). In the northeastern portion of the basin, this general lithofacies pattern is modified by the occurrences of the Shedhorn Sandstone which largely replaces the Park City carbonates (THORNBURG 1990b). This pattern reflects the interplay of paleogeography and sea-level change (Fig. 7.11, 7.13 to 7.16).

The paleogeographic setting of the epeiric Phosphoria sea which covered a pericratonic sag basin (Sublett Basin) and adjacent cratonic areas (WARDLAW & COLLINSON 1986, MAUGHAN 1983, 1984) is again typical for the majority of phosphate basins. The geometry and subsidence history of this basin was modified in its western portion by the Humboldt thrusting, a tectonic event in the framework of the tectonic evolution along the western margin of North America (KETNER 1977). The sea covered substantial portions of Wyoming, Montana, Idaho, Utah, and Nevada (Fig. 7.6). It shallowed gradually toward the east. The paleogeography of the region was modified by the Milk River Uplift in Montana, the Uncompharge Uplift in Utah, and paleotectonic elements of the Ancestral Rocky Mountains (MAUGHAN 1990).

A facies model for the Sublett Basin, based on microfacies or a sedimentological analysis of the phosphate facies and the non-phosphatic sediments and supported by biostratigraphic data from WARDLAW & COLLINSON (1978, 1979a,b), WARDLAW (1979, 1980) and WARDLAW et al. (1979) establishes the genetic and time relationships of the PRC lithofacies association (TRAPPE 1992, 1993). The marine sediments of the PRC represent the eastern half of a very large, but shallow, bowl-like epeiric basin, which in general is better understood as an extremely gentle dipping, semicircular, broad ramp. The sedimentary record of the westernmost portion was mainly destroyed by the post-Permian tectonism and erosion. The preserved eastern portion of the basin reveals an organic C-rich central basin encircled by a fringing, bioclastic carbonate ramp along the marginal portions of the basin (Fig. 7.6, 7.14, 7.16). Detrital sediments supplied by the uplifts in the north and south were trapped in the peritidal zone and accumulated as sand bars and barrier islands (SHELDON 1972, THORNBURG 1990a,b). The dominantly low relief hinterland of broad sabkha flats and playas supplied only dissolved material or fine detritus, both being derived from the continent interior. Trapping in the marginal carbonate belts and the low rate of seaward sediment transport due to the low relief resulted in a relatively starved net sedimentation in the central portion of the marine basin. This deposition was dominated by organic matter and biogenic silica. The abundance of spiculitic sediments triggered diagenetic silicification of major portions of the Phosphoria facies. To the very west, shallow marine sediments, conglomerates, and phosphatic rip-up clasts (KETNER 1977, MARTINDALE 1986) document the presence of the "Humboldt highland belt" and give the large Sublett Basin most probably a silled basin character. The finging carbonate ramp complexes are bioclastic and rich in carbonate mud which suggests protected conditions. The microfacies of the outer carbonate ramps are dominated by brymol biowackestones and biopackstones indicating non-tropical conditions. BRITTENHAM (1973, 1976) describes small bryozoan buildups which were rapidly destroyed and are mostly represented by bioclast accumulations. The development of large carbonate ooid shoals was not achieved in the non-tropical sea. The outer bank sediments interfinger directly with peritidal to lagoonal carbonates (dolomudstones, fenestral limestones, e.g. SIMMONS & SCHOLLE 1990) of this slightly inclined, broad ramp. The influx of dissolved continental material is documented by authigenic glauconite precipitation along the western margin of the basin. Shifting of subsidence centers affected mainly lagoonal to intertidal conditions in the marginal Phosphoria sea during the early PRC deposition in the NW and later in the SW.

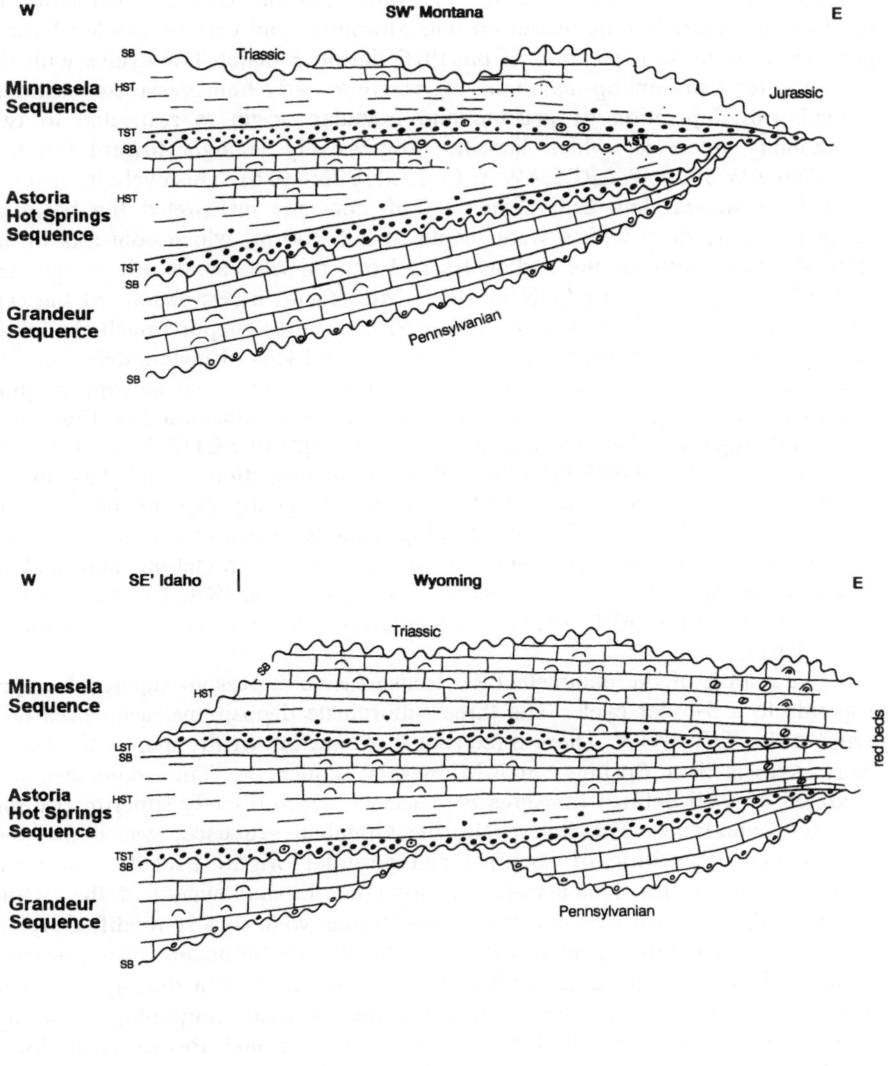

Fig 7.11. Generalized facies architecture in two margin-to-basin center transects with their sequence stratigraphic interpretation.

The interfingering of the PRC lithofacies associations is directly controlled by sea-level changes. SHELDON (1963) proposed deposition of the PRC during two transgressive-regressive (T/R) cycles with phosphorite formation during maximum transgressions. This early concept was later revised by PETERSON (1980a,b, 1984) and THORNBURG (1990b). They introduced a correlation of the lithostratigraphic units based on unconformities and eustatic sea-level changes, which indicate deposition of the PRC during 3 major T/R cycles with the phosphorites representing maximum regressive to early transgressive conditions. Phosphorite deposition of pristine and granular origin is restricted to two sedimentary intervals, which become progressively younger toward the east (WARDLAW 1979, WARDLAW & COLLINSON 1986). This cyclicity is overprinted by several low order cycles. Both concepts interpreted the basin as being relatively deep with a water depth of 200-300 m, which contrasts to the textural composition of the carbonates and phosphorites as already recognized by CRESSMAN & SWANSON in 1964. Microfacies investigations of the carbonates and phosphorites from all major PRC sections support much shallower water depths for the Phosphoria sea of only several tens of meters deep, similar to the modern Persian Gulf (SARNTHEIM 1970). The recent paleogeographic reconstructions support also a protected, semi-enclosed position (see Fig. 7.6). These investigations also confirm the general concepts of PETERSON (1980a,b, 1984) and THORNBURG (1990b) with some modification, which take in account sediment dynamics in a shallow, extremely gently dipping basin floor. A new dynamic facies model was developed on the basis of microfacies, supporting cyclostratigraphic and sequence stratigraphic interpretations and modern ideas concerning early transgressive black shale formation (HALLAM & BRADSHAW 1979, WIGNALL 1991) which addresses the low general topography of the basin.

The duration of the complete PRC transgressive-regressive supercycles corresponds to a 3-order cycle. The three subordinate depositional sequences recveal 4-order T/R cycles. These three sequences are informally named the Grandeur, Astoria Hot Springs, and Minnesela sequences. The sequences are separated by well defined transgressive surfaces which directly superimpose the sequence boundaries. The first cycle, the Grandeur sequence, was dominated by carbonate deposition of lagoonal and peritidal origin. Broadly consistent carbonate microfacies demonstrates the topographic shallowness of the basin. The generally uniform paleogeographic conditions were locally modified by an inherited, low paleotopography of the Ancestral Rocky Mountains. The presence of these elements is documented by areas of non-deposition during Grandeur times. The Grandeur sediments filled up the inherited morphology. Falling sea-level during the late highstand affected hardground formation in local modifications by the retreating sea. Lowstand sediments are usually not preserved. The onset of the Astoria Hot Springs sediments in this very shallow, essentially flat basin coincided with the formation of an extensive transgressive surface. The new transgression uniformly overlapped the paleotopography. This interpretation assigns oxygen deficient deposition to the transgressive systems

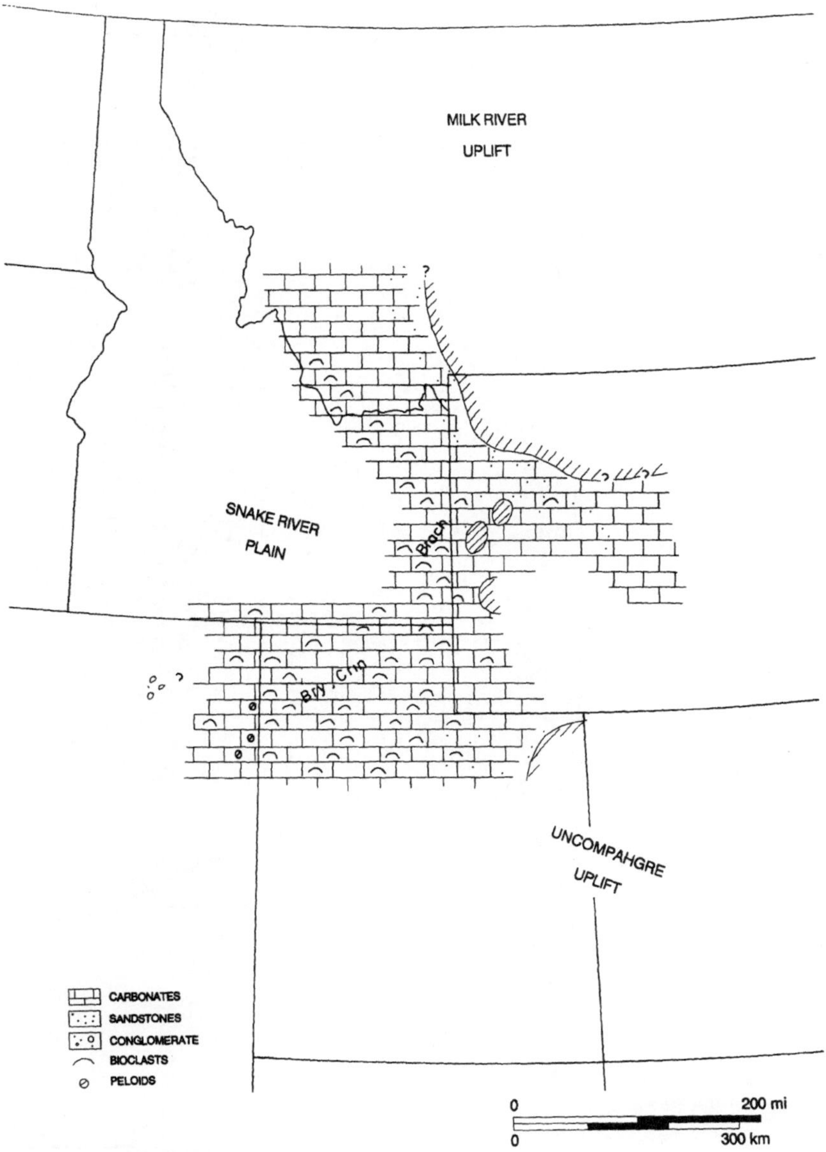

Fig. 7.12. Reconstruction of the lithofacies distribution during the Grandeur sequence based on microfacies investigations (blank areas: no data).

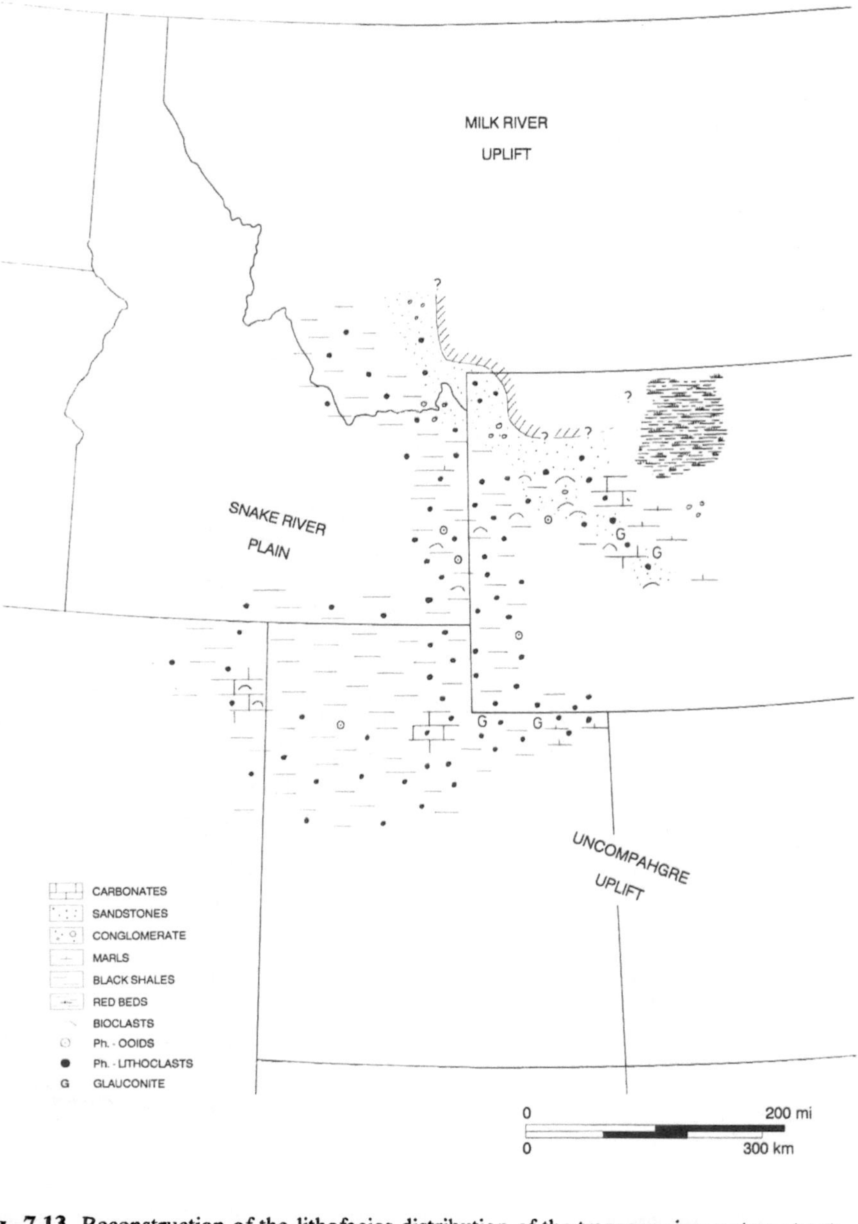

Fig. 7.13. Reconstruction of the lithofacies distribution of the transgressive systems tract of the Astoria Hot Springs sequence (Meade Peak Member and equivalents).

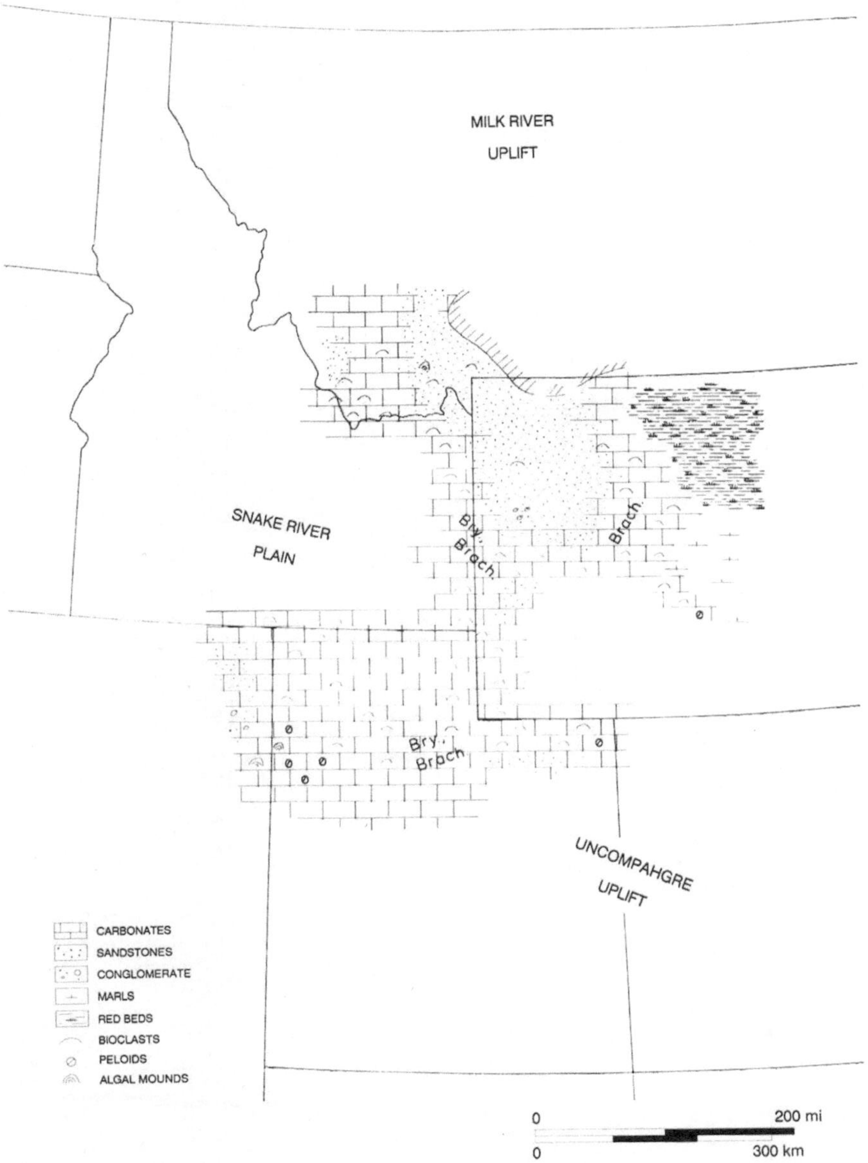

Fig. 7.14. Reconstruction of the lithofacies distribution of the highstand systems tract of the Astoria Hot Springs sequence.

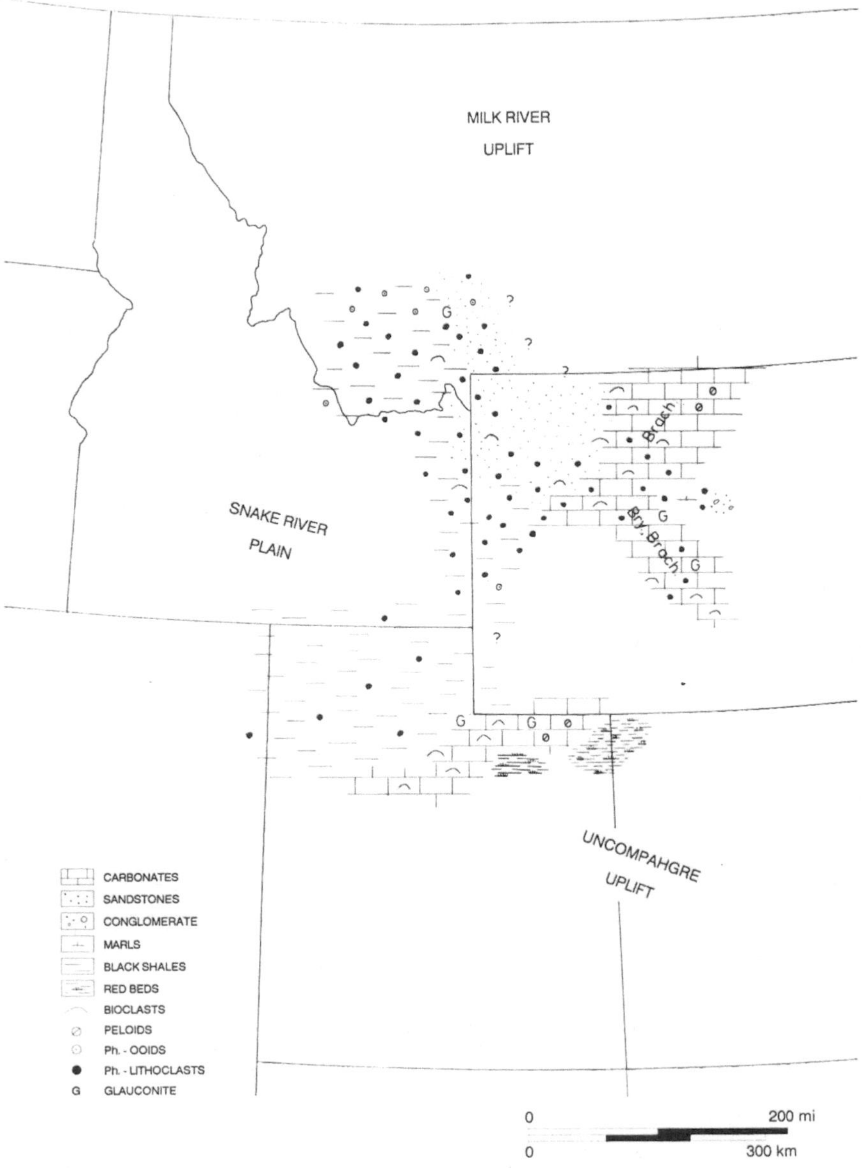

Fig. 7.15. Reconstruction of the lithofacies distribution of the transgressive systems tract of the Minnesela sequence (Retort Phosphate Member and equivalents).

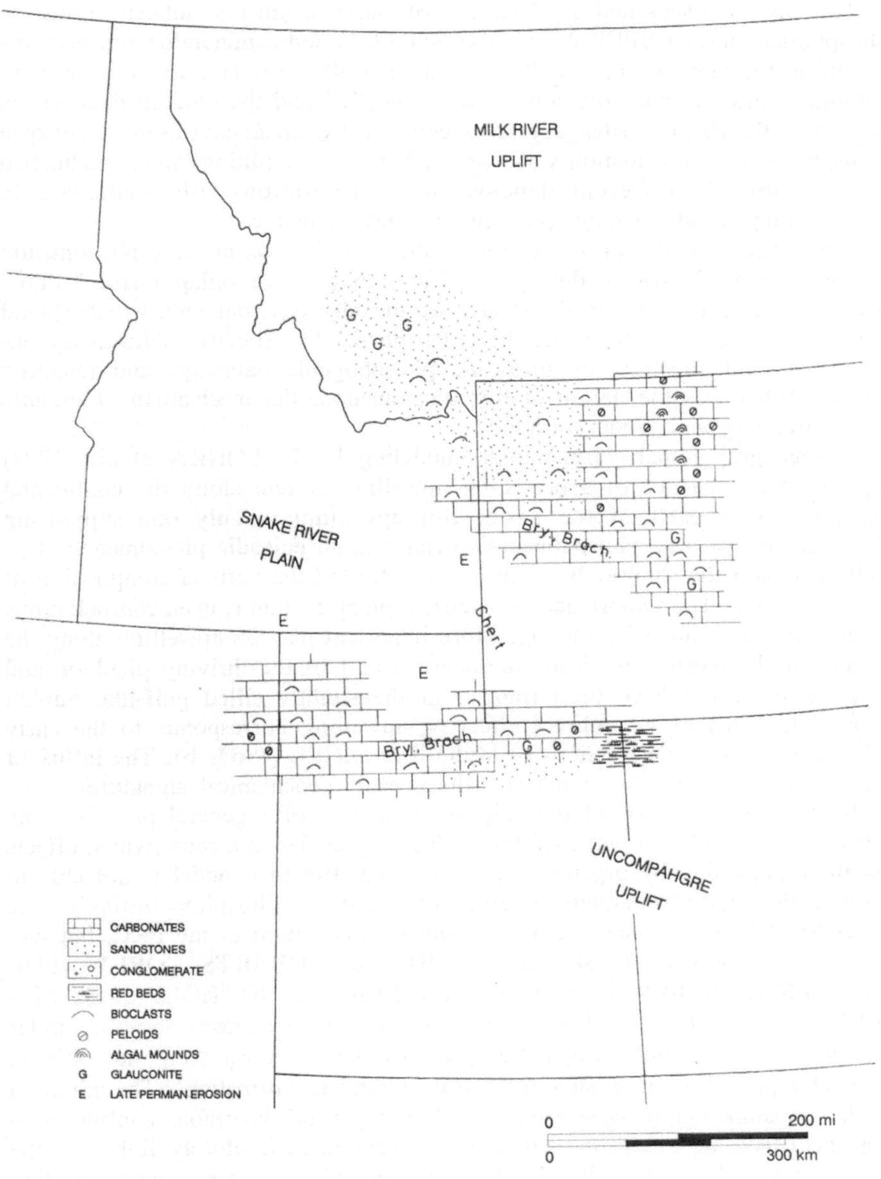

Fig. 7.16. Reconstruction of the lithofacies distribution of the highstand systems tract of the Minnesela sequence.

tract of the Astoria Hot Springs and Minnesela cycles and phosphorite occurrences to its basal portion, which seems to be a stratigraphic paradox.

The trace element and REE-pattern of these sediments indicate a marine phosphorus source (PIPER & MEDRANO 1994) and a mineralization pathway similar to the modern outer shelf sedimentation off Peru. This encouraged these authors to draw a close relation between the PRC and the modern deposystem neglecting the striking paleogeographic contrast. Chemical environment and type of background sedimentation was similar, but these conditions were established in a completely different deposystem configuration with incomparable paleogeography/paleoceanography and sediment dynamics.

It was shown that phosphogenesis in the PRC is a basin wide phenomenon which developed strictly during specific transgressive onlap phases. Phosphogenesis affects at least all subtidal sedimentary environments which spread toward the basin margin during the onlap phase. The specific sedimentary environment and substrate modifies the phosphogenic pathways and resulting fabrics. What was the nature of this phenomenon, the mechanism of organic matter trapping and P-concentration?

Paleogeographic/paleoclimatologic modeling by GOLONKA et al. (1994) supports the existence of a persisting upwelling system along the continental margin and in parts of the basin. But upwelling is only one supporting phenomenon, because phosphogenesis occurs as an episodic phenomenon. Upwelling within the shallow basin may have affected the vertical composition of the water body. The importance as a nutrient pump is minor, even contradictory, because it would intensify mixing. More important may be upwelling along the margin of the North American continent. A part of the thriving plankton and ocean water could have been trapped in the slightly silled gulf-like Sublett basin. This idea of an indirect effect of upwelling corresponds to the early paleogeographic reconstructions of McKEE & ORIEL (1967a,b). The influx of oceanic water would also emphasize the oceanic geochemical signature.

But what is the cause of the episodic formation of a general phosphogenic regime? HIATT (1994) proposed bioproductivity cycles as a cause which affects sea floor poisoning by organic matter overload. But this model is not able to explain the strict sequence stratigraphic control. Phosphogenesis in this pericratonal basin reached a climax during the deposition of the PRC, but was also existent earlier (OBERLINDACHER & ROBERTS TOBEY 1986, NICHOLS & SILBERLING 1990, SILBERLING & NICHOLS 1991, SILBERLING et al. 1995, and own observations). These deposits show a similar stratigraphic orientation. Other transgressive onlap phases were not able to establish a phosphogenic system (e.g. in the Grandeur Formation). The initiation of the phenomenon phosphogenesis within a general favorable configuration must also have an autogenic cause. The development is closely linked to the deposystem evolution and the specific conditions of the phosphogenesis-hosting transgressive environment. In the very early phase, the balanced ecologic condition were shifted to benthos poisoning which generated the preconditions for the establishment of the phosphogenic regime. Phosphogenesis responds differently to environment and substrate which becomes evident by the variation

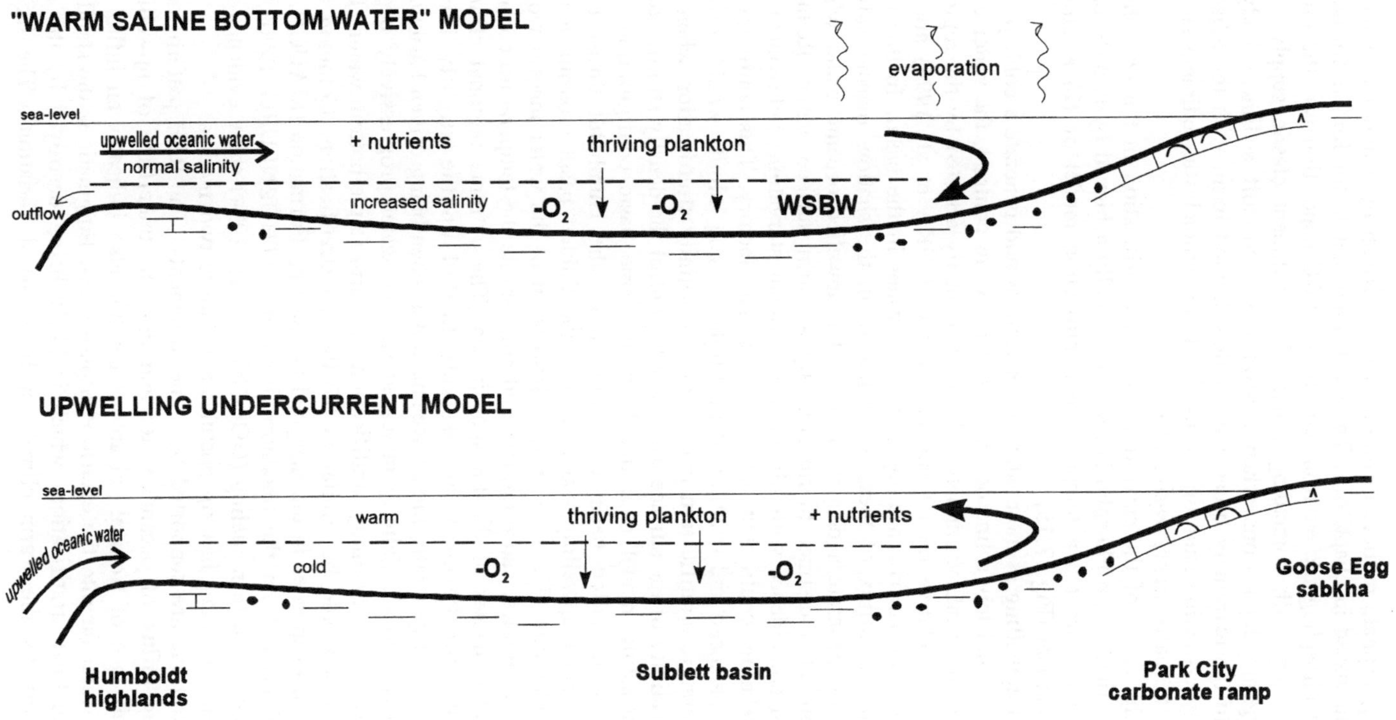

Fig. 7.17. Two contrasting stratification models for the Phosphoria Rock Complex. The basin profile is emphasized in the sketch. WSBW = Warm saline bottom water

in types of phosphate particle formtion. This is driven by a number of factors which are discussed in chapt. 6.1. The most important factor for the formation of a general phosphogenic environment during PRC deposition was the broadness of the ramp with coinciding general low sediment clastic supply. This supports organic depositions when a stratified water/soft sediment body is generated. In the extremely broad and shallow Sublett basin with the adjacent areas, rapidly retreating shorelines caused extreme siliciclastic sediment starvation during the early transgressive stage.

What is the cause of the early transgressive stratification in the Phosphoria sea? Stratification in the Phosphoria sea is potentially achieved by two different scenarios which are (1) a temperature stratification model or (2) a salinity stratification model (Fig. 7.15).

(1) Proposed upwelling at the continental margin would generate a cold, poorly oxygenated, and low saline oceanic water body. In contrast, the water composition in the shallow Sublett basin is dominantly affected by the regional climate. The relative high temperatures (GOLONKA et al. 1994) are expected to cause surface heating of the sea water in the basin. In this configuration, the influx of these water masses in the shallow marine Sublett basin would generate a deeper layer of this upwelled oceanic water which is overlain at a distinct thermocline by a warm surface water. Reduced mixing in coincidence with siliciclastic sediment starvation. Subsequent sea floor poisoning would shut down the carbonate factory. This configuration is a very balanced and friable system. Slightly rising salinity would convert the temperature stratification into an inverse salinity stratification when the elevated salinity of the surface water would control the density stratification.

(2) The contrasting model is initiated by the same basic configuration, but emphazises the silled basin configuration by the Humboldt thrust belt. Nutrient loaded upwelling water flooded the Sublett basin crossing a significant paleohigh. The near surface dislocation of the water masses would rise the temperature, cause mixing, and compensate the temperature contrast to the water masses of the Phosphoria sea. The extreme regional climate during the mid-Permian successively heated up the slightly lower oxygenated, but generally mixed oceanic water when being driven basinward by a current system. Salinity increased and the water progressively began to sink generating a salinity stratification by warm saline bottom water. The general effect would be similar to the thermal stratification. Common occurrences of evaporites in the marginal Goose Egg Formation (MAUGHAN 1965), evidence from the Shedhorn Formation (THORNBURG 1990a,b), and the paleoclimate modeling (GOLONKA et al. 1994) show evidence for general surface water heating postulated by these two models.

Both configurations are supported by some arguments. The REE pattern indicates a strong influx of oceanic water and thus the underflow of upwelling water. The absence of tropical calcareous benthos also indicates an influx of cold oceanic water far into the epeiric Phosphoria sea. Enigmatic is the stability of the thermoclinal stratification, which is likely being destroyed by the increasing salinity by the warm climate or by seasonal variations. The facies

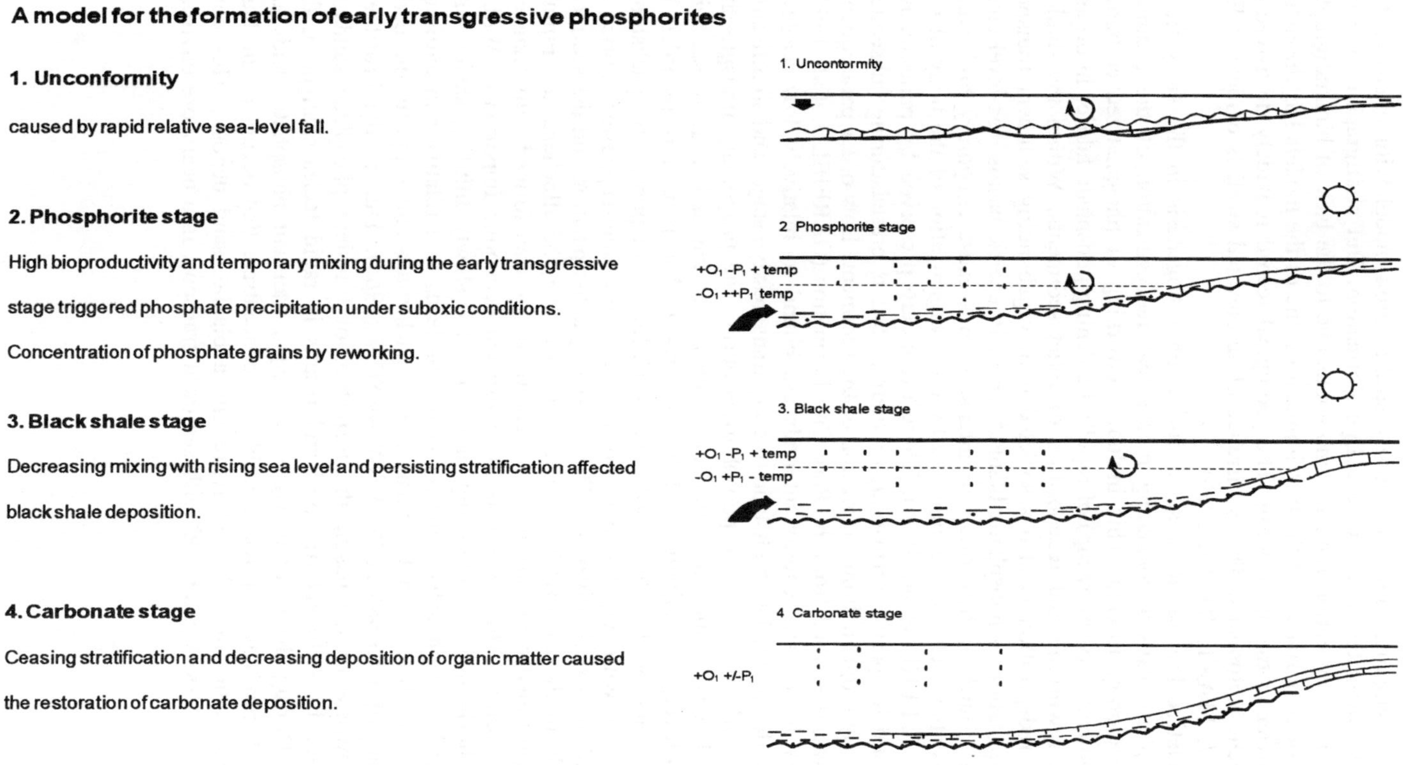
A model for the formation of early transgressive phosphorites
1. Unconformity
caused by rapid relative sea-level fall.
2. Phosphorite stage
High bioproductivity and temporary mixing during the early transgressive stage triggered phosphate precipitation under suboxic conditions.
Concentration of phosphate grains by reworking.
3. Black shale stage
Decreasing mixing with rising sea level and persisting stratification affected black shale deposition.
4. Carbonate stage
Ceasing stratification and decreasing deposition of organic matter caused the restoration of carbonate deposition.
1. Uncontormity
2 Phosphorite stage
+O1 -P1 + temp
-O1 ++P1 temp
3. Black shale stage
+O1 -P1 + temp
-O1 +P1 - temp
4 Carbonate stage
+O1 +/-P1

Fig. 7.18. A dynamic model for granular phosphorite formation in the Phosphoria Rock Complex on basis of the upwelling undercurrent model for phosphogenesis.

distribution supports more the salinity stratification model with intensive formation of evaporites in the Goose Egg Formation. But stratigraphic evidence of maximum evaporation remains ambiguous due to the lack of biostratigraphic data. Major importance for the development of one of the models has the nature of the barrier to the open ocean. The geological record is widely destroyed by the Neogene tectonisms. The occurrence of an elevated swell is evident (KETNER 1977, MARTINDALE 1986).

The relative low sea-level stage during phosphogenesis in the early transgressive phase caused coexisting mechanical reorganization of the sediment during temporary mixing. This includes reworking of phosphatized carbonate rock, exhumation (winnowing) of rapidly hardening phosphate fabrics in organic muds, or reworking and redeposition of relict sediments. Within the shallow sea area, redeposition is also associated with vagabonding sediment transport and the formation of phosphatic blanket sands. In epeiric basins, sea-level changes of the fourth or fifth order or intense storms cause temporary high water energy events which are accompanied by the reorganization of the body of soft sediments (GLENN et al. 1994). These events are preserved by parasequence stacking and tempestite formation, or their record is erased during subsequent events, or by biogenic homogenization of the sediment. Low order parasequence stacking was described from the Park City Formation (GARFIELD et al. 1992). Temporary high water energy intervals are bound to relatively low sea-level conditions during the late highstand to early transgressive stage, and significantly affect the early transgressive phosphorites during the prograding transgressive flooding. Intensity and frequency of sediment reorganization decreases with rising sea-level, respectively is shifted landward. The pattern is recorded by the microfacies of the phosphorites being shifted from phosphatic packstones and rudstones with large particle diversity to more mud-supported fabrics of low phosphate particle diversity. The paleogeographic position and the sedimentation style in the shallow basin determines whether the allochems are rapidly buried, additionally phosphatized/glauconitized, or transformed into transient phosphatic sands. The process is of significant economic importance. Within the Phosphoria facies, winnowing of the relative slowly lithifing shales with rapidly hardening phosphate micro-concretions leads to a relative concentration of phosphate fabrics with the formation of packstones and supports the early assumptions of CRESSMANN & SWANSON (1964). Due to the broadness and shallowness of the basin, the formation of transient phosphate sands is accompanied by sediment transport and results in rapid facies changes in the sediment. During the early transgressive stage, transient phosphate sands are transformed into a widely recognizable lag concentrate that occurs at the base of the phosphatic intervals and represents a blanket sand deposit. This concentration process is combined with sediment transport into nearshore environments.

CASE STUDY 5
Granular platform phosphorites: the Miocene system of the southeastern USA

Anatomy of the deposit

Location: A wide subsurface and surface distribution of the central to northern Florida peninsula; 83° to 81°W and 26°30' to 31°N. Associated deposits occur in the coastal plain of Georgia and the Carolinas (RIGGS & MANHEIM 1988).
Stratigraphy, age: Lithostratigraphic system: established by RIGGS (1979a,b), later extended by SCOTT (1988, 1990), phosphate-bearing succession summarized to the Hawthorn Group, subdivided in several formations with their lateral equivalents of different lithofacies. Age: Miocene.
General paleogeography: Partly drowned shelfal platform with internal paleotopography, influenced by the gulf current system. Possible subaerial exposure of the Ocala Highlands. Paleolatitude: 28° to 32°N (GOLONKA et al. 1994), shallow marine to peritidal, warm (20° C average temperature). To the north: Apalachicola delta. Dynamic upwelling proposed by RIGGS (1984a).
Paleotectonic setting, basin geometry: Stable inner shelfal platform as part of the North American continental margin with subsiding subbasins and platforms.
Depositional system, facies architecture: Mixed carbonate, siliciclastic (terrigenous sediments), phosphorite. Ratios of these components vary with paleotopographic position.
Sea-level control, sequence stratigraphic interpretation: Transgressive/regressive 2-order cycles superimposed by 3- and 4-order cyclicity (COMPTON et al. 1993) and tectonic overprinting, greater resolution in the North Carolina (Pungo River) system (RIGGS 1984a,b, RIGGS et al. 1990). Underlying Suwannee Limestone represents early 2-order sea-level highstand. Partly cycles of transgressive to maximum flooding with organic matter-rich pelitic mudstone to highstand bioclastic carbonate cycles developed. Intensive lateral facies and grain size variations driven by paleogeographic variations.
Type of phosphorite particles: Predominantly phosphoclasts derived from reworking of phosphatized carbonate and sporadically also siliciclastic rock, phosphatized hardgrounds, phosphatized bone fragments and teeth, phosooids, pristine and reworked micro-concretions, rarely microsphorite, phosphatic stromatolites.
Model for the phosphogenic processes: Substrate derived: Replacement, biogenic/skeletal, microbial, authigenic/concretionary. Partly related to sea-level highstand (COMPTON et al. 1993, RIGGS et al. 1990), but dominantly lowstand to early transgressive in association with erosional surfaces, relict grain phosphatization.
Phosphorite formation: Intense reworking during low-order sea-level changes, multiple reworking and composite concentrate formation, coastal-peritidal phosphorite accumulation in agitated water, ?tempestites.
Reserves, resource quality: Identified resources in total Florida 5600 million

tons technical phosphate concentrate with average P_2O_5 contents of 30%, roughly two thirds economic, and one third marginally and subeconomic. Additional 5175 million tons hypothetical reserves (RIGGS & MANHEIM 1988). Mining in two centers: Bartov region in central Florida (W' of Tampa), and the Suwannee district north of Lake City in northern Florida.

The Florida phosphorite deposystem represents best a highly developed composite concentrate deposit The example demonstrates also the development of a large scale economic deposit in a paleogeographic configuration with the absence of equatorial upwelling and an open inner to mid-shelfal setting. The deposition in shallow marine to coastal settings, intensively affected by sea-level fluctuations and storm events, generated a platform deposit within the latest stage of composite phosphorite concentration.

The paleotopography and various subsidence rates on the Neogene Florida platform caused a complex lithofacies development. The lithology of the Miocene Hawthorn Group results from four major components, which are terrigenous clastics, carbonate, organic matter, and authigenic minerals, respectively phosphate. The terrigenous material is mostly supplied from the Appalachian chain in the north. The clastic influx competes with the dominating bioclastic carbonate deposition in the south. In the transition zone, sedimentation is intensively controlled by the paleotopography, which affects siliciclastic sediment bypassing and rapidly changing intensity of synsedimentary erosion. These lateral variations are further affected by sea-level changes of lower order cyclicity. Intervals or areas of low siliciclastic and carbonate deposition are characterized by pelitic and organic matter-rich sediments. The lateral facies relationships correlate also with thickness variations of the Hawthorn Group (SCOTT 1988). The successions in subsidence centers record intensive lithofacies cyclicity which is progressively destroyed by multiple erosion and truncation in approach to positive areas. These lithologic characters significantly affect the type of phosphate fabrics.

(1) Erosional surfaces are phosphatized; carbonate is replaced or sandstone impregnated. These surfaces may further be covered with microbial phosphate crusts.
(2) Rock fragments of various lithology are marginally phosphatized.
(3) Pelitic, organic matter-rich sediments are hosting phosphate micro-concretions. This includes also phosphate precipitation in foraminifera shells.
(4) Primary phosphate-rich skeletal elements are being further phosphatized and silicified.
(5) Microbial coating generates phosphatic ooids.

Elevated water energy in approach to positive areas, or the possibly emerged Ocala High, generated intensive reworking of these fabrics. As a result, most of the Florida phosphate sediments are of granular nature. Pristine fabrics remain exceptional. In contrast to other deposits, the resulting fabrics are broadly of pebble size and include large amounts of phosphatized sediment clasts.

RIGGS (1979a,b) subdivided two major phosphorite facies belts (Fig. 7.19). Intensive drilling of the phosphate industry and the Florida Geological Survey

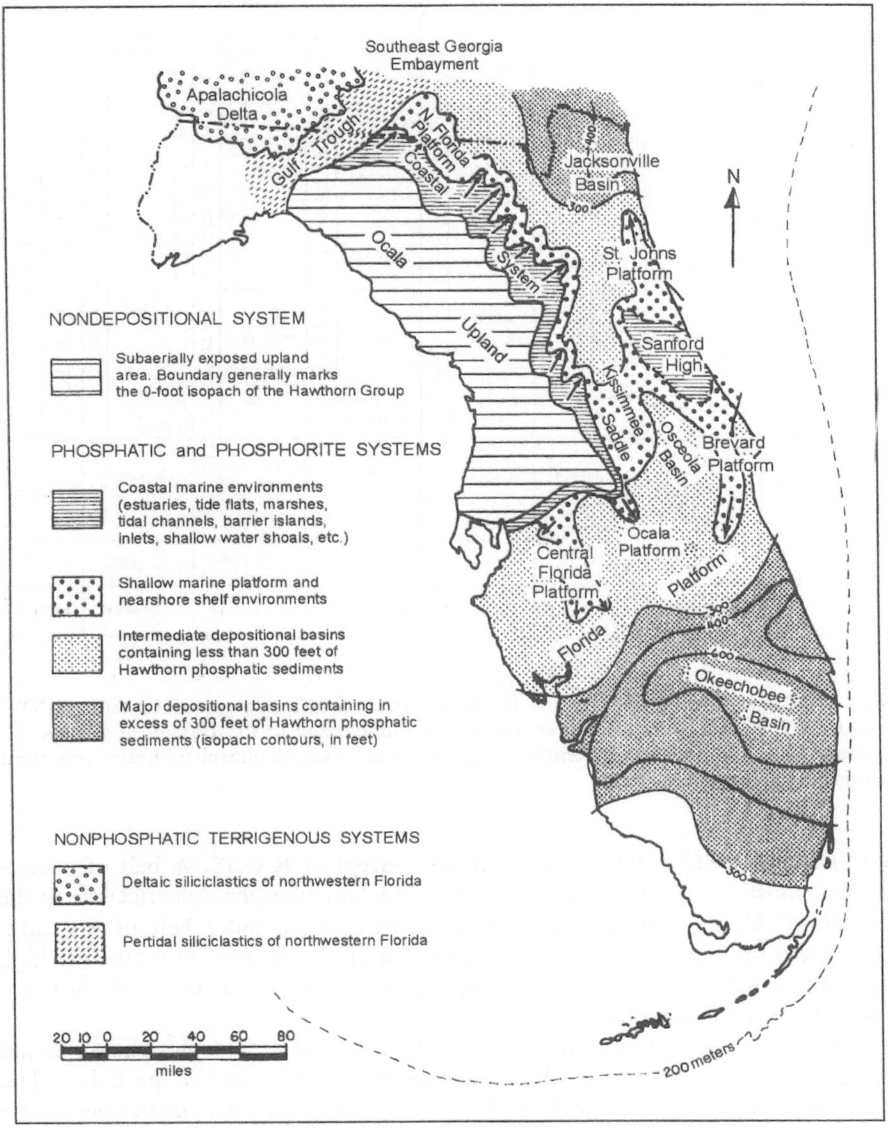

Fig. 7.19. Paleogeographic and structural framework of the Miocene Florida phosphate province (modified from RIGGS 1979b).

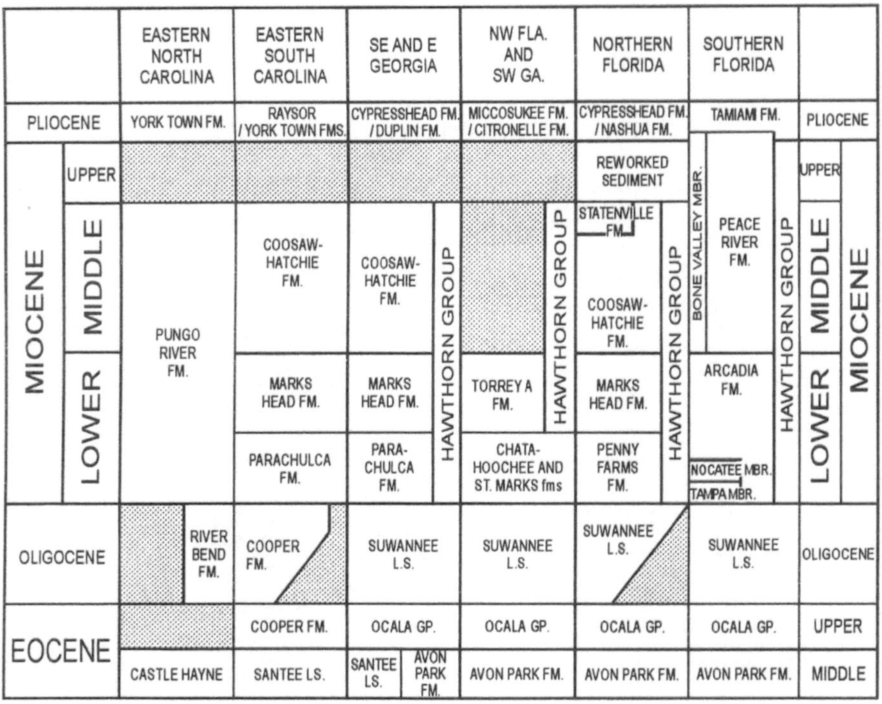

EASTERN NORTH CAROLINA
EASTERN SOUTH CAROLINA
SE AND E GEORGIA
NW FLA. AND SW GA.
NORTHERN FLORIDA
SOUTHERN FLORIDA
PLIOCENE
YORK TOWN FM.
RAYSOR / YORK TOWN FMS.
CYPRESSHEAD FM. / DUPLIN FM.
MICCOSUKEE FM. / CITRONELLE FM.
CYPRESSHEAD FM. / NASHUA FM.
TAMIAMI FM.
PLIOCENE
MIOCENE
UPPER
MIDDLE
LOWER
REWORKED SEDIMENT
STATENVILLE FM.
BONE VALLEY MBR.
PEACE RIVER FM.
PUNGO RIVER FM.
COOSAW-HATCHIE FM.
COOSAW-HATCHIE FM.
COOSAW-HATCHIE FM.
HAWTHORN GROUP
HAWTHORN GROUP
HAWTHORN GROUP
HAWTHORN GROUP
ARCADIA FM.
MARKS HEAD FM.
MARKS HEAD FM.
TORREY A FM.
MARKS HEAD FM.
PARACHULCA FM.
PARA-CHULCA FM.
CHATA-HOOCHEE AND ST. MARKS fms
PENNY FARMS FM.
NOCATEE MBR.
TAMPA MBR.
UPPER
MIDDLE
LOWER
MIOCENE
OLIGOCENE
RIVER BEND FM.
COOPER FM.
SUWANNEE L.S.
SUWANNEE L.S.
SUWANNEE L.S.
SUWANNEE L.S.
OLIGOCENE
EOCENE
COOPER FM.
OCALA GP.
OCALA GP.
OCALA GP.
OCALA GP.
UPPER
CASTLE HAYNE
SANTEE LS.
SANTEE LS.
AVON PARK FM.
AVON PARK FM.
AVON PARK FM.
AVON PARK FM.
MIDDLE

Fig. 7.20. Lithostratigraphy of the Miocene system in Florida (modified from SCOTT 1990). The phosphorite bearing succession is summarized to the Hawthorn Group. The Suwannee Limestone, which includes only sporadically single phosphate grains represents an earlier depositional cycle.

(JOHNSON 1986) widely confirmed the concept of RIGGS. A belt of coarse-clastic, mostly clay rich phosphorite facies (pebble phosphate district) along the postulated Ocala High is further accompanied by an outer belt of a quartz-dominated sand facies. Sediment structures in the latter indicate prevailing high energy conditions. These two lithofacies areas differ also distinctively in phosphate grain composition.

The grain composition in the pebble district is diverse. The phosphate fraction in the sediment is composed of various types of reworked particles. The spectrum consists of structureless phosphate grains, probably micro-concretionary derived or precipitates in foraminifera, and ooids in sand size. This spectrum is supplemented by much coarser phosphate fabrics. These are completely or marginally phosphatic rock fragments. The completely phosphatic grains are intensively pigmented, clay-rich phosphate mudstone, which includes quartz grains, other phosphate grains, and sand-sized rock fragments. These fabrics indicate reworking of detritus-rich sediment layers which were widely phosphatized. These differ from mostly elongated carbonate sediment fragments which are only marginally phosphatized. Most of them represent rip-up clasts from thin carbonate sediment. Large phosphatic skeletal elements are very abundant. All large phosphate fabrics show abrasion and borings to various degree. The phosphate fabrics represent various stages of synsedimentary weathering and organic matter oxidation indicated by different color. The phosphate sediments of this facies belt vary widely in matrix composition, mainly clay, carbonate and quartz ratio. The phosphate fraction differs widely in grain size, sorting and grain type composition in vertical and lateral direction. The distribution lacks a distinct pattern, and documents more a lenticular nature. Hardgrounds occur usually at the base of the sequence, but a distinct stratification is absent.

The quartz-phosphorite facies is much more homogeneous. The grain spectrum is composed of well sorted and well rounded quartz grains, structureless phosphate grains, and ooids in corresponding grain-size. The sediments are

---

Fig. 7.21. Facies of the Florida phosphorites. (A) Two major phosphorite lithofacies types occur in the Florida system: A phosphate pebble facies occurs along the Ocala High and is intensively mined in the Bartov district. The facies differs from that of the second mining area in northern Florida. The latter consists of a fine-grained phosphate fraction and well sorted quartz sand which is cross-bedded and intensively burrowed . (B) Major components of the pebble phosphate facies are large bone fragments and different types of phosphatized lithoclasts. (C) Typical for carbonate lithoclasts are marginal phosphatization and borings.

Fig. 7.22. Microfacies of the Florida phosphorites. (A) A calcareous sandstone includes various phosphate grains of similar grain size. (B) Intensively mixed phosphatic packstone to floatstone. The grain spectrum consists of quartz grains of various size and degree of roundness, further of various large phosphatic lithoclasts. The latter comprise various stages of rounding. The matrix of this concentrate was again phosphatized. (C) Packstone of structureless phosclasts and irregularly shaped bone fragments in a pelitic matrix. (D) Calcareous sandstone with abundant phosclasts, some phosphatized bioclasts, as well as non-phosphatic bioclasts and intraclasts, a composition and texture well known from other deposits. (E) Lithoclasts of phosphatized sandstone. (F) Poorly sorted phosclastic pelitic packstone. The highly mixed phosphate grain spectrum includes different phoslithoclasts, bone fragments and some phosphatized foraminifera shells.

Fig. 7.23. Microfacies of the Florida phosphorites. Hardground related fabrics. (A) The erosional surface on phosclast bearing bioclastic limestone is phosphatized to various degree, indicated by the brown color. Overlying composite concentrate consists of different phosphatic clasts. Clasts of which the lithology corresponds to the reworked limestone have to be assigned to intraclasts. Phosclasts with exoctic lithologic character are considered to be lithoclasts which are supplied from other areas. The matrix, slightly phosphatized, includes peloids (base of photograph 40 mm). (B) Erosional surface which is covered by a phosphatic microbial fixated crust (base of photograph 20 mm). (C) Reworked fragment of slightly phosphatized and bored hardground on carbonate mudstone (diameter 25 mm).

See alos color copies of plate 7.22. and 7.23. in the appendix.

Fig. 7.21.

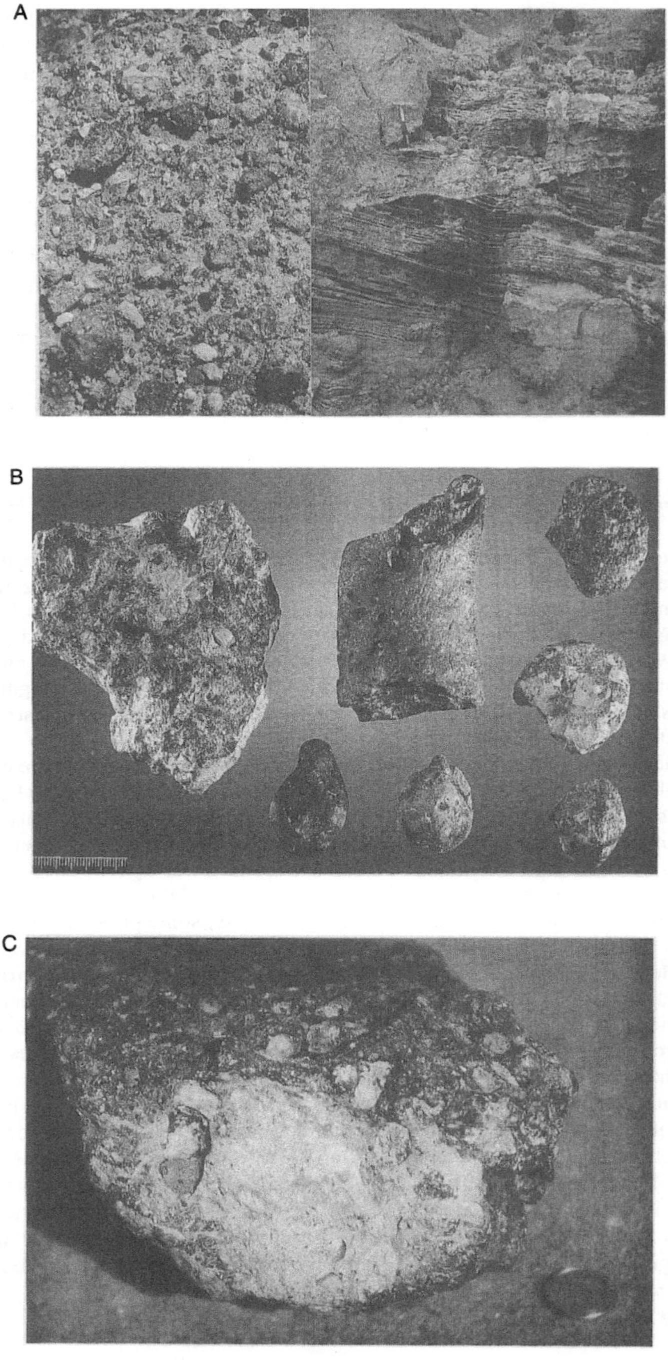

Fig. 7.22.

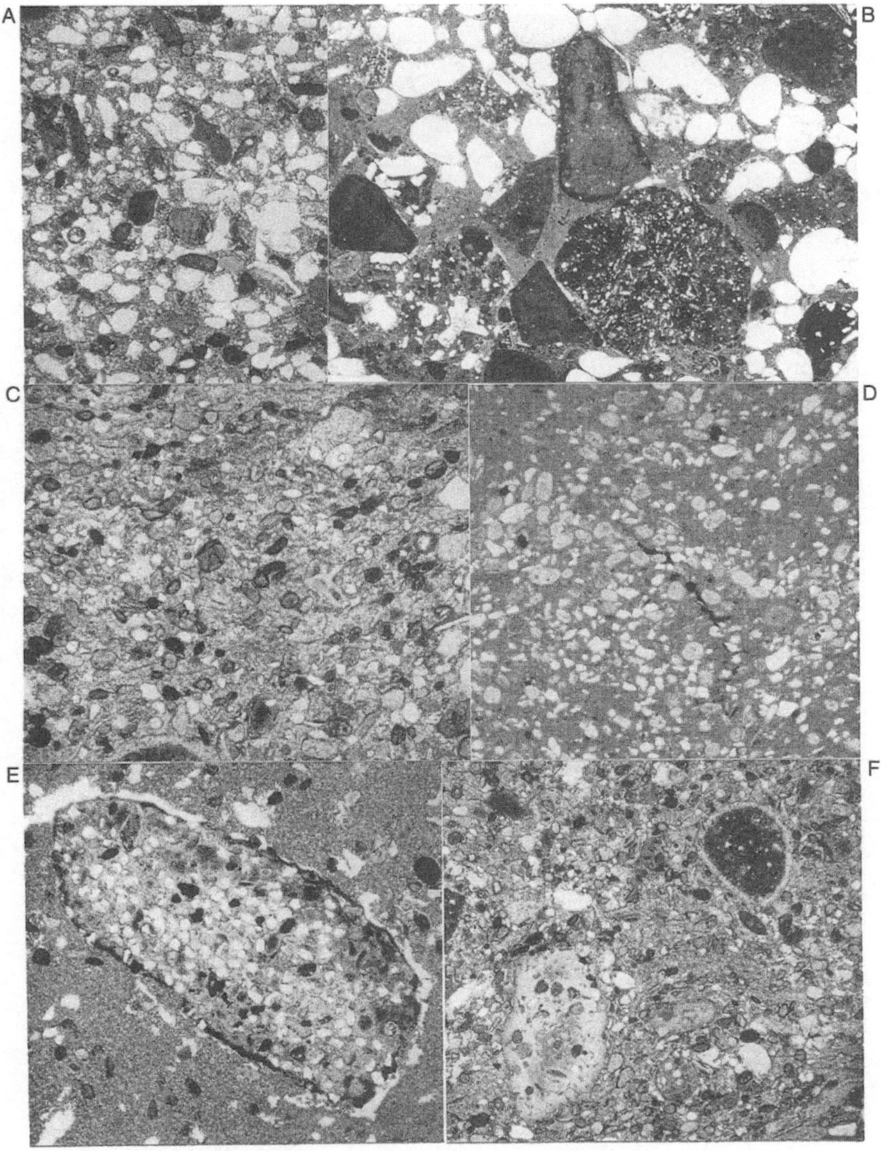

Fig. 7.23.

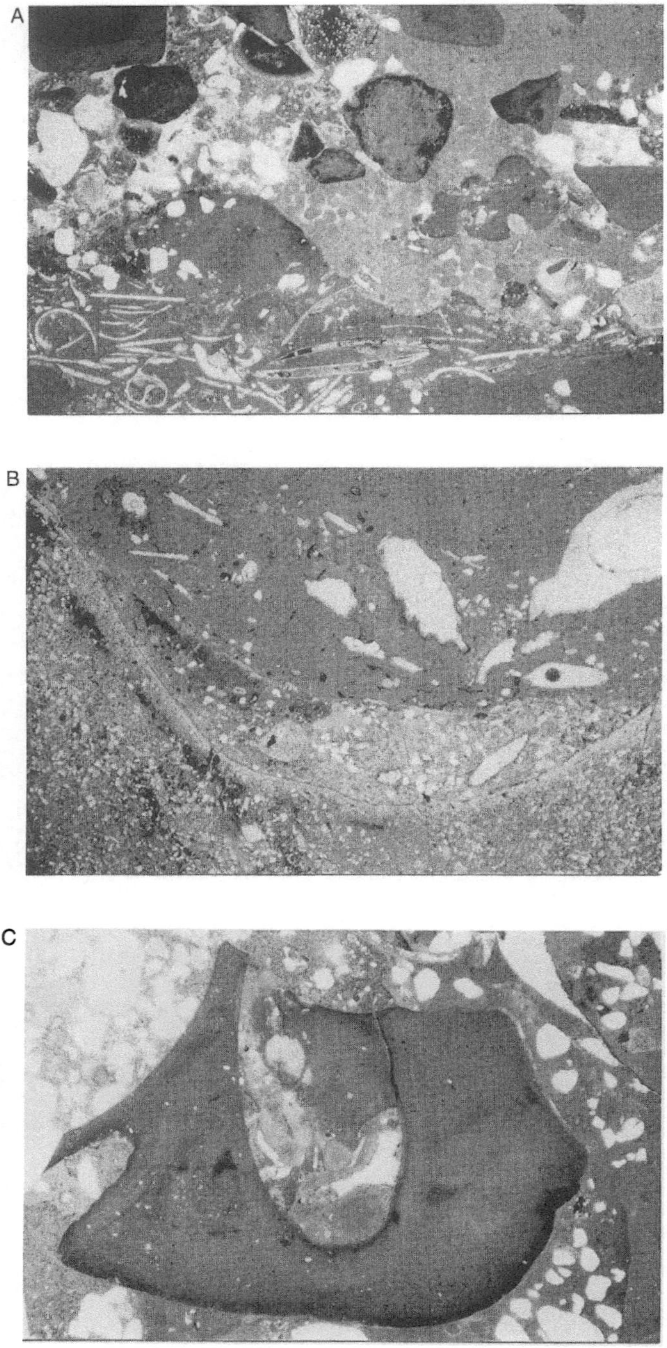

cross-bedded and burrowed (Fig. 7.21). In the Suwannee district, the phosphate content generally decreases to the top.

The transition zone of these facies belts is characterized by an interfingering of the two general phosphate lithologies. In subsidence areas, a cyclicity is documented. Carbonate suites are superimposed by phosphatized hardgrounds, pebbly phosphate and a clayly micro-concretionary phosphate suite. Bioclastic carbonates are completing the cycle upsection (see also COMPTON et al. 1993).

All clastic phosphorites represent abundantly reworked sediments of which the grain spectrum was intensively mixed. The concentrates represent composite concentrates and reveal multiple stages of synsedimentary weathering and relict grain exposure at the sea floor. The latter is indicated by intensive boring activity. The sediments are intensively weathered which erases most primary geochemical signatures. The Ocala High was favored as being a major source area for the clastic phosphate particles. Phosphatic hardgrounds in the present phosphorite deposits document existing phosphogenic conditions also in this area. This is further supported by the occurrences of phosphatic micro-concretions in some shales. But in general, the intensity of sediment reorganization and multiple mixing confines the reconstruction of transport directions and distances.

CASE STUDY 6

Marginal basin blanket phosphorites, Central Morocco, latest Cretaceous to Paleogene

Anatomy of the deposit

Location: Large plateaus in Central Morocco and erosional remnants in the High Atlas region with the adjacent northern and southern basins and plains; 10° to 5°W and 30° to $33^{\circ}30$'N, (Fig. 7.24).
Stratigraphy, Age: Lithostratigraphic system: basinal: informal "Série Phosphatée", "Série Postphosphatée" including lithofacial members (BOUJO 1976), southern and eastern margin: Subatlas Group summarizing several formations and facies areas (HERBIG & TRAPPE 1994). Age: time transgresive base (W to E): Maastrichtian to Thanetian, Top: Lutetian (HERBIG & TRAPPE 1994);(Fig. 7.25).
General paleogeography: Broad gulf of the Atlantic with internal topography. Paleozoic consolidated massifs to the north and south, gulf closes to the east with a narrow embayment. Paleolatitude: $\sim 20^{\circ}$N, warm 20°C average temperature, coastal dry, inland humid, probable offshore summer upwelling site (GOLONKA et al. 1994); depositional environments: peritidal to shallow subtidal (Fig. 7.24, 7.28).
Paleotectonic setting, basin geometry: Inner-shelfal (pericratonal) sag basin, partly above stagnating Atlas rift system (MICHARD 1976, STETS & WURSTER 1981), bowl-like morphology open to the Atlantic.
Depositional system, facies architecture: Mixed carbonate siliciclastic; basinal facies: marl-phosphorite-chert, marginal: fringing bioclastic carbonate ramp complexes, in the final stage rapid progradation of the carbonate ramp, possible sea-level controlled siliciclastic-carbonate ratio (HERBIG 1991, HERBIG & TRAPPE 1994), biofacies zonation of the marginal ramp (TRAPPE 1992a), brymol facies.
Sea-level control, sequence stratigraphic interpretation: Transgressive/regressive supercycle (20 Ma duration) with 4-5 lower order T/R cycles of 1-4 Ma duration and undetermined cyclicity on parasequence level, sea-level record overprinted by subsidence history, sequence stratigraphic concept low resolution due to poor biostratigraphic control, phosphorites in low order ?retrograde portion of a high order general transgressive systems tract.
Type of phosphorite particles: Predominantly phosphoclasts derived from reworking of micro-concretions, phosooids, phosphatized bioclasts, phosphatic coated grains, exceptional phosphatic hardgrounds and crusts (TRAPPE 1989, 1992, PRÉVôT 1990).
Model for the phosphogenic processes: Mostly micro-concretionary growth, phosphatization of bioclasts and carbonate erosional surfaces, minor microbial.
Phosphorite formation: Reworking during low-order sea-level changes and phosphorite blanket sands, phosphorite lags, tempestite shoreward transport during ingression.

Reserves, resource quality: 55 billion tons phosphate rock (BRITISH SULPHUR CORPORATION 1987), mostly in near surface exposures.

The central Moroccan phosphate deposits were selected because this mixed siliciclastic-carbonate-phosphorite deposystem represents a typical example for inner-shelfal, pericratonic sag basin deposit which developed during the Late Cretaceous and the Paleogene in several gulfs along the western and northern margin of the African continent, e.g. Spanish Sahara (MUNOZ CABEZON 1989), Senegal (PASCAL 1989), Togo (SLANSKY 1989), and Tunisia (SASSI 1980). Besides the typical paleogeographic setting, special interest is drawn to the formation and transport of granular phosphorite accumulations which define the economic significance of these deposits. All these phosphorite basins are characterized by a marginal fringing carbonate ramp facies, partly with evaporite sabkhas, e.g. Tunisia, and a basinal marl/phosphorite sedimentation. The phosphorite facies is commonly phosphoclastic with a low grade of particle diversity compared to other phosphate deposits, e.g. the Phosphoria Rock Complex. These granular phosphorites are usually accumulated in seams of several meter thickness which are of significant economic value. Silicification, bedded and nodular, is always present in the phosphorite suite which is generally dominated by marls and dolomites. Some of these deposits, e.g. in Senegal, are altered intensively by weathering (see chapt. 2.2, 3.8).

The Moroccan Phosphate Gulf (Fig.7.24) is an open non-silled basin with a slight internal morphology which causes intensive facies variations (BOUJO 1976, TRAPPE 1991). These rapid vertical and lateral facies changes in the marginal as well as in the internal portion of the basin show a more random pattern with widely lenticular phosphorite bodies. This spatial sediment distribution indicates changing sedimentary conditions, but a lack of a distinct low order sequential pattern (TRAPPE 1989, PRÉVôT 1990, HERBIG 1991). V/(V+Ni) average values of 0.72 for whole rock and 0.69 for phosphorite from fresh samples in the basinal Ganntour deposit indicate typical suboxic to slight anoxic general conditions for the phosphate hosting "Serie phosphatée" (data from PRÉVôT 1990). Microfacies of the fringing carbonate ramp complexes with flourishing benthos indicate well oxygenated conditions in the marginal portions of the basin (HERBIG 1991, TRAPPE 1992a).

This configuration assigns phosphogenesis to the basinal portion of the Central Moroccan Phosphate sea with dominantly organic-rich marls sedimentation (Fig. 7.24). This substrate defines a dominant phosphogenic pathway to micro-concretionary apatite precipitation through rapid burial of organic matter under a stratified water/sediment body (Fig. 7.26). PRÉVôT et al. (1979) and RAUSCHER (1985) propose enhanced organic matter trapping and related phosphogenesis in structural traps. A number of these authigenic grains are nucleated, respectively contain sediment particles from the initial sediment composition (TRAPPE 1989, 1992, PRÉVôT 1990). The presence of a calcareous rock during local formation of erosional surfaces shifted phosphogenic pathways to carbonate replacement (Fig. 7.26). Microbial/ooidal phosphogenesis is present

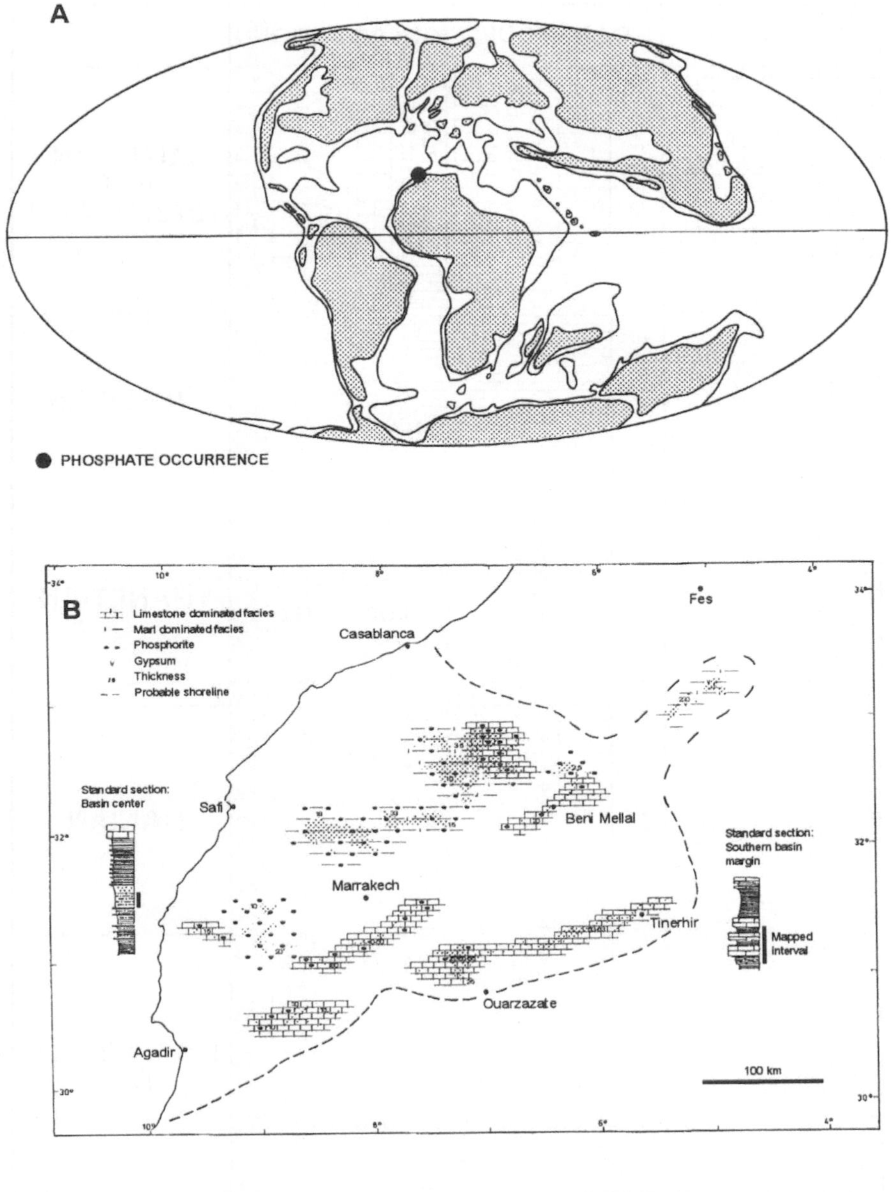

Fig. 7.24. A: Global paleogeographic location of the central Moroccan phosphate basin, based on paleogeographic reconstructions from GOLONKA et al. (1994). B: Facies distribution and paleogeography of the central Moroccan phosphate gulf during the Paleogene (from TRAPPE 1991). Outcrop areas of Paleogene strata are dotted.

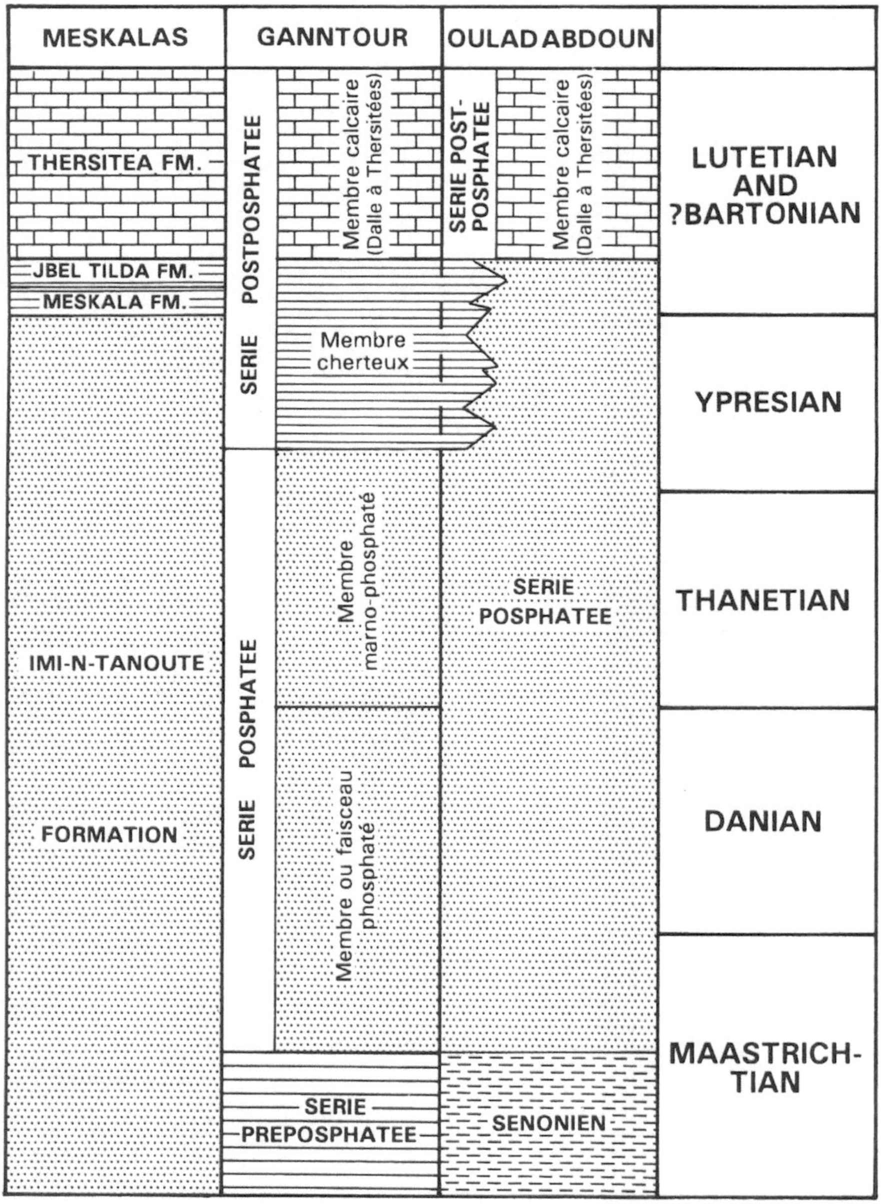

Fig. 7.25. Lithostratigraphic chart of the Subatlas Group and stratigraphic equivalents in Central Morocco (from HERBIG & TRAPPE 1994).

(PRÉVôT 1990) and results most probable from non-burial of phosphate grains, although microbial coating within the sediments cannot be excluded.

Phosphorites generally occur in distinct beds with sharp or rapid changes between phosphate rich and non-phosphatic lithology (AZMANY-FARKHANY et al. 1986, PRÉVôT 1990), implying that initial sedimentation was episodicly affecting numerous levels of phosphate sediments. The degree of reworking and mechanical concentration is controversial. PRÉVôT (1990) postulated only slight reworking for the Ganntour deposits. TRAPPE (1991, 1992a) emphasized the granular character of the phosphorites in the southern portion of the basin from the interpretation of microfacies of the phosphorites and the facies association with well-oxygenated carbonate host rock. The latter confirms the earlier interpretation of BOUJO (1976) for the phosphorites from the central and northern portion of the basin and explains the lenticular, rapidly changing character of the phosphorite seams in submarine large sheets. The microfacies of the phosphorites from the Meskala region documents an intensively mixed phosphorite packstone composition (TRAPPE 1989). The degree of reworking presumably decreases in the central portion of the basin respectively the Ganntour basin, and elucidates partly PRÉVôT's observation. The lateral facies changes, the lenticular nature of the sediment bodies, and the dominantly granular texture of the phosphate accumulation support frequent sediment reorganization by winnowing and reworking. The process is accompanied by vagabonding sediment transport with rapidly forming and decaying phosphate sand sheets and sand waves. Multiple reworking events cause mechanical stratigraphic condensation. The latter interpretation assigns the formation of phosphorite seams to several depositional periods which provide, both, favorable conditions for phosphogenesis and concentration. The typical coupling of both phenomenons is observerable in most Phanerozoic phosphate deposits. Facies and time variations are the result of the internal basin topography.

Fig. 7.26. Major phosphorite microfacies types of the Moroccan phosphorites. (A) Silicified phosbiophosclastpackstone. The highly mixed sediment is composed of bone and teeth fragments of various sizes and darker structureless phosclasts. The dark color of the latter results from enclosed organic matter. (B) Most typical granular phosphate facies type. A packstone composed of various structureless phosphclasts with different degree of pigmentation depending of weathering stage, organic matter preservation, and sediment particle incorporation. (C) A bore hole on an erosional surface is blanketed by a phosphatic microbial crust. The remaining space is filled with a highly mixed, granular phosphorite concentrate. Diameter of the bore hole is about 10 mm. (D) Phosclastpackstone includes numerous large oyster fragments. (E) Polymictic phosclastrudstone. The phosphate grain spectrum includes various phosphatic lithoclasts which are phosphatized carbonate mudstone, phosphatized marl as well as redeposited bioclasts with molds filled with phosphate sediment or complete phosphatized bioclasts. The clasts are of centimeter size. (F) Phosclastpackstone lithoclasts of various sizes and degree of pigmentation. Second major component is quartz detritus of much smaller grain size. The poor sorting documents intensive depositional mixing. Matrix is a micrite.
See also color copy of this plate in the appendix.

Fig. 7.26.

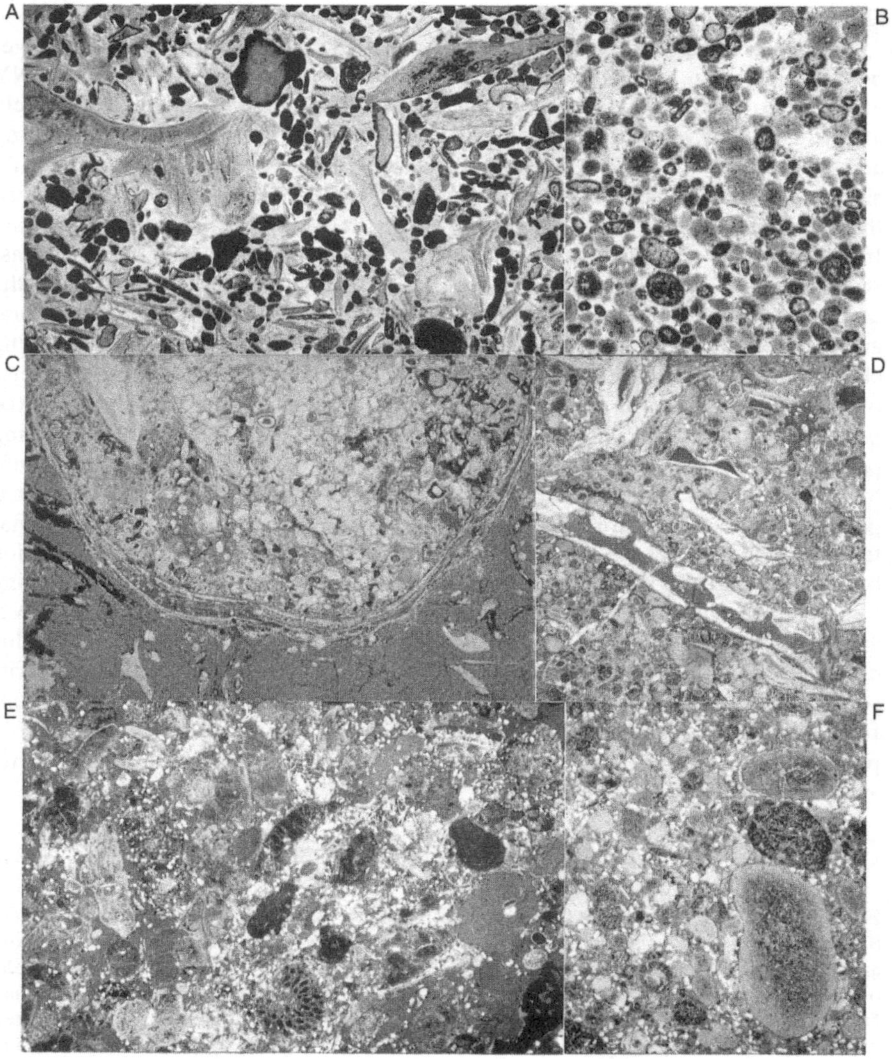

SALVAN (1954) and BOUJO (1976) recognized early the general time transgressive nature of the central Morroccan deposits along the axis of the gulf from the west to the east. Phosphorite deposition is related to the transgressive portion of the Subatlas T/R supercycle (HERBIG & TRAPPE 1994). All phosphorite sediments have to be included in the transgressive systems tract in the scale of this order cyclicity. Some very minor phosphate beds are associated with highstand sediments (e.g. Meskala region, TRAPPE 1989). The mechanical sediment reorganization is associated with significant sediment transport associated with the invasion of the marginal portions of the Moroccan phosphate basin. Granular phosphorites are found in non-phosphogenic, marginal environment (TRAPPE 1991,1992a). These sediment bodies must result from the shift of phosphorite sand sheets toward the basin margin within the shallow basin. The granular phosphorite formation, accumulation and concentration are an episodic phenomenon within the transgressional phase of the Subatlas supercycle. Multiple mechanical sediment reorganization phases destroyed a distinct parasequence pattern in the west. This is affected by the shallowness of the basin and the exposure to the open Atlantic ocean. Only in the very marginal, eastern portion of the gulf, a low order cyclicity is preserved (HERBIG 1991).

The sediment reorganization may result from enhanced storm frequency or permanent elevated water energy, both relating the phosphorite formation to a relative low sea-level stage during low order sea-level fluctuation. Evidence for these periods is documented by hardground formation on the marginal ramp and within the basin (TRAPPE 1989, PRÉVôT 1990, HERBIG 1991). A direct correlation between erosional events on the ramp and phosphorite formation is impossible due to poor biostratigraphic data and the lenticular natur of the phosphorite accumulations. The stratigraphic pattern assigns phosphate beds to two potential sequence stratigraphic positions of low order cyclicity:

(1) the retrograde portions of parasequences or lowest order sequences, or
(2) the early prograde portion of the following (para)sequence.

The absence of significant high energy sediment structures supports reworking by single events, homogenizing the sediment successively. Parts of the matrix are preserved instead of producing cross-bedded sand waves by more persisted high water energy. Magnitude and intensity of reworking remains ambiguous due to the poor biostratigraphic resolution. During these general reworking intervals blanket sands were forming. These features and commonly sharp basal contact support more the early progradational portion of parasequences when sedimentation was renewed during still low sea-level. Morphology and currents caused rapid facies changes.

Sediment transport in this kind of an extreme shallow and broad basin is minor in downslope direction, because of the absence of a significant slope. The overall transgressive trend during phosphorite formation affects cyclic progradation of vagabonding blanket sands in marginal position of the basin (TRAPPE 1991);(Fig. 7.26). Microfacies analysis of these phosphorite hosting rock identify the non-phosphogenic depositional environments in these positions.

The absence of major deltaic or esturine sediment associations and the REE-geochemistry (see Fig. 3.1) indicate a dominantly marine source of phosphorus.

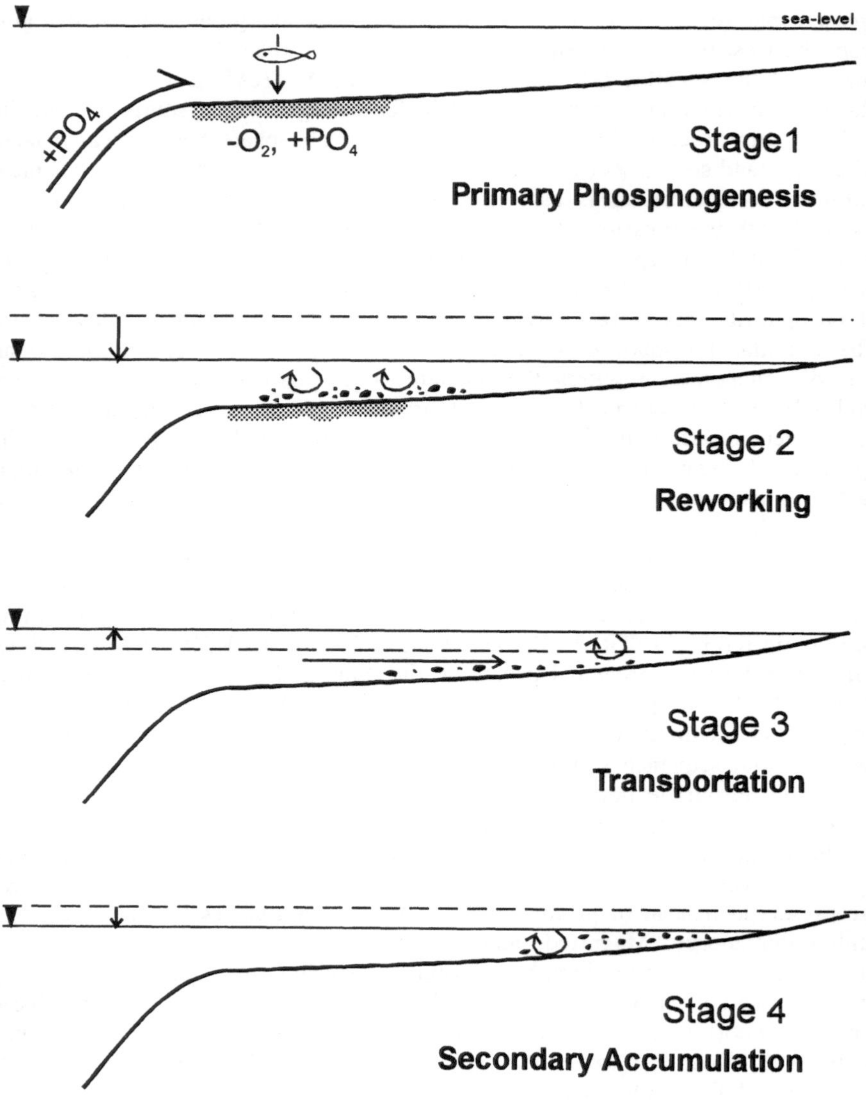

Fig. 7.27. Phases of phosphorite accumulation in a W-E transect through the gulf and the basin center. Within retrograde portions of low order sea-level fluctuations, phosphorites are reworked and transported as a phosphorite sand blanket into marginal portions of the phosphate gulf. Further sea-level changes enhance concentration and mixing. Composite concentrates are formed.

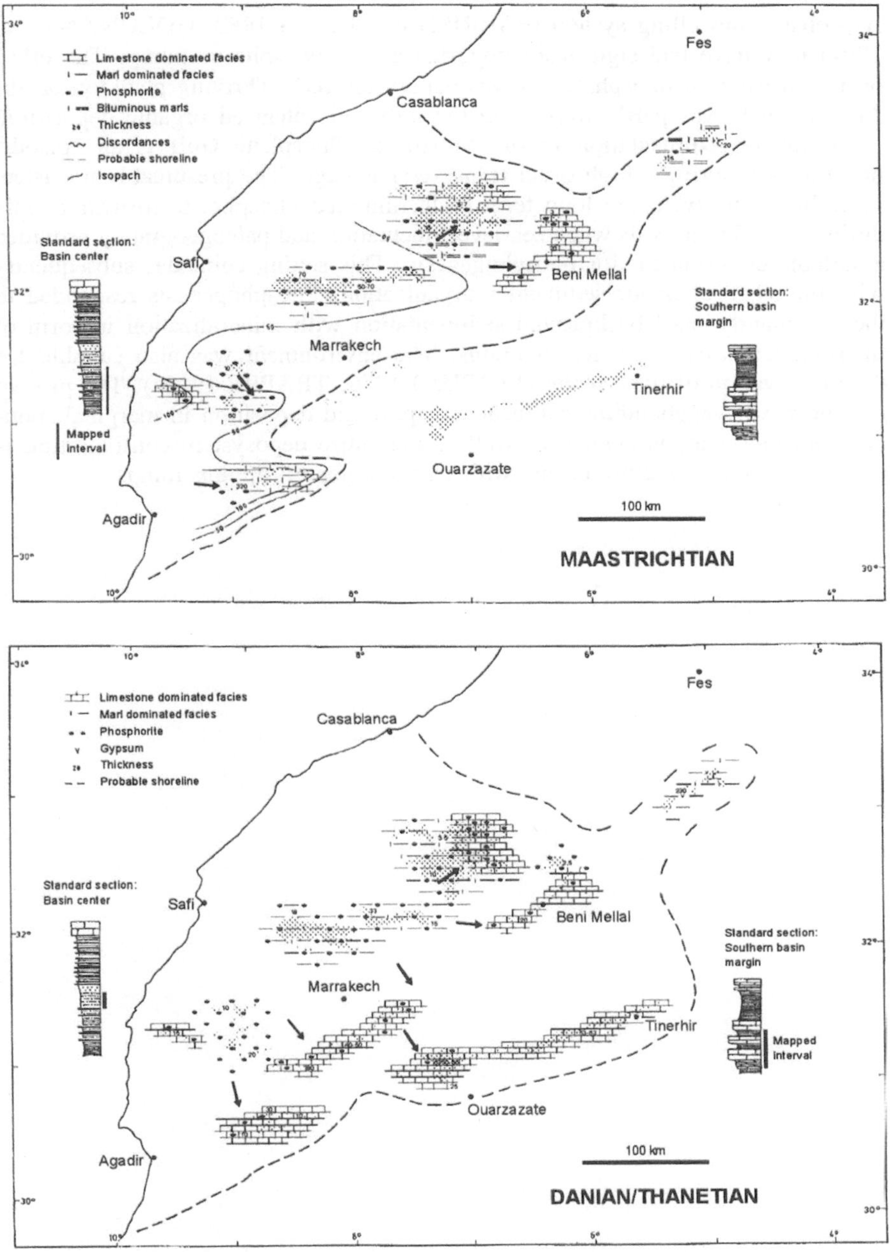

Fig. 7.28. Potential directions of phosphorite sediment transport during the transgressive portion of the Subatlas supercycle. Based on paleogeographic reconstructions of TRAPPE (1991) for the Maastrichtian and Danian to Thanetian.Outcrop areas are dotted.

A probable upwelling system (PARRISH & CURTIS 1982, GOLONKA et al. 1994) may have had significant importance as a phosphorus pump. The effect on the Morroccan phosphate sea was mainly indirect. Thriving plankton on the Atlantic shelf was partly trapped in the gulf and enhanced organic deposition.

Phosphate sedimentation in the Morroccan Phosphate Gulf is an episodic phenomenon during a high order transgressive stage. The presumably persistent upwelling activity in the long term scale, initiated phosphorite formation only during specific intervals when sea-level fluctuation and paleogeography provided a suitable environment for phosphogenesis. This setting coincides subsequently with the stage of major sediment reorganization. Phosphogenesis responded to the dominantly marl background sedimentation with mineralization in form of micro-concretions or nucleated grains. The environment was also suitable for the preservation of bone material (BOUJO 1976, TRAPPE 1992a). The present transgressive, vagabonding sediment transport and deposition in marginal, non-phosphogenic sites is evident as well as the entire deposystem configuration is typical for many shallow basins with extreme gentle dipping ramps.

PART IV
PHOSPHORITE DEPOSYSTEMS

8 Facies and depositional environment of phosphorite occurrences: the first step to phosphorite deposystem recognition

As shown above, the broad spectrum of genetic processes and pathways of phosphorite rock formation are predominantly defined by the depositional environment and the development of the deposystem through time. These two driving parameters find their expression in the lithofacies association and in the facies architecture of the deposit and associated rocks. In the reverse approach, the interpretation of these data reveal depositional processes/pathways, environments, paleogeographic settings, and evolution of ancient deposits which have no modern equivalent. The lithofacies associations and facies models are discussed below. The development through time is outlined in the following chapter 9.

8.1 Lithofacies associations

The phosphorite facies itself, but much more the lithofacies association was commonly considered to be an important attribute for the application of models on genesis and depositional environment. Phosphorite occurrences have been assigned to specific type deposits (e.g. ABED & KRAISHAN 1991). SHELDON (1964a) used lithofacies associations as a tool for exploration strategies. RIGGS (1986) and COOK et al. (1990) listed the following lithofacies associations:

- phosphate - black shale - chert
- phosphate - dolomite-magnesium clays - chert
- phosphate - glauconite
- phosphate - banded iron oxides - chert (restricted to Precambrian)
- phosphate - manganese/ferromanganese oxides

This basic approach for the characterization of the broad spectrum of phosphorite occurrences is not sufficient to address the complexity of their depositional history. Single lithofacies types include genetically different phosphorite facies, respectively genetically similar deposits occur in the different groups. The compilation is also partly driven by lithofacies associations which are mostly known from Neogene and modern systems. These young deposystems are showing a number of parallels, but most ancient phosphorite fabrics are not well represented in modern equivalents (GLENN et al. 1994).

As a consequence, this lithofacies typology has to be extended to address more specifically the facies criteria which are indicative for the reconstruction of their origin, respectively their economic value. This approach has to include the phosphorite facies criteria and stratigraphic relationships. A number of lithofacies association are not well represented by young occurrences. The lithofacies classification needs to reach a preciseness to delimit genetically similar groups. Their identification is a first step to the development of facies models for non-actualistic deposits. According to these requirements, the following lithofacies types are defined for marine, shelfal, or epeiric Phanerozoic deposits (insular and continental phosphorites excluded).

1. concretionary phosphorites - black shales - cherts
2. concretionary phosphorites, mass flow phosphorites - porcelanites
3. micro-concretionary phosphorite, lag phosphorites - black shales - cherts - bioclastic carbonates
4. phosphatic hardgrounds, phosphatic bindstones, granular phosphorites - carbonates
5. granular phosphorites - marls - Mg-clays - cherts
6. phosphatic hardgrounds, phosclasts - glauconites - carbonates
7. phosclasts, phosphate cementations - glauconites - siliciclastics
8. phosbioclasts - glauconites - siliciclastics
9. phosphorite crusts and concretions - Fe/Mn crusts

The lithofacies assemblages 1 to 3 disclose the black shale family, lithofacies assemblages 4 to 5 the carbonate/marl family, and the lithofacies assemblages 6 to 9 the Fe-family.

Nevertheless, the information is not sufficient to determine precisely the phosphate hosting deposystems. Completely uncertain remains the paleogeographic and paleotectonic setting of phosphorite deposystems as well as the facies architecture and time/space relationships, which provide information to sea-level control and event stratigraphy. These informations are only provided by facies models for the specific lithofacies type.

8.2
Facies - facies models - depositional environments

All facies analysis methods are based on two fundamental assumptions which are (1) the existence of a genetic relationship between facies and depositional environment, and (2) that actualistic equivalents can by applied to ancient facies associations.

This method is well established for carbonate rocks (WILSON 1975) and for siliciclastics (REINECK & SINGH 1980). The approach combines physical parameters and observations in modern environments for an interpretative reconstruction of ancient depositional environments. The actualistic approach is extremely limited in ancient phosphorite deposystems, as phosphorite facies and lithofacies do not completely correspond to modern systems. A good example are phosphatic stromatolites, which have not been found today, and their ecologic demands remain enigmatic. The microfacies of phosphorite rocks shows striking similarities to carbonates, and the genetic pathways, which have been identified above, are related to a considerable extent. The few existing studies document the successful application of an extended microfacies analysis for the development of facies models of ancient phosphorite deposystems. Therefore, the development of facies models holds a key position and became possible with the increasing knowledge on processes and the precise reconstruction of depositional environments by microfacies analysis methods. The compilation in Tab. 8.1 summarizes available facies models for the major facies associations.

Existing facies models assign the development of larger phosphorite deposits strictly to the marine realm with a broad spectrum of various depositional environments. Most models localize the ancient phosphorite accumulation in the shallow marine subtidal to peritidal zone. The spectrum of environments shows a distinct preference in various shallow marine settings, which correlate with the maximum phosphorus burial in these environments (s. a. chapt. 1).

Table 8.1. Compilation of the presently available facies models for Phanerozoic phosphorite deposits

PHOSPHORITE FACIES	LITHOFACIES ASSOCIATION	FACIES MODEL	DEPOSIT/AGE	AUTHOR
dominantly phosclastic, concretionary, biogenic, ooidic	black shale-chert, bioclastic and mudstone carbonates, sandstone; lateral: red beds	open shelfal embayment, offshore or deeper water biogenic shale deposition, marginal carbonates and sandstones, broad coastal sabkha	Western Phosphate Field /USA mid-Permian	McKELVEY et al. 1959, McKEE et al. 1967
phosclasts, phosphatic hardgrounds, phosphate bindstones/phoscretes	black shale, carbonate mudstone	shallow epeiric embayment with peritidal to lagoonal pelitic/granular phosphate deposition and basinal carbonate mudstone/phosphatic hardgrounds	Georgina basin/N'Australia Middle Cambrian	DeKEYSER & COOK 1972
phosphatic hardgrounds, phosintraclasts	chalks with nodular cherts	epicontinental carbonate platform	N'France Senonian	JARVIS 1980b, 1992
phosintraclasts, phosbioclasts, phosooids forming pack- and rudstone	bioclastic carbonates, carbonate mudstones, bituminous dolomitic shales	carbonate platform and adjacent upper slope of a cratonic basin	E'Missouri/E'Iowa (USA) Middle to Late Ordovician	BLACK 1985, CAROZZI 1989
"pseudo-oolitic sandy phosphate", "non-detrital" phosphorites (non-differentiated)	marls, bituminous marls & dolomites, cherts, bioclastic carbonates; lateral: evaporites	system of semi-closed gulfs and narrow seaways adjacent to the shelf	Central Morocco Maastrichtian to Eocene	SALVAN 1986
phosphatic hardgrounds, phosintraclasts, phosbindstones/ phoscretes	carbonate mudstones; lateral: black shales	peritidal carbonate platform within pelitic shallow basinal sedimentation	Georgina basin, N'Australia Middle Cambrian	SOUTHGATE 1986a,b, 1988, SOUTHGATE & SHERGOLD 1992

Table 8.1 (cont.). Compilation of the presently available facies models for Phanerozoic phosphorite deposits

PHOSPHORITE FACIES	LITHOFACIES ASSOCIATION	FACIES MODEL	DEPOSIT/AGE	AUTHOR
phosphatic hardgrounds and crusts, phosphate cementations, phosbindstones, phosintraclasts	glauconitic sandstones, sandy lime-stones, calcareous shales	siliciclastic shelfal ramp to continental slope	Helvetian Alps, Switzerland/Austria mid-Cretaceous	FÖLLMI 1989a, 1990
phosclasts, phosphatic micro-concretions	black shales, marls, porcelanites, lateral: variegated shales, sandstones	deltaic inner shelf to mid-shelfal ramp with dominance of siliciclastics	Egypt Late Cretaceous (Campanian)	GLENN & ARTHUR 1990
phoslithoclastic packstones, single phosphatic hardgrounds	marls, bioclastic carbonates, cherts, porcelanites	broad, shallow gulf, bowl-like profile, with marginal biofacial zoned carbonate ramp,allochthonous phosphorite, basinal facies of marls/cherts, granular phosphorites, and porcelanites in the open marine areas	Central Morocco Maastrichtian - Eocene	TRAPPE 1991, 1992a
phosclasts, phosphatic coated grains, phosphate concretions	siliciclastics, shales	slope of continental margin subbasin, bimodal: granular fabrics in peritidal to lagoonal with associated mass flows, basinal phosphate concretions	Baja California (Mexico) Oligocene	SCHWENNICKE 1994

9 Paleogeography and stratigraphy of phosphorites

The facies models for ancient phosphorite depositional environments highlight the contrast between the ancient and modern occurrences according their geographic setting (s. a. chapt. 8). This contrast draws the attention to the potentially different spatial location of ancient phosphorite occurrences. The paleogeographic configuration of a deposystem, which is the sum of paleotectonics, morphology, climate, ocean circulation, and biologic activity, determines not only the geographic context but also the stratigraphic position of phosphorite sediment bodies within a deposystem. As a consequence, characteristic spatial and temporal relationships within phosphorite hosting deposystems directly identify sedimentary preconditions for the development of phosphate deposits and their depositional context, which can be used for the development of exploration strategies in other world regions.

9.1 Paleogeographic settings including climates

The interest in the paleoenvironment of individual deposits and in the global distribution of phosphorite occurrences has increased with the need to improve exploration strategies (SHELDON 1964a,b, 1981, 1984,a,b 1985, FREAS & ECKSTROM 1967, COOK & McELHINNY 1979, SLANSKY 1980a,b, 1986, CHRISTIE 1980, PARRISH & CURTIS 1982, PARRISH 1982, RIGGS 1986, RIGGS & SHELDON 1990). But the application of traditional, commonly misleading interpretations of very generalized facies associations restrained the development of modern reconstructions. This is only tardily changing. A striking example is the interpretation of the black shale-phosphorite-chert facies association, which was broadly assigned to deeper water environments (SHELDON 1987a,b). The modern research could draw a much more differentiated picture (e.g. SANDBERG & GUTSCHICK 1980, 1989 versus NICHOLS & SILBERLING 1990, SILBERLING & NICHOLS 1991, SILBERLING et al. 1995; SCHIEBER 1994, WIGNALL 1991, OSCHMANN 1988, 1990). An unbiased research was furthermore inhibited by the apparent correspondence of a generalized lithofacies of a very few ancient deposits to the modern occurrences. The elegance of the upwelling model and the correspondence to the modern occurrences lead even more to stagnation in this field of research. But an undogmatic investigation of the sedimentology, paleogeography, and location in dynamic stratigraphic models identifies a much broader spectrum and draws the attention to non-upwelling positions (GLENN et al. 1994).

Theoretical studies to sedimentary phosphate system as well as sedimentological characteristics from field studies in ancient phosphorite deposits (see chapt. 4) reveal a broad spectrum of potential paleogeographic positions of phosphorite deposystems. The only precise technique to investigate the paleogeographic setting of the ancient phosphate deposits of which modern equivalents are absent, is the detailed sedimentological analysis of the phosphorites and associated carbonate or siliciclastic rocks. Additional paleoenvironment informations are provided by geochemical methods. These data are the fundament for the earlier discussed facies models. The number of detail paleoenvironment studies remained very limited (see Tabl. 8.1, chapt. 8). But these investigations show that pristine as well as granular phosphorite depositional systems develop in nearly all marine paleogeographic positions, with the emphasis on the shallow marine realm. Despite of a number of sedimentological and geochemical parallels (GLENN et al. 1994), these paleoenvironments reveal a striking contrast to the modern deposits in offshore deeper water, and continental margin position. This fact accentuates the limited application of actualistic concepts for the environmental setting of ancient phosphorite deposystems, and the necessity of reconstructions from the geological record, although steady-state phosphogenic processes and resulting fabrics remain generally the same.

The compilation of modern paleogeographic data from Phanerozoic deposits identifies a distinct distribution pattern (Fig. 9.1). Ancient phosphorite deposystems develop predominantly in epeiric settings:

- open gulfs (e.g. Morocco and other NW African deposits, Late Cretaceous -Paleogene, TRAPPE 1991, WINNOCK 1980)(s.a. Case Study 6),
- shallow carbonate platforms (e.g. northern France, Late Cretaceous (JARVIS 1992; Georgina Basin, Middle Cambrian, SOUTHGATE & SHERGOLD 1991)(s.a. Case Study 2),
- broad, shallow W-E-directed shelfal ramps or platforms (e.g. Tunisia and related North African deposits, Late Cretaceous/Paleogene, WINNOCK 1980)(s.a.Case Study 1, 5),
- semiclosed (silled), shallow, epeiric mega-basins (e.g. Western Phosphate Field/USA, Permian (TRAPPE 1994), Midcontinent/USA, Pennsylvanian, HECKEL 1977)(s.a. Case Study 4),
- seaways (e.g. Ural, Permian TRAPPE 1994; Mali/Niger, Paleogene PASCAL et al. 1989).

Continental margin deposits with general correspondence to the modern phosphorites off Peru are present in the Neogene and Cretaceous system along the westside of the American continent.

Besides the paleogeographic location of a single deposit, a characteristic global distribution was identified early and since then intensively discussed in terms of paleotectonic, climatic and ocean circulation control.

SHELDON (1964b) investigated the paleolatitude distribution of Phanerozoic phosphorite deposits and found a distinct limitation to the zone between 25 degrees latitude north and south. The situation changed with the Neogene, since when the zone was extended to 40 degrees north and south. Phosphorite deposition was also shifted in high paleolatituds during the Permian/Triassic super-

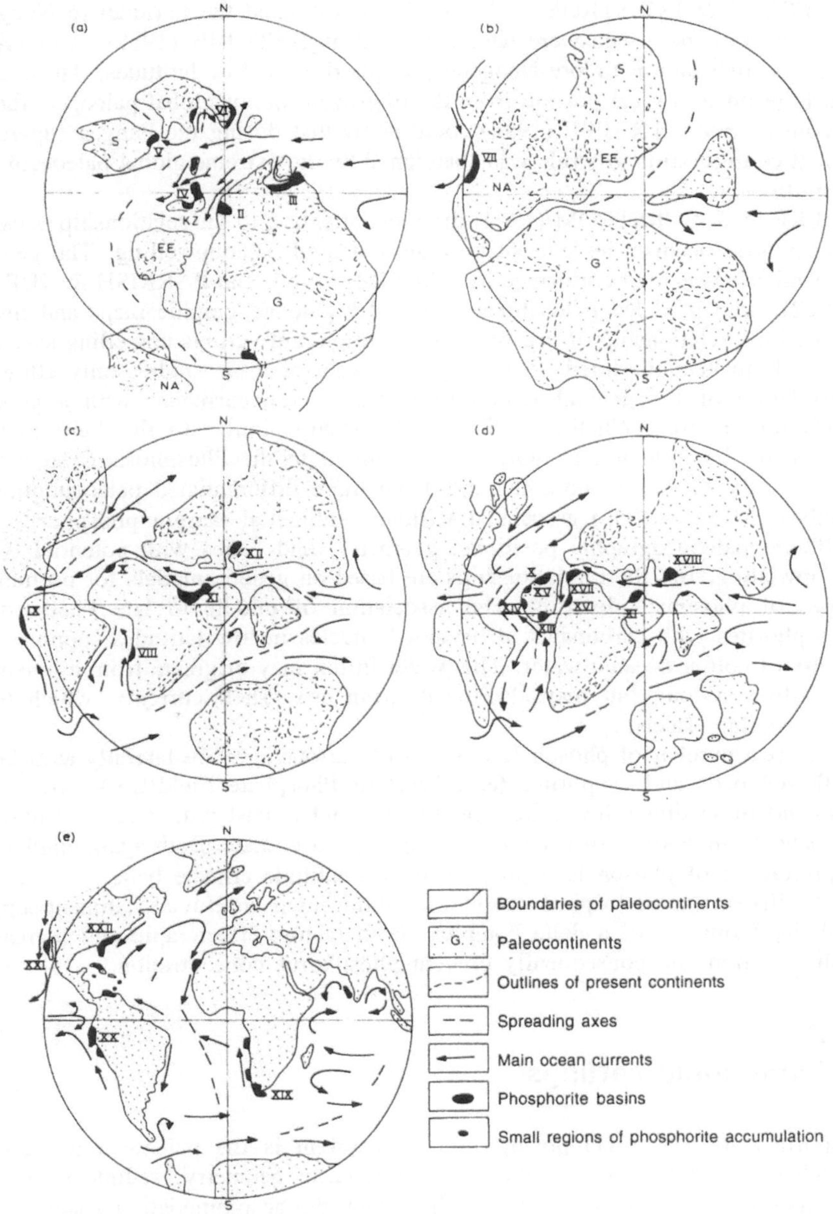

Fig. 9.1. Paleogeographic setting of major phosphate occurrences during important Phanerozoic phosphogenic intervals. From GLENN et al. (1994), after BATURIN (1981).

continent stage (TRAPPE 1994). Generally higher paleolatitudes were also found by FREAS & ECKSTROM (1967,) who investigated the Permian to Neogene periods. But these data were later corrected by BATURIN (1982), who shows again a distribution of pre-Neogene phosphorites in low latitudes. These contrasts result to a great portion from the improvements of global paleogeographic reconstructions. TRAPPE (1994) could show that during the Pangea supercontinent configuration this rule was abandoned by the extreme global paleoecologic conditions.

FREAS & ECKSTROM (1964) draw the attention to the relationship between phosphorite occurrences and sites of the potential paleoupwelling. The general dependence was later confirmed by PARRISH (1982) and PARRISH & CURTIS (1982), who modeled paleoclimate on global paleogeographic maps and investigated the relationship of organic sedimentation and various upwelling systems. It needs to be underlined, that the global scale of these studies only allows a correlation of the general region of phosphorite occurrences with a general upwelling region. Whether a direct relationship similar to the Peru margin exists, or the influence is more passive similar to the Phosphoria System (see Case Study 4), has to be concluded from more differentiated paleogeographic studies. RIGGS (1986) associated various continental margin phosphorites of different paleogeographic positions, geometry, and facies with potential types of upwelling. But these relationships are based on genetic models for prominent types of phosphate deposits. The association of middle to late Phanerozoic phosphorites with carbonates of brymol biofacies in low latitudes supports the influx of colder oceanic water. This water influx may originate from a potential upwelling system, but may also result from cross-shelf currents, which both have similar effects.

A great number of phosphate deposits of various ages are laterally associated with red beds and evaporites (e.g. Western Phosphate Field/USA, Morocco, Tunisia) indicating a low relief and broad sabkha coast with arid or semi-arid climate, or in absence of evaporites, also tropical climate. Both again emphasize a preference of phosphate deposition in low latitude climate belts.

Aridity or intense tropical weathering inhibit intensive fluvial sediment supply and the formation of a delta complex. Both factors effect rapid and increased sedimentation and consequently prevent phosphorus concentration.

9.2 Paleotectonic settings

Important for the development of a deposystem is the paleotectonic setting which significantly controls the topography/basin geometry, sediment supply, subsidence rates, and in relation with eustasy the accommodation space.

Phosphorite deposits generally occur on passive as well as on active continental margins, platforms, and various types of cratonal, orogenic and extensional basins. But their distribution shows distinct preference. The most favorable sites for the development of phosphorite deposystems are passive

Table 9.1. Properties of major paleotectonic settings in terms of phosphogenesis and phosphorite formation with selected, well known ancient examples. In other basin types than listed, larger phosphorite occurrences are commonly absent. CS = Case Study.

Paleotectonic setting, examples	Lithofacies character	Phosphogenesis	Phosphorite genesis
active continental margin (California, Neogene)	black shale, chert	(micro-)concretions, crusts, coated grains	mass-flow; current and event concentrations
passive continental margin (Egypt, Cretaceous)	marl, siliciclastics	micro-concretionary	winnowing, reworking; composite concentrations
inner shelfal sag basins, (Western USA, Permian, CS 1, Morocco, Paleogene, CS 6)	black shale, marl, carbonate, chert	micro-concretions coated grains, biogenic	winnowing, reworking; event, composite and lag concentrations, vagabonding transport
foreland/foredeep basins (Germany, Carboniferous, CS 3, Ural, Permian)	black shale, siliciclastics	concretions	minor; mass flow or current winnowing
platforms (N'-Australia, Cambrian,CS 2, Florida, Neogene, CS 5);including seamounts	(bioclastic) carbonate, siliciclastics; karst-related	carbonate replacements, microbial, coated grains, hardgrounds	reworking, morphology controlled transport; hiatal, composite concentrations
interior sag basins (Midcontinent USA, Pennsylvanian)	black shale, bioclastic carbonate, cyclic	concretions	minor, winnowing
fault related basins (Israel, Cretaceous)	marl, terrigenous clastics	micro-concretions, microbial	winnowing, reworking; event, composite concentrations

continental shelves and adjacent continental sag basins. The vast majority of all phosphate occurrences cover these paleotectonic situations. These subsidence areas are characterized by high bioproductivity, and high phosphorus flux rates in the sediment, furthermore by moderate sediment accumulation rates which, however, are drastically altered by topography and sea-level change. These factors provide best conditions for the initiation of phosphogenesis and subsequent phosphorite genesis. Some of the marginal sag basins are silled (e.g Phosphoria sea) or open to the shelf (Central Moroccan phosphate sea).

Marine platforms, which are characterized by reduced allogenic sediment supply with increasing isolation, represent the second important sites for the development of phosphorite sediment formation with well known Neogene examples. Drowning due to tectonics or sea-level rise shifts sedimentation style further to favorable conditions for phosphorite sedimentation. The opposite effect has intense carbonate deposition in very shallow water depth.

Active margin phosphorites are only known from the Neogene system along the western margin of the American continent. Some ambiguous occurrences are known from the Proterozoic (RIGGS 1986). Phosphorites in backarc basins are unknown.

Only some deposits are reported from foreland basins or syntectonic flysch basins. In general, the high sedimentation rates within a narrow trough-like basin inhibit the development of phosphorite systems. Phosphorites are only deposited in basins with very specific conditions: in seaways with potentially strong current systems through channel effects, e.g. Permian Ural foredeep; in extreme broad basins with relative sediment starvation on the basin shoulder opposite to the orogene, e.g. Mississippian Antler-foreland, or when the subsidence drastically exceeds sedimentation rate (e.g. Lower Carboniferous Rhenish trough). High subsidence rates generally support the phenomenon.

Many phosphorites are known from broad inner cratonal sag basins, most of them related to hiatuses or condensed beds, or are from lacustrine systems.

9.3 Stratigraphic settings of phosphorites and facies architecture

The very specific conditions of phosphogenesis on one side and phosphorite genesis on the other side assign their occurrence to specific stratigraphic positions within phosphorite-hosting deposystems. Consequently, the spatial positions of phosphogenic systems and granular phosphorite accumulation are apparently different, but evidently overlapping in a number of systems. Besides eustasy, significant effects on stratigraphic position and facies architecture of phosphorite sediment bodies have basin geometry, subsidence history, and ratio of autogenic versus allogenic sediment supply. The sequence stratigraphic constellation determines type and intensity of the genetic processes of phosphorite formation and consequently the quality of the resource. Long before sequence stratigraphic concepts have been defined by VAN WAGONER et al. (1988,

1990) and SARG (1988), many researchers observed the transgressive setting of most phosphate deposits, but without differentiating various types (e.g. McKELVEY et al. 1959, CRESSMAN & SWANSON 1964, ARTHUR & JENKYNS 1981, RIGGS & SHELDON 1990, GLENN & ARTHUR 1990, FÖLLMI 1990, HERBIG & TRAPPE 1994). But detailed sequence stratigraphic interpretations of phosphorite deposystems remained exceptional (GLENN 1990a, THORNBURG 1990b, SOUTHGATE & SHERGOLD 1991, GLENN et al. 1994, TRAPPE 1993). One reason for the lack of modern interpretations are the common occurrence of ancient phosphate deposits in epicontinental basins without a prominent slope which complicate the application of traditional sequence stratigraphic concepts. The latter are designed for a continent-shelf-slope-ocean basin profile. The very gentle ramps in epeiric basins are addressed by VAN WAGONER et al. (1990) for siliciclastic systems and by BURCHETTE & WRIGHT (1992) for carbonate ramp and platform systems.

Phosphorite concentrates differ in composition and stratigraphic context which both are indicative for their genesis. Granular phosphate concentrations may have a homogeneous or polymict grain composition. The composition is thought to be a criterion for the frequency of reworking and mixing during transport. The increasing degree of corrosion and amalgamation (microbial or chemical)reveals condensation through long lasting exposure and non-burial. Important are also sedimentary context and bed form. Stratigraphic contacts may be gradational or sharp. The latter comprise omission or erosional surfaces. This pattern and the internal grain size distribution (fining upwards or coarsening upwards) is important for the interpretation of the sequence stratigraphic position and the origin of the concentrates. Similar to shell concentrations (KIDWELL 1991a,b), four major genetic types of phosphorite concentration are distinguished (Fig. 9.2).

Event phosphorite concentrations are commonly thin layers or lenticular bodies with grain compositions of low diversity. Abrasion, corrosion and amalgamation are absent. Host sediment is usually present. These types of concentration occur within the pelitic Phosphoria Formation (see Case Study 4) or within the calcareous Thorntonia Formation (see Case Study 2). The beds have usually a sharp basal contact and a indistinct top.

Composite phosphorite concentrations form larger sediment bodies of a mixed phosphate grain spectrum and include various non-phosphatic grains ofdifferent roundness and size. These beds have usually a sharp basal contact. The top may be gradational or sharp. The beds are a characteristic feature of the Subatlas Group of the Paleogene system in central Morocco (see Case Study 6).

Hiatal phosphorite concentrations are split in two types. Omission or condensation related phosphorite beds are commonly characterized by corrosion and a composite grain spectrum. They result from sediment starvation (e.g. Mid-Cretaceous of the Helvetian Alps, FÖLLMI 1990). But these occurrences may also be erosion-soled. They differ from lag concentrates in the low grade of

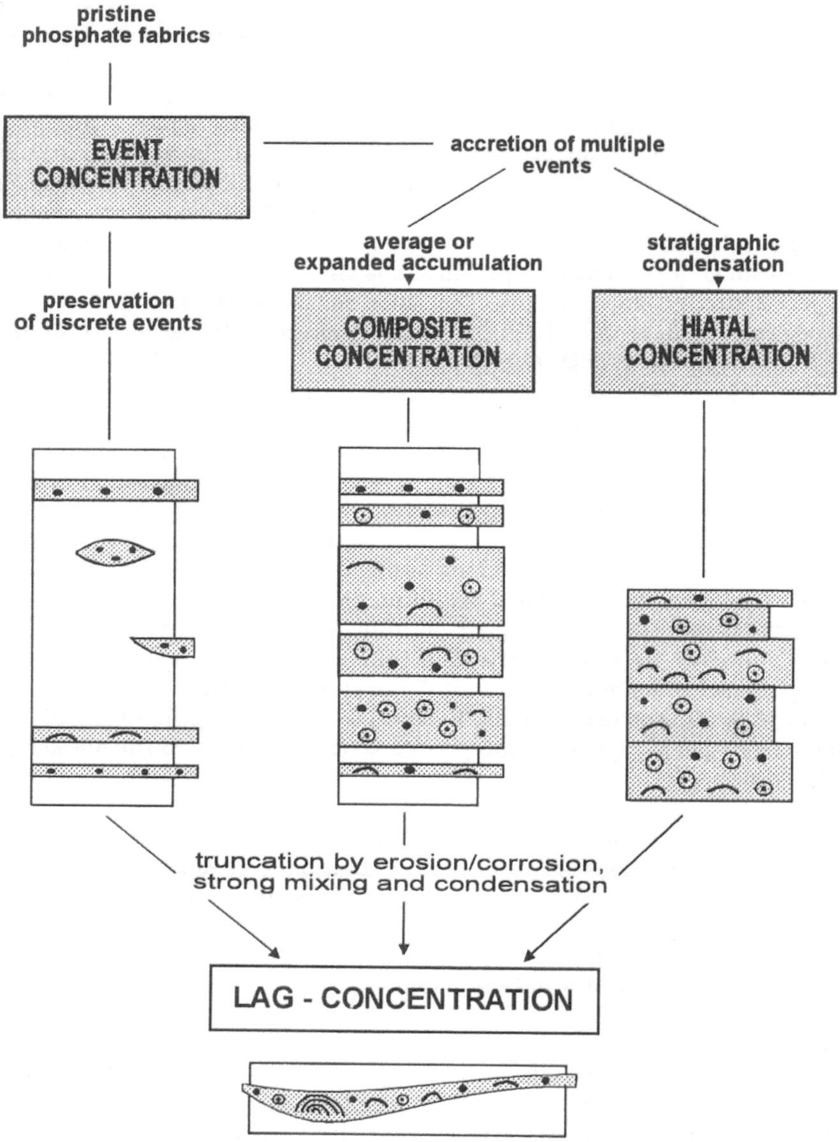

Fig. 9.2. Principal genetic types and stratigraphic appearance of phosphorite concentrations. Lag concentrations are genetically complex beds which combine processes and resulting fabrics of the simple concentration types. Definitions adopted and modified from concepts of shell concentration by KIDWELL (1991a,b).

reworking and mixing. All grains are directly derived from reworking of the erosional surface. Good examples are the hardground successions in the Thorntonia Formation in northern Australia (see Case Study 2).

Lag phosphorite concentrates resemble multiple composite concentrations, partly associated with corrosion and amalgamation, which rest on a distinct truncation surface. These concentrates combine characteristics of both the composite and hiatal concentrations. A well developed example are the basal beds of the Retort Phosphate Member or the Meade Peak Member of the Phosphoria Rock Complex (see Case Study 4).

9.3.1
Fundamentals of sequence stratigraphy in phosphate-siliciclastic deposystems

In siliciclastic systems, facies architecture is dominantly controlled by the allogenic terrigeneous source, the biogenic marine sedimentation, and furthermore by authigenic mineral precipitation to a very minor degree. The interplay of the two major sources in the framework of sea-level change and subsidence defines the development of the various pathways of phosphogenesis. Intensity and type of phosphorite genesis are affected by changes of water energy in the different environments during the T/R cycles.

Phosphogenesis

The important conditions for the initiation of one of the various phosphogenic pathways in siliciclastic dominated deposystems are relative enhanced organic deposition, relative siliciclastic sediment starvation, or more exceptional, rapid burial of large quantities of organic matter. The major prerequisite is further the development of an oxygen deficient water/sediment column. These requirements attribute the development of pristine phosphate fabrics to the transgressive systems tract and the maximum flooding stage, whne allogenic sediment supply is generally reduced in the entire deposystem due to a rising base level. The relative emphasis of organic matter deposition provides a favorable precondition for the authigenesis of apatite, if oxygen deficiency can be established (see chapt. 6). Major control are paleogeography and accommodation space. The phosphate distribution in the Phosphoria Rock Complex shows a distinct preference in the early transgressive stage (see Case Study 4).

Other important locations are stratigraphic intervals of condensed sedimentation. These encompass include two genetically different sites. Phosphorites are observed in the very distal position of a basin/continental margin depositional prism (LOUTIT et al. 1988). During maximum flooding, the source area of allogenic material moves in maximum distance from the basin center or slope causing relative sediment starvation in most distal portions. This sedimentation is associated with phosphate and glauconite authigenesis and was defined in sequence stratigraphic nomenclature as the "condensed section". The sediment body is commonly omission-soled in contrast to onlap (transgressive) condensation which is usually erosion-soled. During maximum flooding of low order cyclicity, potential sediment starvation occurs also in association with downlap and backlap surfaces (Fig. 9.3). These surfaces are in general the trace of the condensed sections of smaller scale cyclicity. These beds usually form thin beds of pavements and are of minor economic importance.

Phosphorite genesis

Fundamentally different are the relative condensation intervals associated with the onlap and toplap (Fig. 9.3).The onlap intervals represent the initial flooding stage of a basin margin with potentially coinciding favorable conditions of phophogenesis (see above) and potential mechanical sediment condensation. In contrast, toplap condensation is related to increased water energy and decreasing accommodation space due to a falling sea-level. In most paleogeographic constellations, falling base level increases sediment supply and phosphogenesis is widely supressed, consequently also granular phosphorite concentrate formation.

During the onlap phase relative sediment starvation by enhanced organic sedimentation during relative low sea-level stage initiates evidently phosphogenesis (see chapt. 6), but favors also the mechanical formation of phosphorite concentrates. This is the preferential site of major phosphorite formation (GLENN 1990a, SOUTHGATE & SHERGOLD 1991, GLENN et al. 1994). The onlap surface occurs in form of an omission surface or an erosional surface. The initial sediment is typically a lag concentrate with phosphatized relict grains, corrosive fabrics and microbial mats. This lag is commonly followed by a composite concentrate which consists of poorly sorted clastic phosphate grains of various types. The composite concentrate results from multiple event concentration and is associated with vagabonding sediment transport. Microbial phosphate formation is reduced. Both sediment bodies are time transgressive to various degree. In the upper sediments of the transgressive systems tract, authigenic phosphate fabrics may be concentrated in individual horizons representing event concentrations. In shallow basins, intensive retrogradational reworking during low order sea-level fluctuation, frequent event concentration, or concentration by cross-shelf current cause intense stratigraphic condensation by mechanical sediment reorganization.

During the late highstand, accommodation space decreases during falling sea-level. This causes again extensive high energy conditions with elevated reworking intensity. But this interval is usually associated also with drastic increase of siliciclastic supply which would accompany reworking with contemporary siliciclastic dilution. Only tropical weathering conditions and a low relief coast would minimize the effect. Enhanced fluvial solution supply of phosphorus sporadically initiates phosphogenesis associated with intensive glauconitization at the presence of Fe, Al, and Mg. These grains are commonly reworked and represent the toplap condensation.

In open shelf configurations, such as the Neogene systems in California (FÖLLMI & GRIMM 1990, FÖLLMI & GARRISON 1991, SCHWENNICKE

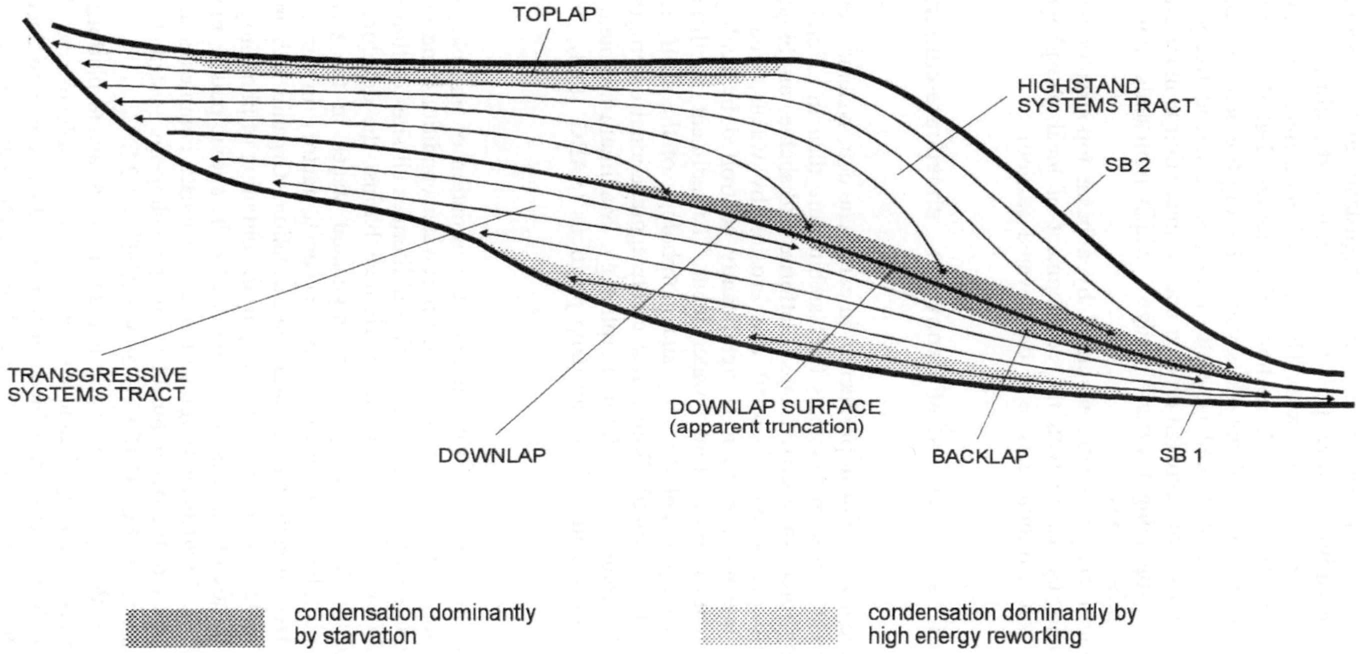

Fig. 9.3. Principle positions of stratigraphic condensation by sediment starvation or erosion as potential sites of phosphogenesis and granular phosphorite formation within a sequence stratigraphic concept for a steepening ramp.

1992) with phosphogenesis persistent also in the highstand interval, sediment mobilization during a falling sea-level produces mass transport sediments (event concentrations) in lowstand systems tracts. The outlined mechanisms are similar in both, continental shelves, as well as homoclinal ramps. But the resulting geometry of facies bodies and systems tracts varies. During lowstand, sediment is mobilized on the entire shelf and generates a distinct lowstand wedge. On a gentle ramp, the contrast of sediment erosion and accumulation is more gradational producing a "prograding lowstand wedge", which represents more the basinward shift of facies zones.

The typical pattern are drastically modified by a basin topography which causes local sediment bypassing with the development of local phosphogenic environments and concentration sites, e.g. on elevated seafloor.

9.3.2
Fundamentals of sequence stratigraphy in phosphate-carbonate deposystems

The dominantly autogenic sediment production in carbonate deposystems causes a significant different facies and systems tract architecture than in siliciclastic systems. Competing forces on epeiric ramps, platforms and shelves are biogenic carbonate production (carbonate factory) and erosion by waves, tides and storms. The carbonate bioproduction is significantly controlled by ecological factors, e.g. water temperature, salinity, water depth. The sediment mobilisation is affected by the paleogeographic setting and morphology of the shelf, ramp or platform (READ 1982, 1985). Both forces create aggradational or progradational systems in the context of seafloor morphology. Maximum sediment accumulation is commonly achieved in mid-ramp positions (SARG 1988).

Phosphogenesis

Rapid carbonate accumulation generally inhibits the initiation of phosphogenic processes in the sediment because the necessary porewater conditions and phosphorus concentrations are not accomplished. This obliterates all sites of flourishing carbonate production and deposition. In rimmed basins, the development of pycnoclines during the transgressive and highstand stages in the basinal position causes a complete shut down of the carbonate factory in the basin center and emphasizes organic background sedimentation. Organic rich muds are deposited as a condensed section giving way to the initiation of phosphogenic processes. This facies architecture is also addressed by the anoxic puddle model of WIGNALL (1994). Centripetal progradation of carbonate complexes along the ramp or slope restricts the development of the phosphogenic system to the transgressive systems tract (BURCHETTE & WRIGHT (1992).

Phosphogenesis on the upper carbonate ramp/shelf or carbonate platform is bound to drowning stages during drastic and rapid relative sea-level rise. The rapid environmental change causes a reduced or terminated carbonate production and a keep-up system is not established again. The relative sediment starvation widely effects intense phosphatization along erosional or omission related sequence boundary. Relict deposits (phosphatic lags) are formed. This phenomenon may also be evident to a much smaller degree in catch-up highstand systems (e.g. GARFIELD et al. 1992). On the basis of sedimentological and biostratigraphic data, SOUTHGATE & SHERGOLD (1991) could determine the retrogradational position on parasequence level within a transgressive systems tract for the occurrence of the phosphorite beds in the Middle Cambrian Georgina Basin deposystem. Increased nutrient supply during highstand in association with suppressed siliciclastic poisoning exceptionally reestablishes again phosphogenesis during late highstand (GLENN 1990).

Phosphorite genesis

Mechanical sediment concentration on ramps or platforms is very much restricted to the peritidal zone or the lowstand phase. In both situations, sediments are reworked in which phosphate mineralization is only exceptionally realized. In relative moderate water depth during the early drowning stages, sediment is mobilized during high energy events. Relict sediments and reworking products from the erosional surface are accumulated downslope or are transformed in vagabonding relict bars, both contribute sediments to the early transgressive systems tract (composite concentrates). The mechanical concentration may result from permanent tides, from current and wave activity, from event reworking during intense storms, or from low-order sea-level change. The latter produces granular phosphorites during the retrogradational portion of the parasequence set. To the marginal portions of the basins, these parasequence sets are successively reworked to a single composite concentrate.

9.4 Phosphorite deposystem development in time and space

The above discussed fundamentals of paleogeography and sequence stratigraphy of phosphorites determine the general sites of potential phosphate resources as well as related facies and geometry of the sediment body. These general features are modified to various degree in adaptation to the particular conditions of the individual deposystem which emphasizes specific genetic pathways and determines duration of favorable conditions for phosphate resources development.

9.4.1 Oxygen-deficient environments and the development of phosphogenic regimes

The important controlling factors for the development of a phosphogenic regime are:

(1) sedimentation style and rate,
(2) bioproduction,

(3) climate/oceanography, and
(4) geometry/subsidence of the basin.

The interplay of these four factors defines the pathway of phosphogenesis and the spatial (paleogeographic and stratigraphic) position within a deposystem. The complex nature of these interrelationships leads to the development of a large number of configurations of which some are evidently more abundant. During the entire Phanerozoic, organic deposition with the initiation of phosphogenesis abundantly developed only in pelitic or mixed carbonate-siliciclastic systems . In pure arenitic or carbonate deposystems, phosphogenic environments canform only during exceptional sedimentation stages. These are extreme condensation intervals due to sediment bypassing or drowning (e.g. FÖLLMI 1989, ILYIN 1994).

In all investigated marine systems, oxygen deficient organic sedimentation on a shelf or homoclinal ramp is mostly a phenomenon of the transgressive systems tract to maximum flooding stage, exceptionally also of the early highstand systems tract (see Case Studies 3,4,6, and GLENN 1990, WIGNALL 1994). Phosphogenesis in the early transgressive phase is gradationally decreasing upsection in most cases. The facies architecture of the Phosphoria Rock Complex seems to be characteristic for many deposits. The phosphatic onlap facies consists in mid-ramp positions of a basal transgressive lag, followed by a succession of phosphorite concentrates (see chapt. 9.4.2) which is finally overlain by a suite of phosphatic shales. The location and magnitude of phosphogenic environments is slightly modified by the paleotopography of the specific basins. The phosphate suite is time transgressive with the amount of time depending on inclination and broadness of the ramp, furthermore of the magnitude of the sea-level change. In the majority of deposits, the time deviation is beyond biostratigraphic resolution. In the Phosphoria Rock Complex, the onlap extends over two conodont zones (see Case Study 4). Preference for the development of the phosphogenic regime is usually the mid-ramp or mid-shelf position, in striking contrast to the outer shelf position of the modern occurrences off Peru. The basinal facies of the Permian Phosphoria deposystem is dominated by black shales instead of phosphorites. During the highstand phase, fringing carbonate bank complexes rapidly prograde into the basin overwhelming the earlier seafloor poisoning. All these systems are characterized by progradational wedges indicating the shallow basinal configuration of most Phanerozoic phosphate basins.

Similar patterns are also reported from other deposits, e.g. Egypt (GLENN 1990), Morocco (TRAPPE 1991), Togo (SLANSKY 1980, 1986, 1987), or Tunisia (SLANSKY 1980, 1987), which are dominantly mixed carbonate-siliciclastic systems.

Phosphogenic pathways and resulting fabrics in oxygen deficient pelitic systems respond to the degree of detrital background sedimentation and most probably also to the sedimentation rate, both being driven by paleogeography and subsidence. Higher detrital sediment proportions, present in most settings, seem to inhibit the growth of coarse phosphatic cements (see Case Study 3) and cause the development of structureless cryptocrystalline micro-concretions (see Case Studies 4 - 6). The sedimentation rates are beyond biostratigraphic resolution, but the texture of the Carboniferous phosphate concretions with well developed cements support growth in a largely pure organic mud. Omission related stratigraphic intervals establish completely different phosphogenic pathways. This configuration supports the development of microbial phosphate fabrics which depend on non-burial or extreme low sedimentation rates, thus leaving open the degree of total microbial phosphate generation in the sediment (see SCHWENNICKE 1992, 1996).

The broadness of a ramp favors the development of a phosphogenic regime. This is becoming very obvious in the configuration of the Mississippian Delle phosphatic interval (SILBERLING et al. 1995). Relative condensation due to the broadness of a shallow shelf versus deep basin configuration was long being in dispute, but the granular character of the phosphorites superimposing erosional surfaces identify the shallow water origin. In deep silled basins, stagnant water conditions may prevail without being significantly affected by minor to medium sea-level changes (e.g. Black Sea).

Both factors, broadness of the ramp and a low relief hinterland, favor relative low siliciclastic sedimentation rates. The development of a stagnant water body is further supported by a silled basin morphology and an evaporitic regime (see chapt. 6). Rare peritidal occurrences, carbonate or siliciclastic, are also bound to transgressions.

The high biologic sediment accumulation rates in carbonates evidently inhibit the development of a phosphogenic regime in most sequence stratigraphic positions. Phosphogenesis is only triggered by the combination of a specific paleogeographic constellation and high-magnitude relative sea-level change. On elevated areas, sediment bypassing, erosion, and omission locally initiate early transgressive phosphatization of preexisting carbonate substrate. The deposits are usually not of economic scale. There seems to be an important relation to non-tropical carbonate deposystems, which are characterized by a carbonate biofacies in brymol association from the late Paleozoic. Coral reef communities and a flourishing foraminifera fauna are commonly absent. Reef builders to a smaller degree are only oysters and bryozoans. Non-tropical carbonate systems are characterized by low carbonate production rates (READ 1995).

Where time control is sufficient, the formation of phosphogenic regimes of economic scale are a phenomenon of 3- to 4-order cyclicity. The persistence of the transgressive oxygen deficiency comprises one or two conodont biozones in the Paleozoic (see Case Studies 3,4). Mesozoic and Cenozoic deposits usually consist also of several pulses of phosphogenesis during an overall transgressive cycle rank within the same scale.

9.4.2 Granular phosphorite deposystems

The economic importance of a phosphate deposit is dominantly determined by style and intensity of the granular phosphate generation. Winnowing, mechanical reworking, transport and concentration are exclusively driven by the water energy. This more simple relationship clearly identifies specific deposystem configurations as favorable for phosphorite occurrences. Different phenomenons of mechanical concentration are present (turbidites, storm events, currents, waves). The formation of granular phosphorites is always bound to an environmental shift of the depositional site in a higher energy regime by base level drop or are the effect of a short-termed energy event (storm, tsunami, earthquake). The spatial distribution of granular phosphorite formations is controlled by the basin geometry and its modification through time, furthermore by eustasy and subsidence, thus shifting a zone of maximum concentration potential for each type through the basin. The controlling factors determine intensity and stratigraphic position. The basin profile also determines the direction and the degree of the transport component.

Sea-level induced environmental shifts to more agitated water conditions in a tectonically stable basin, ramp or platform attribute sediment remobilization to low sea-level stages, respectively to the late highstand, lowstand (where preserved in marginal position) and early transgressive systems tract. The direction and intensity of the transport component are mainly driven by the seafloor morphology. On steep or steepening ramps, rimmed shelves and platform margins, downslope transport during late highstand and lowstand is dominating. Major sediment accumulations would be deposited on the lower ramp or slope, or in the center of a rimmed basin. The resulting sediments are intensively mixed. Stratigraphically isolated beds result from mass transport processes. The beds show a lateral facies variation from proximal to distal (SCHWENNICKE 1992) and are located in a large generally prismatic host sediment body. Intense mixing with non-phosphatic constituents and further the the occurrence of isolated beds in widely unpredictable positions limit the economic importance.

On ramps with low inclination, the late highstand and lowstand interval is commonly not associated with phosphogenesis and the prograding wedge is consequently non-phosphatic. Phosphogenesis and sediment transport are attributed to a following transgression. Transport is vagabonding or slightly upslope directed. The geometry and facies of these concentrates are distinctively different from those of the steep ramps. During the early transgressive stage on shallow ramps or in epeiric basins, several mechanisms count as candidates to generate concentrates: tides, waves, currents, minor sea-level changes or storm events. In many cases, the intensity of mixing diminishes the record of a certain mechanism. JARVIS (1994) favors enhanced current activity for the formation of phosphatic hardgrounds on the Cretaceous epicontinental carbonate platform in France. Also FÖLLMI (1990) claims current activity responsible for the formation of clastic phosphorites of the Helvetian shelf platform. SOUTHGATE & SHERGOLD (1991) and TRAPPE (1991, 1992a) favor low order sea-level fluctuation and storm events for the shallow platforms or basins in Australia and Morocco. The resulting phosphorite hosting sediment body has a sheet-like geometry with a time transgressive onlap toward basin margin. The internal facies variations depend on the basin geometry and intensity of reworking. In shallow basins with intensive sediment reorganization, the phosphorite concentrate can be stratigraphically condensed to a single lag. Lower degree of reworking generates a vertical succession of a basal lag concentrate, a suite of composite concentrates, and event concentrates with a host sediment dominated deposition. This vertical succession correlates with the facies variation toward the basin margin. The allogenic and autogenic sediment supply significantly affect the quality of the concentrate. Persistent benthic carbonate production in the marginal portion of the basin (e.g. Phosphoria Sea) triggers a complementary carbonate allochem component in the concentrates, and the continental influx of Al, Fe, Mg initiates the precipitation of glauconite in these sediments. Composite and event concentrates develop lenticular sediment bodies. Theses "sand waves" generate an irregular stacking pattern of "seams". The common concentrate to host sediment ratio of roughly 1:3 to 1:5 allows economic mining of the several meter thick sediment suite (e.g. Morocco: Case Study 6, BOUJO 1976; Egypt: GLENN & ARTHUR 1990, GLENN 1990a).

PART V
PHOSPHORITE DEPOSYSTEMS THROUGH THE PHANEROZOIC

10 Phosphorite depositional systems and global change

Large scale phosphorite formation during the Phanerozoic is not a persistent, but an episodic phenomenon. Whereas physico-chemical processes of phosphogenesis are apparently static during the Phanerozoic, the distribution and the abundance of phosphogenic environments was changing with the reorganization of the biosphere, with the evolving continents, the biologic evolution, secular changes, and crises. New environments were formed, shifted or destroyed with changes in atmospheric circulation, eustatic sea-level, climate, and land/sea distribution. Certain fundamental factors for the development of phosphogenic environments which are P-flux, bioproductivity, stratification within the water/soft sediment column, detrital sedimentation rate, and the productivity of the carbonate factory are directly affected by the ocean circulation pattern and the paleoclimatic context of the depositional system. Consequently, global changes affect sedimentary systems in all scales of global cyclicity. Long term global changes establish completely unique worlds with different oceans. In these incomparable worlds, pathways for the formation of the phosphogenic environments are different. On the other side, phosphorite formation in particular occurrences is also closely related to short term changes, e.g. eustatic sea-level rises or intensification of current systems due to climatic changes.

10.1 Episodicity of phosphate giants

The broad spectrum of genetic pathways and phosphogenic environments gives way for the assumption, that phosphogenic systems are able to respond successfully to the global changes during the Phanerozoic Earth history. But the compilation of major Phanerozoic phosphorite deposits by COOK & McELHINNY (1979) identifies an episodic occurrence of major phosphorite deposition rather than being a persistent phenomenon (Fig. 10.1).

Some progress in dating of some of the early Phanerozoic phosphorites as well as some new calculations to resource sizes in the last decades have only very minor effect on this relative old compilation of COOK & McELHINNY (1979). The review of the periodically updated global phosphorite resource data base by the British Sulphur Corporation (1971, 1987) shows that the number of new discoveries are relative small. Also the reserve and resource data have only slightly changed. Some improvement was achieved by dating the old oc-

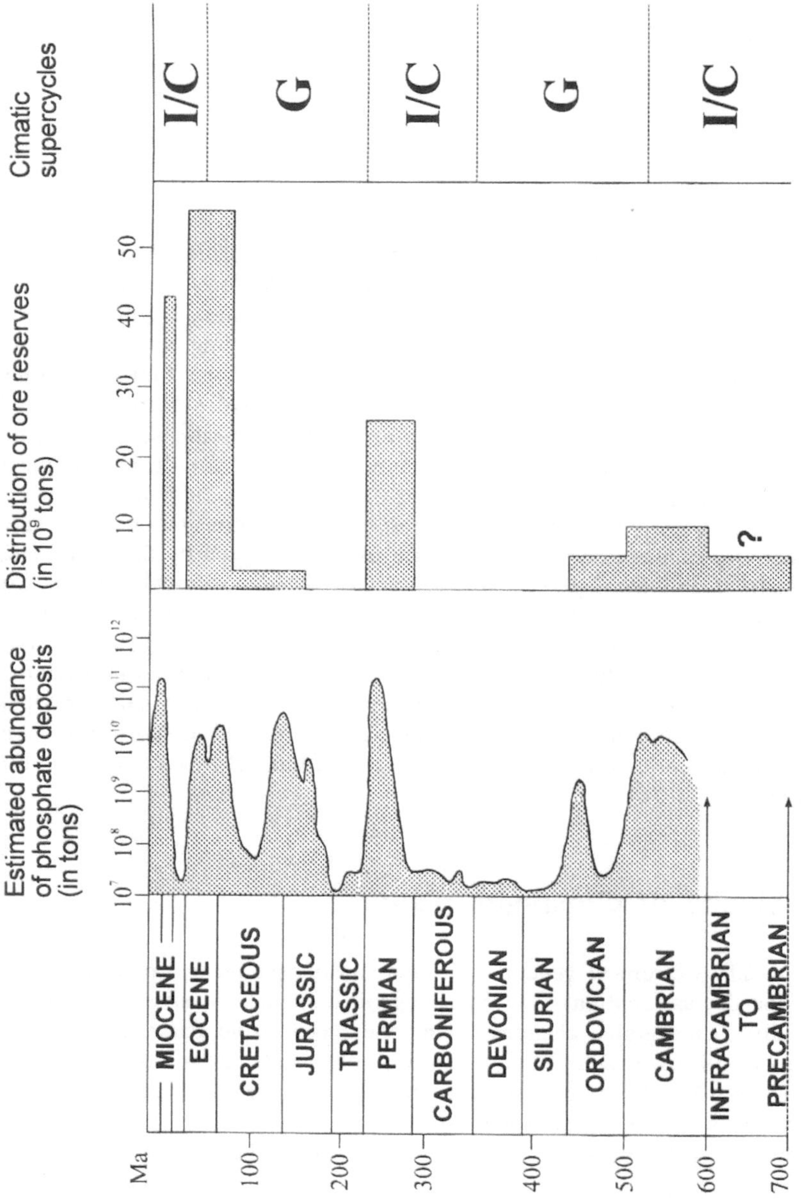

Fig. 10.1. The occurrence of phosphate giants during the Phanerozoic after COOK & McELHINNY (1979) in relation to global tectonic-climatic supercycles (VEEVERS 1990).

currences which would emphasize the elevation of the Cambrian peak to some degree (see discussion in COOK 1992).

Since the publication of this outstanding paper, a large number of contributions speculated about the reasons for this distinct episodic temporal distribution. COOK & McELHINNY (1979) examined the relationship of the temporal distribution and the global plate tectonic configuration. They concluded that major phosphorite deposits are forming when the global plate tectonic configuration provides a large number of preferential sites for upwelling. The authors emphasize, that climate, volcanism and sedimentary changes due to plate tectonics have only secondary effects. But this approach leaves open why phosphogenesis is also an episodic phenomenon in sites which remain in a favorable upwelling position over a longer time interval. Furthermore, the occurrence of many phosphorite deposits is not closely linked to upwelling systems (see chapt. 6 and GLENN et al. 1994). The control on the formation of epeiric phosphorites continues to be ambiguous.

A large number of researches investigated the relationship of phosphate occurrences and biologic changes or crises. COOK & SHERGOLD (1984, 1986), DONNELLY et al. (1990), and COOK (1992) suggested that the abundance of phosphate deposits around the Proterozoic-Cambrian transition was related to a prior oceanic anoxia and the successive supply of the stored phosphorus to the shallow marine areas. This "excess phosphate" generates an intense bioproductivity and the deposition of organic sediments including phosphorites (COOK 1992, BRASIER 1990a,b). This is linked with evolutionary trends, and modelled pO and pCO_2 variations during the Cambrian phosphatic event (DONNELLY et al. 1990). The model is supported by isotopic excursions at the Proterozoic-Cambrian transition (COOK & SHERGOLD 1984, DONNELLY 1988, COOK 1992). COOK & COOK (1985) postulated that oceanic overturn and a resulting high phosphorus concentration caused the intense phosphorite deposition and the biological changes at the Cretaceous-Tertiary transition. A contrasting model was proposed by HERRING (1995), in which the intense phosphorus deposition during the mid-Permian affects an oceanic P-depletion and subsequently a biologic crisis. These models only explain the cause of the specific phosphogenic interval, and imply a polygenetic origin. TRAPPE (1994) found a latest Permian to earliest Triassic phosphate gap, which correlates with a crisis of other organic sedimentation and a δC^{13} isotope excursion (ERWIN 1993, 1994, FLÜGEL 1994, FAURE et al. 1995, RETLACK et al. 1996).

An astronomic cause for specific intervals is also suggested by SHELDON (1984a, 1985) by his ice-ring or tectite ring hypothesis, which should cause temporary low latitude oceanic overturn.

STRAKHOV (1960) highlighted the correlation of phosphorite occurrences and major transgressions. This dependence was later confirmed by several authors (BURNETT & VEEH 1977, ARTHUR & JENKYNS 1981, RIGGS 1984a,b, SHERGOLD & BRASIER 1986, RIGGS & SHELDON 1990). This dependence is also evident in the deposition of organic matter (SCHLANGER & JENKYNS 1976).

Influenced by the general hypothesis of FISCHER & ARTHUR (1977) on the origin of global anoxia, a group of researchers have considered a multi-factor cause for the occurrence of phosphogenic events . They postulate expanded oxygen minimum zones during episodes of high sea-level and warm global climate (polytaxic episodes). Oceanic turnover and diminished upwelling would lead to an increased phosphorus storage in the deeper parts of the ocean. The shallow water origin of phosphate deposition remained difficult to explain. SHELDON (1980) further developed the ideas of FISCHER & ARTHUR. He has advanced the idea to correlate periods of phosphogenesis with episodes of vertical oceanic mixing during polytaxic stages and to intervals following the climax of sea-level with beginning cooling by glaciations. SHELDON (1981) and later BATURIN et al. (1995) pointed out, that in this framework, episodicity appears to be the result of paleogeographic, eustatic, oceanographic, and climatic factors.

SHELDON (1980, 1984b) and RIGGS & SHELDON (1990) have drawn the attention to the temporal relation with global glaciation intervals in which global cooling enhanced oceanic circulation and thus the nutrient supply (COMPTON et al. 1990, 1993). COOK (1992) discussed the stratigraphic problems for the relation of glaciation and phosphogenesis in the Proterozoic/Cambrian transition.

Several studies related phosphorus burial, dispersed or concentrated, with climatic cycles and related weathering intensity (ARTHUR & JENKYINS 1981, BARRON & FRAKES 1990, COMPTON et al. 1990, 1993, FÖLLMI et al. 1993, FÖLLMI 1995, 1996, FILIPPELLI & DELANEY 1994a,b,). These studies document some linearities between weathering and climate evolution, but the complex relationships between weathering, response of oceanic P-cycle, type of P-burial (organic matter, dispersed apatite, phosphorites), and of P-burial rates are still poorly understood. Further research is necessary (FILIPPELLI & DELANEY 1994a,b, GLENN et al. 1994, FILIPPELLI 1997).

PIPER & CODISPOTI (1975) attribute episodicity in phosphogenesis to the variation in global denitrification intensity in response to the extent of the oceanic oxygen minimum layers. The temporal distribution of phosphate giants lacks direct correlation with oceanic anoxic events (ARTHUR & JENKYNS 1981) or stages of general anoxic oceans (WRIGHT et al. 1984, 1987, WIGNALL & HALLAM 1992, FÖLLMI 1996).

The direct or indirect effect of tectonism and volcanism on the global phosphorite formation was frequently discusses (BRODSKAYA 1974, COOK & McELHINNY 1979, YANSHIN 1986, SCHLANGER & JENKYNS 1981), but final evidence was not found.

Despite of these different approaches, the episodicity of phosphate giants remains an apparent enigma. The major reason for the enigmatic relationship is the complexity of the phosphogenic systems. The development of phosphogenic systems in all stages of the box model (Fig. 5.5) is controlled by bulk environmental conditions which are determined by a bundle of ecologic factors. The limited availability of phosphorus, the short residence time of the element in most environments, as well as the narrow range of the apatite precipitation window allow phosphogenesis of a larger scale only if the entire chain of events for one of the phosphogenic pathways is successfully complete. This requires that all environmental criteria generate favorable conditions during all stages. Because of the number of various mechanisms and pathways within the box model, phosphogenic pathways develop and respond multifarious to the local environmental conditions. As a result, phosphogenesis occurs in various settings and facies, but the narrow range and complexity of parameters causes a very rare appearance. These specific P-burial conditions are superimposed by global processes of the P-cycle (BATURIN et al. 1995, FÖLLMI 1996). This multi-factor control is further complicated by dynamic global changes in the organization of the biosphere. These relationships are investigated in the following chapter.

10.2 Spatial distribution of phosphorite deposition during Phanerozoic phosphate giants

The comparison of two major intervals of phosphorite formation which are formed during different global supercycle stages (VEEVERS 1990) reveals striking variations. The mid-Permian to Triassic coldhouse interval roughly corresponds in time and amount of phosphate rock formation to the Late Cretaceous to mid-Paleogene greenhouse interval. The temporal and spatial distribution during these two phases differ significantly.

During the Permian/Triassic interval the majority of the buried phosphorus (non-dispersed) was accumulated during a short interval in two areas in North America. All other deposits are only of minor amount (TRAPPE 1994). These two dominating occurrences are situated in a west coast position of the North American continent. Relative phosphorus flux rates in the sediment are in the range of other phosphate deposits (FILIPPELLI & DELANEY 1992, FILIPPELLI 1997). Peculiar is the enormous lateral extent of the phosphatic environment. This configurations requires a large, similar depositional environment, which is provided by the typical paleogeographic constellation of the stable North American continent with enormous marginal flooding areas. Furthermore, a stable and intensive basinwide stratification mechanism is required (see Case Study 4), which may result from the extreme climate in the Permian and Triassic, as well as from a phosphorus pump of some kind which charges large quantities of phosphorus into the basins. For both of these deposits, an oceanic upwelling pump is most likely, further intensified by the enhanced oceanic and atmospheric circulation during the coldhouse stage (TRAPPE 1994).

The Late Cretaceous to mid-Paleogene interval is characterized by phosphogenesis in numerous medium sized deposits. Phosphate deposition in general is a long lasting phenomenon in some individual deposits comprising durations in the scale of chronostratigraphic age intervals instead of the duration of single conodont biozones in the Permian. The Late Cretaceous/Paleogene deposits are located along the western margins of the continents as well as along the northern and southern margins. In this course, it is important to note that all deposits

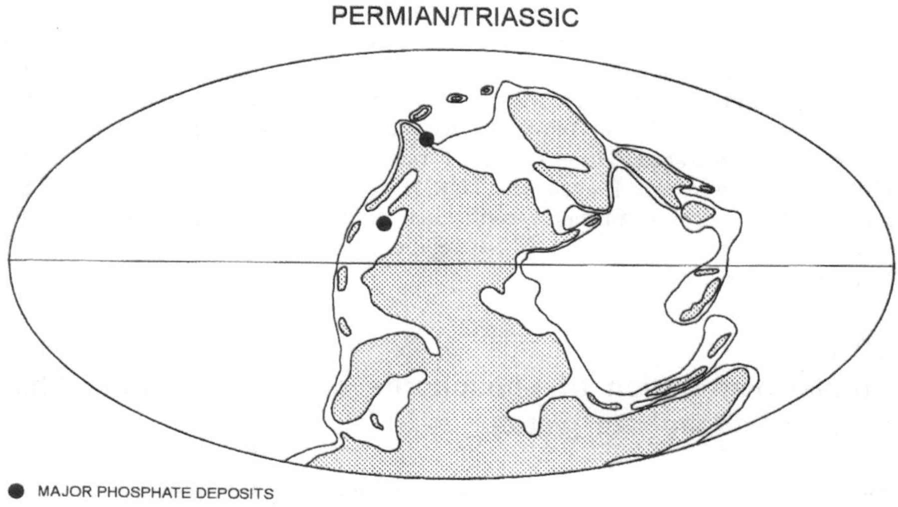

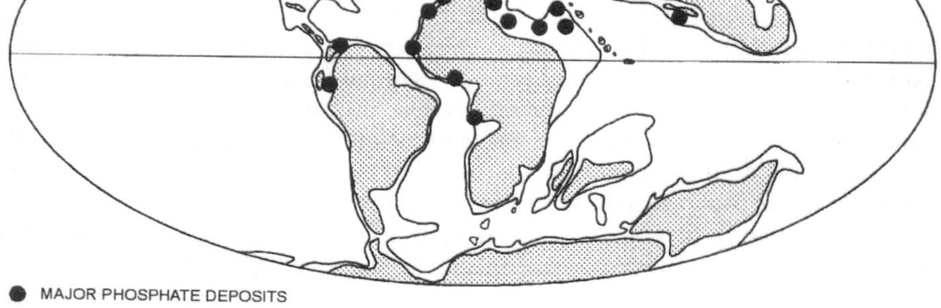

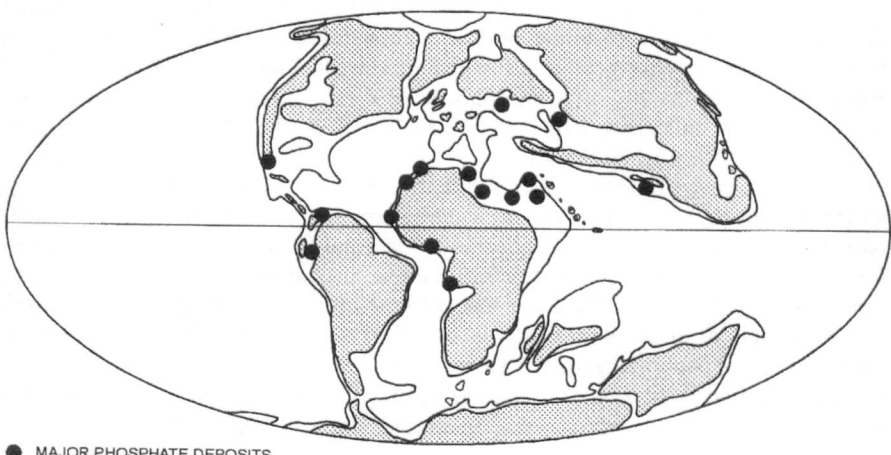

Fig. 10.2. The distribution of phosphorite occurrences during the mid-Permian to Triassic coldhouse interval and the Late Cretaceous to mid-Paleogene greenhouse interval.

are situated in an inner shelf position or are related to epeiric sea areas, where the role of upwelling currents are being an indirect effect. The presence of an upwelling situation as a cause of the development of the Tethys occurrences is purely hypothetical and only concluded from the presence of phosphorites (see chapt. 6.3). The large number of deposits in inner shelfal environments is difficult to explain during greenhouse intervals with weak oceanic and atmospheric circulation. Other mechanisms have to be taken in account which address more the specific environmental conditions during the time interval. Rising chemoclines such as discussed for the formation of Cretaceous black shales (SCHLANGER & JENKYNS 1976) have to be taken in account as well as continental phosphorus supply (GERMANN et al. 1985, 1987, FÖLLMI 1996; s.a. chapt. 10.3). The flooding of continental margins by oxygen deficient waters which generate favorable conditions for the development of phosphogenic systems explains the wide occurrence of medium-sized deposits much better than a bioproductivity derived sea-floor poisoning model through upwelling (see chapt. 10.3).

10.3 The effect of long-term changes: the global supercycles

Long-term global changes, for example climatic supercycles, are affected by the tectonic plate constellation. During a greenhouse and a dispersed continent configuration, an equal and warm-humid global climate intensified chemical weathering and consequently the P-flux into the ocean. Rising chemoclines in sluggish oceans coupled with a relative high sea-level caused widespread development of seawater stratification and oxygen-minimum zones (OMZ) in westcoast upwelling positions as well as in all other shallow marine shelfal areas, where local upwelling by Ekman transport may have played a role of uncertain significance. During these intervals, broadness of the shelfal areas or marine epicontinental basins in spatial relationship with dominantly chemical weathering reduced the detrital net sedimentation rate and enhanced the P-flux into the world ocean. Both favored again phosphogenesis by emphasizing organic deposition. The climatic and paleogeographic conditions during greenhouse intervals resulted in a widespread epicontinental basin and shallow shelf phosphogenesis.

Coldhouse/icehouse supercontinent stages favored the development of a small number of large deposits (Fig. 10.2). Vigorous ocean circulation due to extreme climatic contrast, and possible mega-monsoonal atmospheric circulation enhanced strong upwelling along the westcoasts of the supercontinents. Hot climate caused extreme heating of the shallow marine surface water as well as salinity stratification. Cold water bottom currents along the westcoasts of the continents intensified the process. The extreme atmospheric circulation during coldhouse/icehouse intervals enhanced Ekman transport especially within the trade wind belts. Dominantly physical weathering may have reduced the terrestrial P-flux into the ocean. Phosphorite formation was shifted into high latitudes, and was

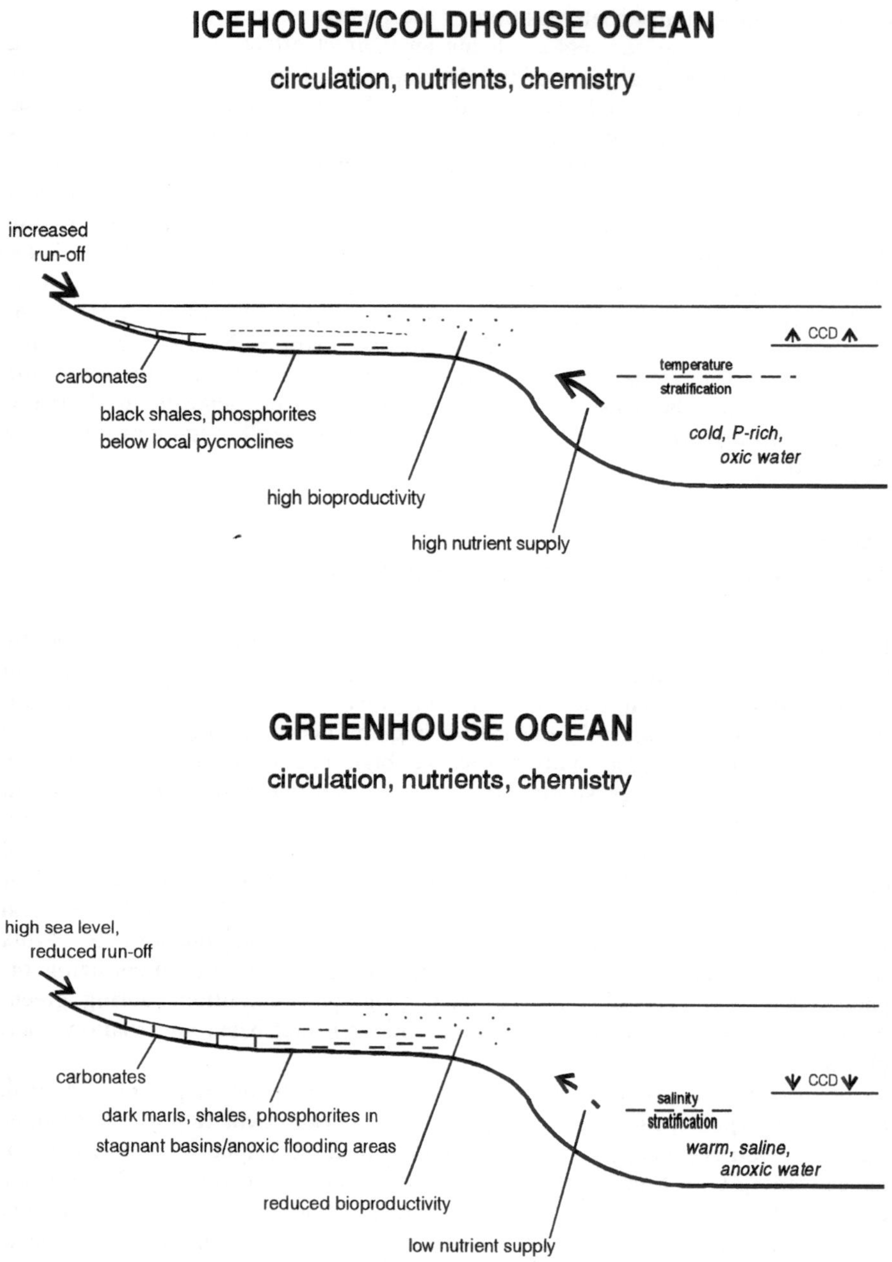

Fig. 10.3. The influence of different global climatic configurations on the development of phosphogenic systems.

very much restricted to westcoasts of the supercontinent with extreme circulation pattern.

10.4
The effect of short-term changes

Short-term global changes were effective within the scale of eustatic sea-level changes. Current activity, but also the shelfal carbonate factory were strictly controlled by the temperature distribution as well as by salinity, and responded rapidly to changes e.g. interglacial stages. In most occurrences, the process of phosphogenesis itself is a short-term phenomenon (e.g. see Case Study 4). Relative starvation intervals in shelfal or epicontinental sea areas were related to early transgressive sea-level stages or to platform drowning. Upwelling enhanced net organic deposition, and hence the phosphorus flux into the sediment via increased bioproductivity. Subsequently, phosphorites can be reworked into rock sequences which represent much longer time intervals. The timing of phosphorite genesis within eustatic sea-level cycles depends on the specific paleogeographic and oceanographic positon of the particular occurrence.

10.5
The impact of global change and global crisis on phosphorite deposystems

Some authors discussed the impact of global climatic changes on the global marine phosphorus cycle (SHELDON 1980, 1981, ZANIN 1981, FÖLLMI 1995, 1996). Altered climatic configuration affects magnitude of phosphate sink on land by coal formation and type of weathering, further the oceanic circulation system which consequently modifies the P-flux and the latitudinal distribution. Most authors emphazised the effects of weathering and the increase of the P-flux into the world ocean, respectively into epicontinental sea areas and lakes. Enhanced nutrient supply would increase the bioproduction in most environments and subsequently the P-sink into the sediments. This relationship is based on the presumption that in long term the P content in the world ocean remains constant. But the flux of P into the ocean through weathering as well as the total sink accounts only for 2 % of the fluxes between ocean water and the organisms. In this global approach, the effect on the increase of continental fluxes would be minor in the short term and probably being absorbed by production of living biomass. In long term, the influx and sink must be balanced to prevent depleation or replenishment of the world ocean. More significant for organic sedimentation and formation of phosphate deposits are temporary disturbance of the P-cycling in the entire biosphere or in specific areas. All quantitative approaches offer only very simplified models. The very important shifts of the phosphorus distribution with climate changes in ancient sea, especially in shallow marginal sea areas, are beyond quantitative modelling of phosphorus

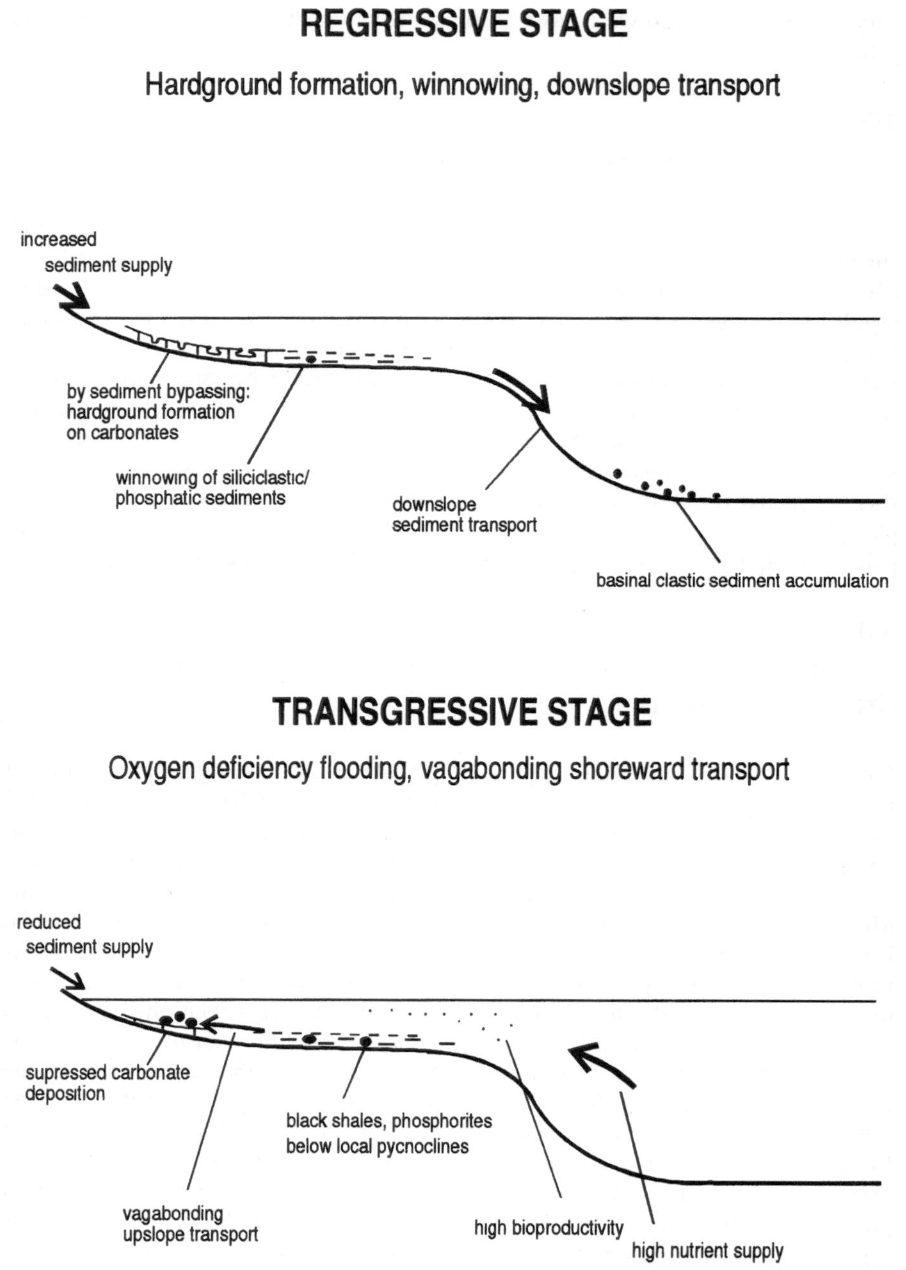

Fig. 10.4. The effect of short-term changes on the development of phosphorite deposits.

fluxes in specific environments and on global scale. These changes in flux rates are most probably very significant for the formation of phosphate deposits. The second important effects are rising chemoclines in all marine environments or mixing of water bodies with enhanced nutrient regeneration during intervals of global changes. Intense stratification and the formation of oxygen-depleted water columns would increase the organic matter deposition and consequently the P-sink into the sediment independently from the global P-cycle. The phenomenon is a more regional effect rather than a global one. Only during "oceanic anoxic events", global consequence would be likely. Stagnation within the ocean and distinct and shallow chemoclines should increase the P-sink in ocean water and shelfal areas (see SCHLANGER & JENKYNS 1976, also controversial discussion in FÖLLMI 1996, INGALL & JAHNKE 1997). SHELDON (1984b), HIATT (1994) and PIPER & MEDRANO (1995) related bioproductivity cycles and subsequently phosphorite forming settings to glaciation intervals.

11 Factors controlling the formation of phosphorite deposystems in time and space (conclusions)

In the course of this study, it was shown that "the phosphorite deposystem" is not existent. Phosphatic sediments and phosphorites are manifold in origin, facies, sedimentary history, and depositional setting. This diversity results from the development of various chains of events to phosphate sediment formation in response to substrate, environment/paleogeography and global environmental changes.

Most phosphorite deposits owe their origin from a fundamental 2-stage development: (1) the process of phosphogenesis (phosphate mineralization in sediments) and (2) the subsequent phosphorite genesis (mechanical alteration).

The depositional pathways of phosphogenesis are determined by a chain of basic requirements: P-supply from organic matter, P-trapping, P-concentration, and phosphate mineral precipitation. These fixed conditions are defined by the steady-state chemo-physical parameters for apatite precipitation, the "Apatite Precipitation Window". Nonetheless the narrow range of conditions for apatite precipitation, the mineralization in sedimentary systems follows not a single process, but are manifold and assemble a great variety of pathways. These find their expressions in facies and geochemistry. The only exception from this rule may be represented by skeletal and microbial phosphate mineralizations, which remain widely enigmatic.

This variety of processes and resulting facies is not only evident during phosphogenesis, but is also characterizing the subsequent sedimentological processes of phosphorite genesis. These mechanical alterations lead to concentration and sediment transport. Intensity and type of phosphorite concentrate formation is always a direct response to paleogeography and sea-level change. These factors are even more important during sediment transport and finally determine the location, quality, and geometry of an economic phosphorite sediment body within a deposystem.

The genetic complexity of phosphate deposits delayed the early identification of the driving factors of phosphogenesis and the development of a general genetic model. A second inhibiting factor in the research on the genesis of phosphate deposits is the common occurrence within sedimentary intervals, which record breaks in sedimentation or exceptional stages. The multiplicity of processes in all stages of phosphogenesis and phosphorite genesis is best described by a modular box model, which was outlined in the course of this study.

The box model identifies the preconditions for a deposystem being potentially phosphogenic. The evident importance of relative elevated organic deposition assigns favorable locations in both, carbonate and siliciclastic systems, to sites of low sedimentation rates. These may result from erosion, sediment bypassing, or starvation. The conditions have to remain stable for some time until a sufficient P-concentration is reached. This requirement links the spatial occurrence of phosphogenesis also with sea-level changes. The positions of phosphogenic sites depend furthermore on the paleogeography of the basin or shelf. Besides these autogenic (i.e. basin internal) control factors, global phenomenons trigger phosphogenesis in specific deposystems. Climate and ocean circulation on regional and global scale has significant impact on the nutrient cycle, and thus the bioproductivity and potential organic matter deposition. But these allogenic factors are only supporting. Primary importance has the occurrence of suitable conditions within a deposystem. The complex interrelationships of global processes with their secular variations of nutrient mobility with resulting bioproductivity and the local factors of the specific deposystem evolution leave causes of the Phanerozoic episodicity of elevated phosphorite formation wide open and ambiguous.

Much better understood, but in many cases underestimated, are the sedimentary processes of phosphorite formation. The interpretation of the mechanical processes benefited from the investigation of other clastic deposystems. The regional parameters basin profile, tectonic activity, and relative sea-level change (as the sum of subsidence and eustasy) have overwhelming importance. Global factors like climate and magnitude of sea-level change have only secondary influence on the sediment reorganization. The latter affect the intensity of marine sediment remobilization as well as terrigenous sediment supply and productivity of the carbonate factory, both potentially inhibiting concentration of phosphate particles.

The application of microfacies techniques and sequence stratigraphy for paleoenvironment reconstructions and the identification of time and space relationships of the different occurrences of phosphate sediments within a deposystem improved the understanding furthermore. These techniques could identify various environments of phosphogenesis and subsequent phosphorite formation. Broad passive shelves and epeiric basins with relative low net sedimentation rates, which are further reduced during early transgressive intervals, clearly favor phosphogenesis, even though phosphorites develop also in other sites, but more exceptionally. Fine-clastic deposystems with relative high organic matter deposition rates are emphasized. Phosphogenesis is dominatly initiated under suboxic conditions within the denitrification zone of early and rapid organic matter degradation. In carbonate deposystems, which are commonly characterized by high sediment accumulation rates and rapid lithification, phosphogenesis usually develops only during discontinuous deposition.

During Phanerozoic earth history, some time intervals reveal a conspicuously elevated phosphate deposition. The major peaks are the Early to Middle Cambrian, the mid-Permian, the late Cretaceous to Paleogene and the Miocene. Besides these major intervals, the Ordovician and Early Carboniferous seem to be also very important. The Silurian, Devonian and Triassic are extremely poor in phosphate occurrences. This episodicity shows no distinct correlation to global environmental or paleotectonic stages, changes, or crisis, nonetheless causal relationships which can be drawn to global environmental phenomenons and single intervals. The multifarious chains of events during phosphogenesis allow largely a response to different environments. The interpretation on global scale is further complicated by the subsequent sediment reorganization, which destroyed primary signatures to a large extent.

Phosphorite deposystems with their broad spectrum of facies, depositional setting, and paleogeographic distribution are an example of a depositional resource system, which shows with conspicuous clarity a dynamic response to the ecosystem Earth. The focus on the variations and their environmental causes enabled the development of a general model for the genesis of this resource, a concept, of course, which has to be a dynamic. This is addressed by a modular genetic model for phosphogenesis and phosphorite genesis, which defines genetic pathways including all their variations, resulting sediments, and settings.

References

ABED, A M, FAKHOURI, K (1990) Role of microbial processes in the genesis of Jordanian Upper Cretaceous phosphorites. Geol Soc London, Spec Publ, 52: 193-203

ABED, A M (1989) On the genesis of the phosphorite-chert association of the Amman Formation in the Tel Es Sur area, Ruseifa, Jordan. Sci Géol, Bull, 42: 141-153

ABED, A M (1994) Shallow marine phosphorite-chert-palygorskite assocition, Upper Cretaceous Ammon Formation Formation, Jordan. In: ILJIMA, A, ABED, A M, GARRISON, R E (eds) Siliceous, phosphatic and glauconitic sediments of the Tertiary and Mesozoic, Proc 29th Intl Geol Congess, C: 205-224

ABED, A M, KRAISHAN, G M (1991) Evidence for shallow-marine origin of a "Montery-Formation Type" Chert-Phosphorite-Dolomite Sequence: Amman Formation (Late Cretaceous) Central Jordan. Facies, 24: 25-38

AHARON, P, VEEH, H H (1984) Isotope studies of insular phosphates explain atoll phosphatization. Nature, 309: 614-617

ALTSCHULER, Z S (1980) The geochemistry of trace elements in marine phosphorites, Part I Characteristic abundance andenrichment. Soc Econ Paleont Miner, Spec Publ, 29: 19-30

ALTSCHULER, Z S, CATHCART, J B, YOUNG, E J (1964):Geology and geochemistry of the Bone Valley Formation and its phosphate deposits, west-central Florida. Geol Soc America, Ann Meeting Miami Beach, Field Guidebook, 6: 68 pp

ALTSCHULER, Z S, BERMAN, S, CUTTITTA, F (1967) Rare earths in elements - Geochemistry and potential recovery. Intermountain Association of Geologists, Guidebook of the 15th Annual Field Conference: 125-135

ALTSCHULER, Z S, CLARKE, R S, Jr, YOUNG, E J (1958) Geochemistry of uranium in apatite and phosphorites. US Geol Surv, Prof Pap, 314-D: 87 pp

AMES, L L (1959) The genesis of carbonate apatites. Econ Geol, 54: 829-841

AMES, L L (1960) Some cation substitutions during the formation of phosphorite from calcite. Econ Geol, 55: 354-362

ARAMBOURG, C (1952) Les Vertébrés fossiles des gisements de Phosphates (Maroc, Algérie, Tunisie). Notes & Mém Serv géol Maroc, 92: 372pp

ARTHUR, M A, JENKYNS, H C (1981) Phosphorites andpaleooceanography. Oceanol Acta, 1981, Proceed 26th Intern Geol Congr, Paris, 83-96

ARTHUR, M A, SAGEMAN, B B (1994) Marine black shales: depositional mechanisms and environments of ancientdeposits. Annu Rev Earth Planet Sci, 22: 499-551

ATLAS, E L, PYTKOWICZ, R M (1977) Solubility behaviour of apatite in seawater. Limnol Oceanogr, 22: 290-300

AVITAL, Y, STARINSKY, A, KOLODNEY, Y (1983) Uranium geochemistry and fission track mapping of phosphorites, Zefa Field, Isreal. Econ Geol, 78: 121-131

AZMANY-FARKHANY, M, BOUJO, A, SALVAN, H M (1986) Description des disements et depôts phosphates Marocains. In: Geologie des gites mineraux Marocains, Vol 2 (2nd ed), Notes & Memoires Serv Geol Maroc, 276: 153-264

BAKER, K B, BURNETT, W C (1988) Distribution, texture and composition of modern phosphate pellets in Peru shelf muds. Mar Geol, 80: 195-214

BALSON, P S (1990) Episods of phosphogenesis and phosphorite concretion formation in the North Sea Tertiary. Geol SocLondon, Spec Publ, 52: 125-137

BANERJEE, D M (1971) Precambrian stromatolitic phosphorite of Udaipur, Rajasthan, India. Geol Soc America, Bull 82: 2319-2330

BANERJEE, D M, SCHIDLOWSKI, M, ARNETH, J D (1986) Genesis of Upper Proterozoic-Cambrian phosphorite deposits of India: Isotopic inference from carbonate fluorapatite, carbonate and organic carbon. Precambrian Res, 33: 239-253

BARRON, E J, FRAKES, L A (1990) Climate model evidence for variable continental precipitation and its significance for phosphorite formation. In: BURNETT, W C, RIGGS, S R(eds) Phosphate deposits of the world Vol 3: Neogene to modern phosphorites, 260-272

BARRON, E J, MOORE, G T (1994) Climate model application in paleoenvironmental analysis. Soc Econ Paleont Miner, Short Course, 33: 339pp

BASHYAL, R P (1984) Stromatolitic phosphorite occurrences in the lesser Himalaya of far western Nepal. Geol Surv India, Spec Publ, 17: 197-200

BATURIN, G N (1969) Authigenic phosphate concretions in recent sediments of the southwest African shelf. Doklady AkadNauk SSSR, 189: 227-230

BATURIN, G N (1971a) Formation of phosphate sediments and water dynamics. Oceanology, 11: 372-376

BATURIN, G N (1971b) Stages of phosphorite formation on the sea floor. Nature, 232: 61-62

BATURIN, G N (1982) Phosphorites on the sea floor. Dev Sed, 33: 343 pp

BATURIN, G N (1983) Some unique sedimentological and geochemical features of deposits in coastal upwelling regions. In: THIEDE, J, SUESS, E (eds) Coastal upwelling, its sediment record Part B: Sedimentary record of ancient coastal upwelling: 11-27

BATURIN, G N, BEZRUKOV, P L (1979) Phosphorites on the sea floor and their origin. Mar Geol, 31: 317-332

BATURIN, G N, LUCAS, J, PRÉVôT-LUCAS, L (1995) Phosphorus behavior in marine sedimentation. CR Acad Sci Paris, 321,IIa: 263-278

BATURIN, G N, MERKULOVA, K I, CHALOV, P I (1972) Radiometric evidence for recent formation of phosphatic nodules in marine shelf sediments. Mar Geol, 13: 37-41

BATURIN, G N, ORESHKIN, V N (1984) Behavior of cadnium in ocean-floor bone phosphate. Geochem Int, 35: 69-74

BEIN, A, AMIT, A (1982) Depositional environment of the Senonian cherts, phosphorite and oil shale sequencein Israel deduced from their organic matter composition. Sedimentology, 29: 81-90

BELAYOUNI, H, TRICHERT, J (1981) Preliminary data on the origin of the organic matter in the phosphate basin of Gafsa (Tunisia). In: BJOROY, M (ed) Advances in organic geochemistry: 328-355

BELAYOUNI, H, TRICHET, J (1984) Hydrocarbons in phosphatized and nonphosphatized sediments from the phosphate basin of Gafsa. Org Geochem, 6: 741-754

BENMORE, R A, COLEMAN, M L, McARTHUR, J M (1983) Origin of sedimentary francolite from its sulphur and carbon isotope composition. Nature, 302: 516-518

BENTOR, Y K (1953) Relations entre la tectonique et les dépots de phosphates dans le Negev Israelien. 19th Int Geol Cong, Algier, 1952, 11: 93-101

BENTOR, Y K (ed)(1980) Marine Phosphorites - Occurrence, Genesis. Soc Econ Paleont Miner, Spec Publ, 29: 249 pp

BERTRAM, C J, ELDERFIELD, H (1993) The geochemical balanceof the rare earth elements and neodymium isotopes in the ocean. Geochim Cosmochim Acta, 57: 1957-1986

BIDAUT, H (1953) Note préliminaire sur un mode de formation possible des phosphates dinantiens des Pyrenées. C R 19 Congr Géol Intern, Fasc 11: 185-190

BIRCH, G F (1977) Surficial sediments on the continental margin off the west coast of South Africa. Mar Geol, 23: 305-337

BIRCH, G F (1979a) Phosphatic rocks on the western margin of South Africa. J Sed Petrol, 49: 93-110

BIRCH, G F (1979b) Phosphorite pellets and rock from the western continental margin and adjacent coastal terrace of South Africa. Mar Geol, 33: 91-116

BIRCH, G F (1979c) The nature and origin of mixed glauconite/apatite pellets from the continental margin of South Africa. Mar Geol, 29: 313-334

BIRCH, G F (1980) A model of penecontemporaneous phosphatization by diagenetic and authigenic mechanisms from the western margin of Southern Africa. Soc Econ Paleont Miner, Spec Pub, 29: 79-100

BLACK, N R (1985) Petrography and diagenesis of the Galena (Middle Orovizian) - Maquoketa (Late Ordovician) contact and the basal Maquoketa phosphorites in eastern Missouri and eastern Iowa, USA. Unpubl MS thesis, Univ of Illinois at Urbana-Champaign, 133pp

BLISKOVSKII, V Z (1967) Phosphorite-bearing weathered crust in the Bol'shie Dzhebarty (Eastern Sayan). Lith Miner Resour, 1967,4: 428-435

BLISKOVSKIY, V Z (1976) On kurskite and francolite. Lithol Miner Resour, 11: 332-341

BOCK, W-D (1987) Geochemie und Genese der oberkretazischen Phosphorite Ägyptens. Berliner geowiss Abh (A), 82: 138 pp

BÖGER, H (1962) Zur Stratigraphie des Unterkarbons im Velberter Sattel. Decheniana, 114: 133-170

BOUJO, A (1976) Contribution à l'étude géologique du gisement de phosphate crétacé - éocene des Ganntour (Maroc occide tal). Notes & Mém Serv géol Maroc, 262: 227pp

BOURROUILH-LE JAN, F (1980) Phosphates, sols bauxitiques et karsts dolomitiques du Centre et Sud-Ouest Pacifique Comparaisons sédimentologiques et géochimiques. BRGM, Doc, 24: 113-140

BOURROUILH-LeJAN, F G, CARSIN, J-L, NIAUSSAT, P-M, THOMMERRET, Y (1985) Sédimentation phosphatée actuelle dans le lagon confiné de l'Ile de Clipperton (Océan Pacifique) Datations, sédimentologie et géochimie. Sci Géol, Mém, 77: 109-124

BOWEN, H J M (1966) Trace elements in biochemistry. pp

BOYER, B W (1982) Green River laminites: Does the playa-lake model really invalidate the stratified-lake model. Geology, 10: 321-324

BRAID, G C (1978) Pebbly phosphorites in shale: A key to recognition of a widespread submarine discontinuity in the Middle Devonian of New York. J Sediment Petrol, 48: 545-555

BRAITHWAITE, C J R (1980) The petrology of oolitic phosphorites from Esprit (Aldabra) Western Indian Ocean. Phil Trans R Soc Lond B, 288: 511-540

BRASIER, M D (1990a) Nutrients in the Early Cambrian. Nature, 347: 521-522

BRASIER, M D (1990b) Phosphogenic events and skeletal preservation across the Precambrian-Cambrian boundary interval. Geol Soc London, Spec Publ, 52: 289-303

BRASS, G W, SOUTHAM, J R, PETERSON, W H (1982) Warm saline bottom water in the ancient ocean. Nature, 296: 620 -623

BRAUN, A, GURSKY, H-J (1991) Kieselige Sedimentgesteine des Unter-Karbons im Phenoherzynikum - eine Bestandsaufnahme. Geologica et Palaeontologica, 25: 57-77

BREMMER, J M (1980) Concretionary phosphorites from Southwest Africa. J Geol Soc London, 137: 773-786

BRENNER, R L (1978) Sussex sandstone of Wyoming: an example of Cretaceous offshore sedimentation. Amer Assoc Petrol Geol, Bull, 62: 181-200

BRENNER, R L, DAVIS, D K (1974) Oxfordian sedimentation in Western Interior United States. Amer Assoc Petrol Geol, Bull, 58: 444-467

BRITISH SULPHUR CORPORATION (1971) World survey of phosphate deposits. 3rd Ed, 180 pp

BRITISH SULPHUR CORPORATION (1987) World survey of phosphate deposits. 5th Ed, 274 pp

BRITTENHAM, M D (1973) Permian Phosphoria bioherms and related facies, southeastern Idaho. Master Thesis, Univ Montana: 213 pp; Missoula

BRITTENHAM, M D (1976) Permian Phosphoria carbonate banks, Idaho- Wyoming thrust belt. Rocky Mountains Association of Geologists, 1976 Symposium, 173-191

BRODSKAYA, N G (1974) Role of volcanism in phosphorite formation. Acad Sci USSR, Transactions, 258: 199 pp

BROMLEY, R G (1975) Trace fossils at omission surfaces. In: FREY, R W: The study of trace fossils, 562 pp

BUCKLAND, W (1829) On the discovery of coprolithes, or fossil faeces, in the Lias at Lyme Regis, and in other formations . Geol Soc London Trans, 3: 223-238

BUCKLAND, W (1843) On the causes of the general presence of phosphates in the strata of the earth and inall fertile soils. Jour Royal Agricult Soc, 10: 520-525

BURCHETTE, T P, WRIGHT, V P (1992) Carbonate ramp depositional systems. Sediment Geol, 79: 3-57

BURNETT, W C (1974) Phosphorite deposits from the sea floor off Peru and Chile: radiometrical and geochemical investigations concerning their origin. Hawaii University, Hawaii Institute of Geophysics, Report, HIG 74-3: 163 pp

BURNETT, W C (1977) Geochemistry and origin of phosphorite deposits from off Peru and Chile. Geol Soc America, Bull, 813-823

BURNETT, W C (1980a) Apatite-glauconite associations off Peru and Chile: palaeo-oceanographic implications. J geol Soc London, 137: 757-764

BURNETT, W C (1980b) Oceanic phosphate deposits. In: SHELDON, R P, BURNETT, W C: Fertilizer Mineral Potential in Asia and the Pacific, East-West Resource Systems Institute, 119-144

BURNETT, W C, BAKER, K B, CHIN, P A, McCABE, W, DITCHBURN, R (1988) Uranium-series and AMS 14C studies of modern phosphatic pellets from Peru shelf muds. Marine Geology, 88: 215-230

BURNETT, W C, GOMBERG, D N (1977) Uranium oxidation and pro bable subaerial weathering of phosphatized limestone from Pourtales Terrace. Sedimentology, 24: 291-302

BURNETT, W C, LANDING, W M, LYONS, W B, OREM, W (1989) Jellyfish Lake, Palau, A model anoxic environment for geochemical studies. EOS 70: 777-779, 783

BURNETT, W C, RIGGS, S R (eds)(1990) Phosphate deposits of the world, Vol 3, Neogene to Modern phosphorites. 464 pp

BURNETT, W C, VEEH, H H (1977) Uranium series disequilibrium series in phosphorite nodules from the west coast of South America. Geochim et Cosmochim Acta, 41: 755-764

BURNETT, W C, VEEH, H H, SOUTAR, A (1980) U-series, oceanographic and sedimentary evidence in support of recent formation of phosphate nodules off Peru. Soc Econ Paleont Miner Spec Pub, 29: 61-71

BUSHINSKI, G I (1935) Structure and origin of the phosphorites of the USSR. J Sediment Petrol, 5: 81-92

BUSHINSKI, G I (1964) On shallow water origin of phosphorite sediments. In: STRAATEN, L M van: Deltaic and shallow marine deposits. Dev Sed, 1: 62-70

BUSHINSKI, G I (1966) The origin of marine phosphorites. Lith, Min Res, 1966/3: 292-311

BUSHINSKI, G I (1969) Old phosphorites of Asia and their origin. Akad Nauk SSSR, Transaction, 149: 266 pp

CAROZZI, A V (1989) Carbonate rock depositional models. 604 pp

CARSON, G A, CROWLEY, S F (1993) The glauconite-phosphate association in hardgrounds: examples from the Cenomanian of devon, southwest England. Cretaceous Res, 14: 69-89

CATHCART, J B (1978) Uranium in phosphate rock. US Geol Surv, Prof Pap, 988-A: A1-A6

CATHCART, J B (1989) The phosphate deposits of Tennessee, USA. In:NOTHOLT, A J G, SHELDON, R P, DAVIDSON, D F: Phosphate deposits of the world, Vol 2: 6-13

CATHCART, J B (1991) Phosphate deposits of the United States - discovery, development; Economic geology and outlook for the future. In: GLUSCOTER, H J, RICE, D D, TAYLOR, R B (eds) The geology of North America, Vol P-2 Economic Geology, US, Geol Soc America: 153-164

CATHCART, J B (1992) Uranium in phosphate rock with special references to the Central Florida deposits. US Geol Surv, Circ, 1069: 32-35

CATHCART, J B, McGREEVY, K J (1959) Results of geologic exploration by core drilling, 1953, land-pebble phosphate district, Florida. US Geol Surv, Bull, 1046-K: K221-K298

CAYEUX, M L (1936) Existence de nombreuses bactéries dans les phosphates sédimentaires de tout âge. C R Acad Sci, 203: 1198-1200

CAYEUX, M. L (1950) Les phosphates de chaux sédimentaires de France, III. Etude des gîtes minéraux de la France, Serv carte Géol Fr: 458 pp

CHALYSHEV, V I (1968) The phosphorite assemblage of the Permian and Triassic deposits of the northern pre-Ural downwrap. Lith Miner Res, 1968/2: 174-183

CHAUDAN, D S (1979) Phosphorite bearing stromatolites of the Precambrian Aravalli phosphorite deposits of Udaipur region, their environmental significance and genesis of phosphorite. Precambrian Res, 8: 95-126

CHAUHAN, D S, SISODIA, M S (1984) Nature of Udaipur phosphorite and its genetic implications. Geol Surv India, Spec Publ, 17: 79-104

CHRISTIE, R L (1980) Paleolatitudes and potential for phosphorite deposition in Canada. In: Current research, part B. Geol Surv Canada, Pap, 80-1B: 241-248

CLAUSEN, C-D, LEUTERITZ, K, ZIEGLER, W (1989) Ausgewählte Profile an der Devon/Karbon-Grenze im Sauerland (Rheinisches Schiefergebirge). Fortschr Geol Rheinl u Westf 35: 161-226

COMPTON, J S, HODELL, D A, GARRIDO, J R, MALLINSON, D J (1993) Origin and age of phosphorite from the south-central Florida Platform: Relation of Phosphogenesis to sea-level fluctuations and δ13C excursions. Geochim Cosmochim Act: 57: 131-146

COMPTON, J S, SNYDER, S W, HODELL, D A (1990) Phosphogenesis and weathering of shelf sediments from the south-eastern United States: Implications for Miocene δ13C excursions and global cooling. Geology, 18: 1227-1230

CONKIN, J E, CONKIN, B M (1975) The Devonian-Mississippian and Kinderhookian-Oseagean boundaries in the East-Central United States and paracontinuities. University of Louisville Studies in Paleontology and Stratigraphy, 4: 54 pp

COOK, P J (1969) The petrology and geochemistry of the Meade Peak Member of the Phosphoria Formation. PhD Thesis, University of Colorado: 204 pp

COOK, P J (1970) Repeated diagenetic calcitization, phosphatization, and silicification in the Phosphoria Formation. Geol Soc America, Bull, 81: 2107-2116

COOK, P J (1972a) Petrology and geochemistry of the phosphate deposits of northwest Queensland, Australia. Econ Geol, 67: 1193-1213

COOK, P J (1972b) Sedimentological studies on the Stairway Sandstone of central Australia. Austr Bur Min Res, Bull, 95: 73 pp

COOK, P J (1976a) Georgina Basin phosphatic province, Queensland and Northern Territory - Regional overview. In: KNIGHT, C L (ed) Economic geology of Australia and Papua New Guinea, 4Industrial minerals and rocks. Australian Institute of Mining and Metallurgy, Monograph, 8: 245-250

COOK, P J (1976b) Sedimentary phosphate deposits. In: WOLF, K H (ed) Handbook of strata-bound and stratiform ore deposits: 505-535

COOK, P J (1989) Phosphate deposits of the Georgina Basin, northern Australia. In: NOTHOLT, A J G, SHELDON, R P, DAVIDSON, D F: Phosphate deposits of the world, Vol 2: 533-544

COOK, P J (1992) Phosphogenesis around the Proterozoic-Phanerozoic transition. J Geol Soc London, 149: 615-620

COOK, P J, COOK, J R (1985) Marine biological changes and phosphogenesis around the Cretaceous-Tertiary boundary. Sci Géol, Mém, 77: 105-108

COOK, P J, ELGUETA, S A (1986) Proterozoic and Cambrian phosphorites. deposits: Lady Annie, Queensland, Australia . In: COOK, P J, SHERGOLD, J H: Phosphate deposits of the world, Vol 1: Proterozoic and Cambrian phosphorites: 132-148

COOK, P J, McELHINNY, M W (1979) A reevaluation of the spatial and temporal distribution of sedimentary phosphate deposits in the light of plate tectonics. Econ Geol, 74: 315-330

COOK, P J, SHERGOLD, J H (1984) Late Proterozoic-Cambrian phosphorites and phosphogenesis. Proc 27th Intern Geol Congr, Vol 15: 397-444

COOK, P J, SHERGOLD, J H (1986) Proterozoic and Cambrian phosphorites - nature and origin. In: COOK, P j, SHERGOLD, J H (eds), Phosphate deposits of the world, Vol 1: Proterozoic and Cambrian phosphorites, 369-386

COOK, P J, SHERGOLD, J H (eds)(1986) Phosphate deposits of the world, Vol I: Proterozoic and Cambrian phosphorites. 386 pp

COOK, P J (1967) Winnowing - an important process in the concentration of the Stairway Sandstone (Ordovician) phosphorites of Central Australia. J Sed Petrol, 37: 818-828

COOK, P J, SHERGOLD, J H, BURNETT, W C,, RIGGS, S R (1990) Phosphorite research: a historical overview. Geol Soc London, Spec Publ, 52: 1-22

CRESSMAN, E R, SWANSON, R W (1964) Stratigraphy and Petrology of the Permian Rocks of Southwestern Montana. U S Geol Surv, Prof Pap, 313-C: 275-569

CULLEN, D J (1980) Distribution, composition and age of submarine phosphorites on the Chatham Rise, east of New Zea land. Soc Econ Paleont Miner, Spec Publ, 29: 139-148

CULLEN, D J (1988) Mineralogy of nitrogenous guano on the Bounty Islands, SW Pacific Ocean. Sedimentology, 93: 421-428

CULLEN, D J (1989) The Chatham Rise phosphorites of New Zealand. In: NOTHOLT, A J G, SHELDON, R P, DAVIDSON, D F: Phosphate deposits of the world, Vol 2: 528-532

CULLEN, D J, BURNETT, W C (1986) Phosphorite association on seamounts in the tropical southwest Pacific Ocean. Mar Geol 71: 215-236

CULLEN, D J, CHALLIS, G A, DRUMMOND, G W (1990) Late Holocene estuarine phosphogenesis in Raglan Harbour, New Zealand. Sedimentology, 37: 847-857

D'ANGLEJAN, B F (1967) Origin of marine phosphorites off Baja California, Mexico. Mar Geol, 5: 15-44

DA ROCHA ARAUJO, P R, FLICOTEAUX, R, PARRON, C, TROMPETTE, R (1992) Phosphorites of Rochina Mine - Patos de Minas (Minas Gerais, Brazil) Genesis and evolution of a Middle Proterozoic deposit tectonized by the Brazilian orogeny. Econ Geol, 87: 332-351

DAPPLES, E C (1955) General lithofacies relationship of St. Peter Sandstone and Simpson Group. Amer Assoc Petrol Geol, Bull, 39: 444-467

DEGENS, E T, STOFFERS, P (1976) Stratified waters as a key to the past. Nature, 263: 22-27

DEKEYSER, F (1969) On the genesis of the Georgina Basin phosphorites, northwestern Queensland. BMR, Record, 1969/79: 20 pp + 49 fig

DEKEYSER, F, COOK, P J (1972) Geology of the Middle Cambrian phosphorites and associated sediments of northwestern Queensland. Austr Bur Miner Resour, Bull, 138: 79 pp

DEMAISON, G J, MOORE, G T (1980) Anoxic environments and oil source beds. Am Assoc Petrol Geol, Bull, 64: 1179-1209

DENISON, R E, KOEPNICK, R B, BURKE, W H, HETHERINGTON, E A, FLETCHER, A (1994) Construction of the Mississippian, Pennsylvanian and Permian seawater 87Sr/86Sr curve. Chem Geol, 112: 145-167

DENISON, R E, KOEPNICK, R B, FLETCHER, A, HOWELL, M W, GALLAWAY, W S (1994) Criteria for the retention of original seawater 87Sr/86Sr in ancient shelf limestones. Chem Geol, 112: 131-143

DEOLIVEIRA, N P (1980) Mineralogie und Geochemie der phosphatführenden Laterite von Itacupim und Trauria, Nordbrasilien. Diss Thesis, Univ Erlangen-Nürnberg, 149pp

DIX, G R (1988) Late Holocene, insular phosphorite from western Australia. Econ Geol, 83: 1279-1284

DONNELLY, T H, SHERGOLD, J H, SOUTHGATE, P N (1988) Ano malous geochemical signals from phosphatic Middle Cambrian rocks in the southern Georgina Basin, Australia. Sedimen tology, 35: 549-570

DONNELLY, T H, SHERGOLD, J H, SOUTHGATE, P N, BARNES, C J (1990) Events leading to global phosphogenesis around the Proterozoic/Cambrian boundary. Geol Soc London, Spec Publ, 52: 273-287

DOTT, R H, jr, ROSHARDT, M A (1972) Analysis of crossstratification orientation in the St. Peter Sandstone in southwestern Wisconsin. Geol Soc America, Bull, 83: 2589-2596

DOYLE, L J, BLAKE, N J, WOO, C C, YEVICH, P (1978) Recent biogenic phosphorites: Concretions in mollusk kidneys . Science, 199: 1431-1433

DUNHAM, R J (1962) Classification of carbonate rocks according to depositional texture. AmerAssoc Petrol Geol, Mem, 1: 108-121

EGANOV, E A (1988) Phosphate deposition and stromatolites. Inst Geol, Geophys, Akad Nauk SSSR: 89pp

EGANOV, E A, SOVETOV, Y K, YANSHIN, A L (1986) Proterozoic and Cambrian phosphorites deposits: Karatau, southern Kazakhstan, USSR. In: COOK, PJ, SHERGOLD, JH (eds), Phosphate deposits of the world, Vol 1, Proterozoic and Cambrian phosphorites, 175-189

EICKHOFF, H-G (1962) Zur Stratigraphie und Tektonik des Oberdevons nördlich Lautenthal/Harz. unpubl Dipl Thesis Univ Göttingen: 82 pp

EKMAN, V W (1905) On the influence of the earth's rotation on ocena currents. Royal Swed Acad Sci, Arkiv för matematik och fysik, 2/11: 1-53

EL FALEH, E M (1988) Les méchanismes de synthèse de l'apatite par activité bactérienne Rôle et comportement de quelques éléments minéaux Application aux phosphates sédimentaires . PhD Thesis, Université Louis Pasteur, Strasbourg

ELDERFIED, H (1986) Strontium isotope stratigraphy. Palaeogeog Palaeoclimatol Palaeoecol, 57: 71-90

ELDERFIELD, H, GREAVES, M J (1982) The rare earth elements in seawater. Nature, 296: 214-219

ELIOTT, J C (1985) Controlled crystallization. Nature, 317: 387-388

EMBRY, A F, KLOVAN, J E (1971) A Late Devonian reef tract on northeastern Banks Island, NWT. Bull Canad Petrol Geol, 19: 730-781

ERWIN, D H (1993) The Great Paleozoic Crisis - Life and death in the Permian. 327pp

ERWIN, D H (1995) The End-Permian mass extinction. In: SCHOLLE, P A, PERYT, T M, ULMER-SCHOLLE, D S (eds) The Permian of northern Pangea, Vol 1: Paleogeography, paleoclimates, stratigraphy: 20-34

EUGSTER, H P, HARDIE, A H (1975) Sedimentation in an ancient playa-lake complex: The Wilkens Peak Member of the Green River Formation in Wyoming. Geol Soc America, Bull, 86: 319-334

EUGSTER, H P, SURDAM, R C (1973) Depositional environment of the Green River Formation of Wyoming: A preliminary report. Geol Soc America, Bull, 84: 1115-1120

FAURE, K, DE WIT, M, WILLIS, J P (1995) Late Permian global coal hiatus linked to 13C-depleated CO2 flux into the atmosphere during final consolidation of Pangea. Geology, 23: 507-510

FIKRI, A, LAMBOY, M, BENALIOULHAJ, S TRICHET, J, BELAYOUNI, H (1989) Contribution àl'étude pétrologique de la matière organique dans les phosphates naturels Novelles approches méthodologiques. Bull Soc géol France, 8,5: 979-987

FILIPPELLI, G M (1997) Controls on phosphorus concentration and accumulation in oceanic sediments. Mar Geol, 139: 231-239

FILIPPELLI, G M, DELANEY, M L (1992) Similar phosphorus flux in ancient phosphorite deposits and a modern phospho genetic environment. Geology, 20: 709-712

FILIPPELLI, G M, DELANEY, M L (1994a) Phosphorus geochemistry, diagenesis and mass balance of the Miocene Monterey Formation at Shell Beach, California. US Geol Surv, Bull, 1995: G1-G11

FILIPPELLI, G M, DELANEY, M L (1994b) The oceanic phosphorus cycle and continental waethering during the Neogene. Paleoceanography, 9: 643-652

FILIPPELLI, G M, DELANEY, M L, GARRISON, R E, OMARZAI, S K, BEHL, R J (1994) Phosphorus accumulation rates in a Miocene low oxygen basin: The Monterey Formation (Pismo Basin), California. Mar Geol 116: 419-430

FISCHER, A G, ARTHUR, M A (1977) Secular variations in the pelagic realm. Soc Econ Paleont Miner Spec Pub, 25: 19-50

FLEMING, P J G (1974) Origin of some Cambrian bedded cherts, and some other aspects of silicification in the Georgina Basin, Queensland. Geol Surv Qld, Publ, 358: 9 pp+ 6 pl

FLEMING, P J G (1977) Faunas, lithologies, and the origin of phosphorites in parts of the Middle Cambrian Beetle Creek Formation of northwestern Queensland. Geol Surv Qld, Publ, 364: 21 pp + 7 pl

FLICOTEAUX, R (1982) Genèse des phosphates alumineaux du Sénégal occidental - Étapes et guides de l'altération. Sci Géol, Mem, 67: 229 pp+IXpl

FLICOTEAUX, R, LUCAS, J (1984) Weathering of phosphate minerals. In: NRIAGU, J O, MOORE, P B (eds) Phosphate minerals, 292-317

FLICOTEAUX, R, NAHON, d, PAQUET, H (1977) Genese des phosphates alumineux a partier des sédiments argilo-phosphatés du Tertiaire de Lam-Lam (Sénégal) Suite minéralogique Permanences etchangements de structures. Sci Géol, Bull, 30: 153-174

FLÜGEL, E (1982) Microfacies analysis of limestones. 633 pp (Springer)

FLÜGEL, E (1994) Pangean shelf carbonates: Controls and paleoclimatic significance of Permian and Triassic reefs. In: KLEIN, G D (ed), Pangea: paleoclimate, tectonics, and sedimentation during accretion, zenith and breakup of a supercontinent, Geol Soc America, Spec Pap, 288: 247-266

FÖLLMI, K B (1989a) Evolution of the Mid-Cretaceous triad. Lecture Notes of Earth Science, 23: 153 pp

FÖLLMI, K B (1989b) Mid-Cretaceous platform drowning, current-induced condensation and phosphogenesis, and pelagic sedimentation along the eastern Helvetic shelf (northern Tethys margin). In: WIEDMAN, J (Ed), Cretaceous of the western Tethys, 585-606

FÖLLMI, K B (1990) Condensation and phosphogenesis: example of the Helvetic mid-Cretaceous (northern Tethyan margin). Geol Soc London, Spec Publ, 52: 237-252

FÖLLMI, K B (1995) 160 my record of marine sedimentary phosphorus burial: Coupling of climate and continental weathering under greenhouse and icehouse conditions. Geology, 23: 503-506

FÖLLMI, K B (1996) The phosphorus cycle, phosphogenesis and marine phosphate-rich deposits. Earth-Science Review, 40: 55-124

FÖLLMI, K B, GARRISON, R E (1991) Phosphatic sediments, ordinary or extraordinary deposits? The example of the Miocene Monterey Formation (California). In: MÜLLER, D W, McKENZIE, J A, WEISSERT, H: Controversies in modern Geology: 55-84

FÖLLMI, K B, GRIMM, K A (1990) Doomed pioneers: Gravity-flow deposition and bioturbation in marine oxygen-deficient environments. Geology, 18: 1069-1072

FÖLLMI, K B, GARRISON, R E, GRIMM, K A (1991) Stratification in phosphatic sediments: Illustrations from the Neogene of California. In: EINSELE, G, RICKEN, W, SEILACHER, A: Cycles and events in stratigraphy: 492-507

FÖLLMI, K B, GARRISON, R E, RAMIREZ, P C, ZAMBRANO-ORTIZ, F KENNEDY, W J, LEHNER, B L (1992) Cyclic phosphate-rich successions in the upper Cretaceous of Columbia. Palaeogeogr Palaeoclimatol Palaeoecol, 93: 151-182

FÖLLMI, K B, WEISSERT, H, LINI, A (1993) Nonlinearities in phosphogenesis and phosphorus-carbon coupling and their implications for global change. In: WOLLAST, R, MACKENZIE, F T, CHOU, L (eds) Interactions of C, N, P and S, biogeochemical cycles and global change, NATO ASI Series, I4: 447-474

FÖLLMI, K B, WEISSERT, H, BISPING, M, FUNK, H (1994) Phosphogenesis, carbon isotope stratigraphy, and carbonate platform evolution on the Lower Cretaceous northern Tethyan margin. Geol Soc America, Bull, 106: 729-746

FOLK, R L (1962) Spectral subdivision of limestone types. Amer Assoc Petrol Geol,Mem, 1: 20-32

FRAZIER, F R (ed)(1981) Coastal upwelling. Coastal and Estuarine Science, 1: 529 pp

FREAS, D H, ECKSTROM, C L (1967) Areas of potential upwelling and phosphorite deposition during Tertiary, Mesozoic, and late Paleozoic time. In: Proceedings of the Seminar on sources of mineral raw materials for fertilizer industry in Asia and the Far East, Mineral Resour Dev Ser, 32, UNECAFE: 228-238

FREAS, D L, RIGGS, S R (1965) Environments of phosphorite deposition in Central Florida phosphate district. In: BROWN, L F (ed) 4th Forum on Geology of Industrial Minerals, Univ of Texas at Austin, Bur Econ Geol, 117-128

FRIEDMAN, G M (1962) On sorting coeffizients and lognormality of grainsize distribution of sandstonesJour Geol, 70: 737-753

FROELICH, P N, ARTHUR, M A, BURNETT, W C, DEAKIN, M, HENSLEY, V, JAHNKE, R, KAUL, L, KIM, K-H, ROE, K, SOUTAR, A, VATHAKANON, C (1988) Early diagenesis of organic matter in Peru continental margin sediments: phosphorite precipitation. Mar Geol, 80: 309-343

FROELICH, P N, KLINKHAMMER, G P, BENDER, M L, LUEDTKE, N A, HEATH, G R, CULLEN, D, DAUPHIN, P, HAMMOND, D, HARTMAN, B, MAYNARD, V (1979) Early oxidation of organic matter in pelagic sediments of the eastern equatorial Atlantic: suboxic diagenesis. Geochim Cosmochim Acta, 43: 1075-1090

FÜRSICH, F T (1979) Genesis, environments, and ecology of Jurassic hardgrounds. N Jb Geol Paläont Abh, 158: 1-63

GALLI-OLIVER, C, GARDUÑO, G, GAMIÑO, J (1990) Phosphorite deposits in the Upper Oligocene, San Gregorio Formation at San Juan de la Costa, Baja California Sur, Mexico. In: BURNETT, W C:, RIGGS, S R (eds) Phosphate deposits of the world, Vol 3: Neogene to Modern Phosphorites, 122-126

GARFIELD, T R, HURLEY, N F, BUDD, D A (1992) Little Sand Draw oil field, Bighorn Basin, Wyoming: A hybrid dual-porosity and single porosity reservoir in the Phosphoria For mation. Amer Assoc Petrol Geol, Bull, 76: 371-391

GARRELS, R M, MACKENZIE, F T, HUNT, C (1975) Chemical cycles and the global environment. 206 pp

GARRISON, R E, KASTNER, M (1990) Phosphatic sediments and rocks recovered from the Peru margin during ODP leg 112. In: SUESS, E, VON HUENE, et al (eds), Proc ODP, Sci Results, 112: 111-134

GARRISON, R E, KASTNER, M, REIMERS, C E (1990) Miocene phosphogenesis in California. In: BURNETT, W C, RIGGS, S R, Phosphate deposits of the world, Vol 3: Neogene to modern phosphorites, 285-299

GERDES, G, KRUMBEIN, W E (1987) Biolaminated deposits. Lecture Notes in Earth Sciences, 9: 183 pp

GERMANN, K, BOCK, W-D, SCHRÖTER, T (1984) Facies development of upper cretaceous phosphorites in Egypt: sedimentological and geochemical aspects. Berliner geowiss Abh, (A),50: 345-361

GERMANN, K, BOCK, W-D, SCHRÖTER, T (1985) Properties and origin of Upper Campanian phosphorites in Egypt. Sci Géol, Mém, 77: 23-33

GERMANN, K, BOCK, W-D, GANZ, H, SCHRÖTER, T, TRÖGER, U (1987) Depositional conditions of Late Cretaceous phosphorites and black shales in Egypt. Berliner geowiss Abh, (A), 753: 629-668

GERMANN, K, PAGEL, J-M, PAREKH, P P (1979) Iodine in karst type phosphorites from the Lahn region, Germany. Chem Geol, 25: 305-316

GERMANN, K, PAGEL, J-M, PAREKH, P P (1981) Eigenschaften und Entstehung der "Lahn-Phosphorite". Z dt geol Ges, 132: 305-323

GERVY, R (1973) Les phosphate et l'agriculture. 298 pp

GILINSKAYA, L G (1991) A new type of PO43-centre in apatite. J Struct Chem, 31: 892-898

GILINSKAYA, L G (1993) Stable paramagnetic Pb3+(2S172) centres in natural apatite. Phy Solid State, 35: 35-73

GILINSKAYA, L G, ZANIN, Y N (1993) EPR investigation of isomorphic impurity in the form of VO2+ in apatite from phosphorite. Dokl Akad Nauk SSSR, 273: 1463-1467 (in Russian)

GILINSKAYA, L G, ZANIN, Y N, KNUBOVETS, R G, KORNEVA, T A, FADEEVA, V P (1993) Organophosphorus radicals in naturalapatites (Ca5(PO4)3(F,OH). J Struct Chem, 33: 1463-1467

GIMMELFARB, B M, KRASILNIKOVA, N A, TUSHINA, A M (1959) Classification of phosphorites. Dokl(Proc) Acad Sci USSR, 128: 1024-1026

GINSBURG, R N (1991) Controversies about stromatolites: Vices and virtues. In: MÜLLER, D W, McKENZIE, J A, WEISSERT, H (eds) Controversies in modern Geology: 25-36

GLENN, C R (1990a) Depositional sequences of the Duwi, Sibâîya and Phosphate Formations, Egypt: phosphogenesis and glauconitization in a Late Cretaceous epeiric sea. Geol Soc London, Spec Publ, 52: 205-222

GLENN, C R (1990b) Pore water, petrologic and stable carbon isotopic data bearing on the origin of Modern Peru margin phosphorites and associated authigenic phases -in: BURNETT, W C, RIGGS, S R: Phosphate deposits of the world, Vol 3: Neogene to modern phosphorites 46-61

GLENN, C R, ARTHUR, M A (1985) Sedimentary and geochemical indicators of productivity and oxygen contents in modern and ancient basins: The Holocene Black Sea as the "type" anoxic basin. Chem Geol, 48: 325-354

GLENN, C R, ARTHUR, M A (1988) Petrology and major element geochemistry of Peru margin phosphorites and associated diagenetic minerals: Authigenesis in modern organic rich sediments. Marine Geology, 80: 231-268

GLENN, C R, ARTHUR, M A (1990) Anatomy and origin of a Cretaceous phosphorite-greensand giant, Egypt. Sedimentology, 37: 123-154

GLENN, C R, ARTHUR, M A, RESIG, M J, BURNETT, W C, DEAN, W E, JAHNKE, R A (1994) Are modern and ancient phosphorites really so different? In: IIJIMA et al (eds), Siliceaous , phosphatic and glauconitic sediments of the Tertiary and Mesozoic, 159-188

GLENN, C R, ARTHUR, M A, YEH, H-W, BURNETT, W C (1988) Carbon isotopic composition and lattice-bound carbonate of Peru-Chile margin phosphorites. Marine Geology, 80: 287-308

GLENN, G R, KRONEN, J D (1993) Origin and significance of late Pliocene phosphatic hardgrounds on the Queensland Plateau, northeastern Australian margin. In: McKENZIE, J A, DAVIS, P J, PALMER-JULSON, A (eds) Proc ODP, Sci Results, 133: 525-432

GLENN, C R, FÖLLMI, K B, RIGGS, S R, BATURIN, G N, GRIMM, K A, TRAPPE, J, ABED, A M, GALLI-OLIVER, C, GARRISON, R E, ILYIN, A V, JEHL, C, ROEHRLICH, V, SADAQAH, R M Y, SCHIDLOWSKI, M, SHELDON, R E, SIEGMUND, H (1994) Phosphorus and phosphorites: sedimentology and environment of formation. Eclogea geol Helv, 87: 747-788

GOLDBERG, E D, KOIDE, M, SCHMITT, R A, SMITH, H V (1963) Rare earth distribution in the marine environment. J Geophys Res, 68: 4209-4217

GOLONKA, J, ROSS, M I, SCOTESE, C R (1994) Phanerozoic paleogeographic and paleoclimatic modeling maps. In: EMBRY, A F, BEAUCHAMP, B, GLASS, D J (eds) Pangea: Global environments and resources. Canad Soc Petrol Geol, Mem, 17: 1-48

GRANDJEAN, P, ALBARÉDE, F (1989) Ion probe measurement of rare earth elements in biogenic phosphates. Geochim Cosmochim Acta, 53: 3179-3183

GRANDJEAN-LÉCUYER, P, FEIST, R, ALBARÉDE, F (1993) Rare earth elements in old biogenic apatites. Geochim Cosmochim Acta, 57: 2507-2514

GRIMM, K A (1992) The sedimentology of coastal upwelling systems. PhD Thesis, Univ Calif Santa Cruz

GULBRANDSEN, R A (1960) A method of x-ray analysis for determining the ratio of calcite to dolomite in mineral mixtures. US Geol Surv, Bull, 1111-D: 147-152

GULBRANDSEN, R A (1966) Chemical composition of phosphorites of the Phosphoria Formation. Geochim et Cosmochim Acta, 30: 769-778

GULBRANDSEN, R A (1969) Physical and chemical factors in the formation of marine apatite. Econ Geol, 64: 365-382

GULBRANDSEN, R A (1970) Relation of carbon dioxide content of apatite of the Phosphoria Formation to regional facies. US Geol Surv, Prof Pap, 700-B: B9-B13

GULBRANDSEN, R A, ROBERSON, C E (1973) Inorganic phosphorus in seawater. In: Environmental phosphorus handbook, 77-93

GUSEV, G M, ZANIN, Y N, KRIVOPUTSKAYA, L M, LEMINA, N M, IUSUPOV, T S (1976) Transformation of apatite composition under conditions of weathering and leaching. Dokl Acad Nauk SSSR, 229: 971-973

HALLAM, A, BRADSHAW, M J (1979) Bitumunous shales and oolitic ironstones as indicators of transgressions and regressions. J Geol Soc Lond, 136: 157-164

HARDIE, L A (ed)(1977) Sedimentation on the modern carbonate tidal flats of northwest Andros Island, Bahamas. John Hopkins University Studies in Geology, 22: 202 pp

HAY, W W, BROCK, J C (1992) Temporal variation in intensity of upwelling off southwest Africa. In: SUMMERHAYS, C P et al (eds), Upwelling systems: evolution since the Early Miocene, Geol Soc London, Spec Publ, 63: 463-497

HECKEL, P H (1977) Origin of phosphatic black shale facies in Pennsylvanian cyclothems of mid-continent of North America . Amer Assoc Petrol Geol, Bull, 61: 1045-1068

HEIN, J R, YEH, H-W, GUNN, S H, SLITER, W V, BENNINGER, L M, WANG, C-H (1993) Two major Cenozoic episodes of phosphogenesis recorded in equatorial pacific seamount deposits. Paleoceanography, 8: 293-311

HEINSALU, H, RAUDSEP, R (1993) Lithostratigraphic subdivision of the phosphate bearing (C3-O1kl) strata in the Rakvere area of northern Estonia. Bull Geol Surv Estonia, 3/1: 4-12

HEINSALU, H (1990) Tremadoc phosphate-bearing rocks of North Estonia and shelly phosphorite. In: KALJO, D, NESTOR,H: Field meeting Estonia 1990, an excursion guidebook, 37-39

HEINSALU, H, VIIRA, V, RAUDSEP, R (in press) Environmental conditions of shelly phosphorite accumulation in the Rakvere phosphorite region, northern Estonia. Bull Geol Surv Estonia

HENDERSON, R A, CUFF, C, SOUTHGATE, P N (1979) A brine-leaching mechanism of phosphorite genesis. In: COOK, P J, SHERGOLD, J H (eds) Proterozoic and Cambrian Phospho rites, p22

HENIN, S (1982) Le phosphore et al vie - Introduction. C R Séance Acad Agric Fr, 68 (4) 277-281

HERBIG, H-G (1991) Das Paläogen am Südrand des Hohen Atlas und im Mittleren Atlas MarokkosStratigraphie, Fazies, Paläogeographie und Paläotektonik. Berliner geowiss Abh (A), 135: 289 pp

HERBIG, H-G, BENDER, P (1992) A eustatically driven calciturbidite sequence from the Dinantian II of the Eastern Rheinisches Schiefergebirge. Facies, 27: 245-262

HERBIG, H-G, TRAPPE, J (1994) Stratigraphy of the Subatlas Group (Maastrichtian - Middle Eocene, Morocco). Newsl Stratigraphy, 30: 1-42

HERRING, J R (1995) Permian phosphorites: a paradox of phosphogenesis. In: SCHOLLE, P A, PERYT, T M, ULMER-SCHOLLE, D S The Permian of the northern Pangea, Vol 2, Sedimentary basins and economic resources, 292-312

HEWITT, R A (1980) Microstuctural contrast between some sedimentary francolites. J geol Soc London, 137: 661-667

HIATT, E E (1993) Productivity cycles within the Permian Phosphoria sea: chemostratigraphic analysis of an ancient upwelling deposit, SE Idaho. Carboniferous to Jurrassic Pangea, Ann Meeting, Canad Soc Petrol Geol, Program with Abstracts: 312

HILLER, N (1993) A modern analogue for the Lower Ordovician Obolus conglomerate of Estonia. Geol Mag 130: 265-267

HIRSCHLER, A (1990) Étude de l'intervention des microorganismes dans la formation de l'apatite. PhD Thesis, Université Louis Pasteur, Strasbourg

HIRSCHLER, A, LUCAS, J, HUBERT, J-C (1990a) Bacterial involvement in apatite genesis. FEMS Microbiology Ecology, 73: 211-220

HIRSCHLER, A, LUCAS, J, HUBERT, J-G (1990b) Apatite genesis: A biologically induced or biologically controlled mineral formation process? Geomicrobiol J, 7: 47-57

HITE, R J (1978) Possible genetic relationships between evaporites, and iron-rich sediments. The Mountain Geologist, 14: 97-107

HODELL, D A, MUELLER, P A, GARRIDO, J R (1991) Variations in strontium isotopic composition of seawater during the Neogene. Geology, 19: 24-27

HOFMANN, H J (1975) Bolopora not a cryozoan, but an Ordovician phosphatic, oncolitic accretion. Geol Mag 112: 523-526

HORTON, A, IVIMEY-COOK, H C, HARRION, R K, YOUNG, B R (1980) Phosphatic öoids in the Upper Lias (Lower Jurassic) of central England. J geol Soc London, 137: 731-740

HOWARD, P F (1986) Proterozoic and Cambrian phosphorites - regional review: Australia. In: COOK, P J, SHERGOLD, J H: Phosphate deposits of the world, Vol 1, Proterozoic and Cambrian phosphorites: 20-4

HOWARD, P F (1989a) The D Tree phosphate deposit, Georgina Basin, Australia. In: NOTHOLT, A J G, SHELDON, R P, DAVIDSON, D F: Phosphate deposits of the world, Vol 2: 551-557

HOWARD, P F (1989b) The Wonarah phosphate deposits, Georgina Basin, Australia. In: NOTHOLT, A J G, SHELDON, R P, DAVIDSON, D F: Phosphate deposits of the world, Vol 2: 545-550

HOWARD, P F (1990) The distribution of phosphatic facies in the Georgina, Wiso and Daly River Basins, Northern Australia. Geol Soc London, Spec Publ, 52: 261-272

HOWARD, P F, HOUGH, M J (1979) On the geochemistry and origin of the D Tree, Wonarah, and Sherrin Creek phosphorite deposits of the Geogina basin, Northern Australia. Econ Geol, 74: 260-284

HUTCHINSON, G E (1950) Survey of contemporary konwledge of biochemistry 3 The biochemistry of vertebrate excretion. Bull Am Mus Hist 96: 553 pp

HUTCHINSON, G E (1957) A treatise on limnology Vol 1: Geography, physics, and chemistry. 1015 pp

ILYIN, A V (1994) Cenomanian phosphorites in the former Soviet Union. Sediment Geol, 94: 109-127

ILYIN, A V, HEINSALU, H N (1990) Early Ordovician shelly phosphorites of the Baltic Phosphate Basin. Geol Soc London, Spec Publ, 52: 253-259

ILYIN, A V, RATNIKOVA, G I (1976) Rare-earth element distributionin the Hobso Gol phosphorites (Mongolia). Geochem Int, 13: 53-56

ILYIN, A V, RATNIKOVA, G I (1981) Primary, bedded, structureless Phosphorite of the Khubsugul Basin, Mongolia. Jour Sed Petrology, 51: 1215-1222

INGALL, E, JAHNKE, R (1997) Influence of water-column anoxia on the element fractionation of carbon and phosphorus during sediment diagenesis. Mar Geol, 139: 219-229

IRWIN, H, CURTIS, C, COLEMAN, M (1977) Isotopic evidence for source of diagenetic carbonates formed during burial of organic-rich sediments. Nature, 269: 209-213

JAHNKE, R A, EMERSON, S R, ROE, K K, BURNETT, W C (1983) The present day formation of apatite in Mexican continental margin sediments. Geochim Cosmochim Acta, 47: 259-266

JARVIS, I (1980a) Geochemistry of phosphatic chalks and hardgrounds from the Santonian to Early Campanian (Cretaceous) of northern France. J Geol Soc London, 137: 705-721

JARVIS, I (1980b) The initiation of phosphatic chalk sedimentation. the Senonian (Cretaceous) of the Anglo-Paris Basin. Soc Econ Paleont Min, Spec Pub, 29: 167-192

JARVIS, I (1992) Sedimentology, geochemistry and origin of phosphatic calks: The upper Cretaceous deposits of NW Europe. Sedimentology, 39: 55-97

JARVIS, I, BURNETT, W C, NATHAN, Y, ALMBAYDIN, F S M, ATTIA, A K M, CASTRO, L N, FLICOTEAUX, R, HILMY, M E, HUSAIN, V, QUTAWNAH, A A, SERJANI, A, ZANIN, Y N (1994) Phosphorite geochemistry: state-of-the-art and environmental concerns. Eclogae geol Helv, 87: 643-700

JOHNSON, R A (1986) Shallow stratigraphic core tests on file at the Florida Geological Survey. Fla Geol Survey, Inf Circ, 103: 431 pp

KALJO, D (1990) An introduction to the geology of Estonia. In: KALJO, D, NESTOR, H (eds) Field Meeting Estonia 1990, an excursion Guidebook, 6-10

KALJO, D, BOROVKO, N, HEINSALU, H, KHAUANOVICH, K, MENS, K, POPOV, L, SERGEYEVA, S, SOBOLEVSKAYA, R, VIIRA, V (1986) The Cambrian-Ordovician boundary in the Baltic-Ladoga Clint area (North Estonia and Leningrad region, USSR). Proc Acad Sci Estonian SSR, Geology, 35: 97-108

KALJO, D, HEINSALU, H, MENS, K, PUURA, I, VIIRA, V (1988) Cambrian-Ordovician boundary beds at Tônismägi, Tallinn, North Estonia. Geol Mag, 125: 457-463

KAZAKOV, A V (1937) The phosphorite facies and the genesis of phosphorites. In: HIMMELFARB, B M, KAZAKOV, A V, KURMAN, I M: Geological investigations of aricultural ores USSR. Trans Sci Inst Fertil, Insectofungicides, 142: 95-113

KELTS, K (1988) Environments of deposition of lacustrine petroleum source rocks: an introduction. In: FLEET et al (eds), Lacustrine petroleum source rocks. Geol Soc London, Spec Publ, 40: 3-26

KEMPE, S, DEGENS, E T (1985) An early soda ocean. Chem Geol, 53: 95-108

KEMPE, S, KAZMIERCZAK, J (1994) The role of alkalinity in the evolution of ocean chemistry, organization of living systems, and biocalcification processes. In: DOUMENGE, F, ALLEMAND, D, TOULEMONT, A (eds) Past and present biomineralization processes, Bull de l'Insitut océanographique Monaco, nspécial 13: 61-117

KENDALL, C G, SCHLAGER, W (1981) Carbonates and relative changes in sea level. Mar Geol, 44: 181-212

KENNEDY, W J, GARRISON, R E (1975) Morphology and genesis of nodular phosphates in the Cenomanian Glauconitic Marl of south-east England. Lethaia, 8: 339-360

KESTER, D R, PYTKOWICZ, R M (1967) Determination of the apparent dissociation constants of phosphoric acid in sea water. Limnol Oceanogr, 12

KETNER, K B (1977) Late Paleozoic orogeny and sedimentation, southern California, Nevada, Idaho, and Montana. In: STE WART, J H, STEVENS, C H, FRISCHE, A E: Paleozoic paleogeography of the western United States: 363-369

KIBALCZYC, W, CHRISTOFFERSEN, J, CHRISTOFFERSEN, M R, ZIELENKIEWICZ, A, ZIENKIEWICZ, W (1990) The effect of magnesium ions on the precipitation of calcium phosphates. J Crysal Growth, 106: 355-366

KIDDER, D L (1985) Petrology and origin of phosphate nodules from the Midcontinent Pennsylvanian sea. J Sediment Petrol, 55: 809-816 KIDDER, D L, EDDY-DILEK, C A (1994) Rare earth variations in phosphate nodules from the Midcontinent Pennsylvanian cyclothems. J Sediment Res, A64: 584-592

KIDWELL, S M (1991a) Condensed deposits in siliciclastic sequences: expected and observed features. In: EINSELE et al (eds) Cycles and events in stratigraphy: 682-695

KIDWELL, S M (1991b) The stratigraphy of shell concentrations . In: ALLISON, P A, BRIGGS, D E G: Taphonomy: Releasing data locked in the fossil record: 211-290

KIM, K H, BURNETT, W C (1988) Accumulation and biological mixing of Peru margin sediments. Marine Geology, 80: 181-194

KIRKLAND, D W, EVANS, R (1981) Source-rock potential of evaporitic environments. Amer Assoc Petrol Geol, Bull, 65: 181-190

KIZEVETTER, I V (1973) Biochemistry of resources of aquatic origin. Pishchevaya Promyshlennost, Moscow

KOLODNY, Y (1969) Are marine phosphorites forming today? Nature, 224: 1017-1018

KOLODNY, Y, GARRISON, R E (1994) Sedimentation and diagenesis in paleo-upwelling zones of epeiric sea and basinal settings. In: ILJIMA, A, ABED, A M, GARRISON, R E (eds) Siliceous, phosphatic and glauconitic sediments of the Tertiary and Mesozoic, Proc 29th Intl Geol Congess, C: 205-224

KOLODNY, Y, KAPLAN, I R (1970) Uranium isotopes in sea floor phosphorites. Geochim Cosmochim Acta, 34: 3-24

KRAJEWSKI, K P (1981) Pelagic stromatolites from the High-Tatric Albian limestones in the Tatra Mtns. Kwart Geol 25: 731-759

KRAJEWSKI, K P (1983) Albian pelagic phosphate-rich macrooncoids from the Tatra Mts (Poland). In: PERYT, T M (ed) Coated grains: 344-357

KRAJEWSKI, K P (1984) Early diagenetic phosphate cements in the Albian condensed glauconitic limestone of the Tatra Mountains, Western Carparthians. Sedimentology, 31: 443-470

KRAJEWSKI, K P (1989) Organic geochemistry of a phosphorite to black shale transgressive succession: Wilhelmöya and Janusfjellet Formations (Rhaetian to Jurassic) in central Spitsbergen, Arctic Ocean. Chem Geol, 74: 249-263

KRAJEWSKI, K P, VAN CAPPELLEN, P, TRICHET, J, KUHN, O, LUCAS, J, MARTIN-ALGARRA, A, PREVOT, L, TEWARI, V C, GASPAR, L, KNIGHT, R I, LAMBOY, M (1994) Biological processes and apatite formation in sedimentary envrionments. Eclogae geol Helv, 87: 701-745

KRAMER, J R (1964) Seawater, saturation with apatite and carbonate. Science, 146: 637-638

KRASIL'NIKOVA, N A, PAUL, R K (1984) Stromatolitic phosphorites of the Gornaya Shoria region, USSR. Geol Surv India, Spec Publ, 17: 191-196

KRASILNIKOVA, N A, ILYIN, A V (1989) The Ordovician Baltic phosphorite Basin,USSR. In: NOTHOLT, A J G, SHELDON, R P, DAVIDSON, D F: Phosphate deposits of the world, Vol 2: 494-496

KREJCI-GRAF, K (1966) Geochemische Faziesdiagnostik. Freiberger Forschungshefte, C224: 80pp

KRESS, A G, VEEH, H H (1980) Geochemistry and radiometric ages of phosphatic nodules from the continental margin of northern New South Wales, Australia. Mar Geol, 36: 143-157

KRUMBEIN, W E (1983) Stromatolites - the challange of a term in space and time. Precambrian Res, 20: 493-531

KUDRASS, H R (1982) Submarine phosphorite nodules on the Chatham Rise/New Zealand: geological aspects and a preliminary estimation of reserves. In: HALBACH, P, WINTER, P (eds) Marine mineral deposits, 45-59

KUDRASS, H R, CULLEN, D J (1982) Submarine phosphorite nodules from the Central Chatham Rise off New Zealand - Compostion, distribution, and reserves (VALDIVIA-Cruise). Geol Jb, D51: 3-41

KUDRASS, H R, RAD, U von (1984) Geology and some mining aspects of the Chatham Rise phosphorite: a synthesis of SONNE-17 results. Geol Jb, 65: 233-252

KUMAR, S, MULLER, G (1988) Geochemistry, mineralogy and genesis of phosphatic stromatolites, Gangolihat Dolomite (Riphean), Kumaun Himalaya, India. Heidelberger Geowiss Abh, 20: 87-126

LAMBOY, M (1982a) Importance des pelotes fécales comme origine des grains de phosphate; l'exemple du gisement de Gafsa (Tunisie). C R Acad Sci Paris, (2)295: 595-600

LAMBOY, M (1982b) La phosphatisation en micromilieu granulaire d'apres un exemple Tunisien. C R Acad Sci Paris (2) 295: 799-802

LAMBOY, M (1986) Relations entre propriétés optiques et nannostructures des grains de phosphate Implications génétiques. Rev Géol dyn Géogr phy, 27: 311-318

LAMBOY, M (1987a) Genèse de grains de phosphate à partir de débris de squelette d'échinodermes: les processus et leur signification. Bull Soc géol France, 8,3: 979-987

LAMBOY, M (1987b) Genèse des phosphates granulaires, Enseignements des grains centrés sur des foraminifères Importance et modalités de la precipitation. C R Acad Sc Paris, 304: 435-440
LAMBOY, M (1990a) Microbial mediation in phosphogenesis: new data from the Cretaceous phosphatic chalks of northern France. Geol Soc London, Spec Publ, 52: 157-167
LAMBOY, M (1990b) Microstructures of a phosphatic crust from the Peruvian continental margin: phosphatized bacteria and associated phenomena. Oceanologica Acta, 13: 439-451
LAMBOY, M (1993) Phosphatization of calcium carbonate in phosphorites: microstructure and importance. Sedimentology, 40: 53-62
LEANZA, H A, SPIEGELMAN, A T, HUGO, C A, MASTANDREA,, OBLITAS, C J (1989) Phanerozoic sedimentary phosphatic rocks of Argentina. In: NOTHOLT, A J G, SHELDON, R P, DAVIDSON, D F: Phosphate deposits of the world, Vol 2, Phosphate rock resources: 147-158
LEHR, J R, McCLELLAN, G H, SMITH, J P, FRAZIER, A W (1967) Characterization of apatites in commercial phosphate rocks. In: Proceedings of International Colloquium of Solid Inorganic Phosphates, Toulouse, May 16-20, Société Chimique de France, 29-44
LEWY, Z (1990) Pebbly phosphate and granular phosphorite (Late Cretaceous, southern Israel) and their bearing on phosphati zation processes. Geol Soc London, Spec Publ, 52: 169-178
LOUTIT, T S, HARDENBOL, J, VAIL, P R, BAUM, G R (1988) Condensed sections: The key to age dating and correlation of continental margin sequences. Soc Econ Paleont Miner, Spec Publ, 42: 183-213
LOVE, J D (1964) Uraniferous phosphatic lake beds of Eocene age in intermontane basins of Wyoming and Utah. US Geol Surv, Prof Pap, 474-E: E1-E66
LOWE, D R (1972) The relationship between silicic volcanism and the formation of some sedimentary phosphorites. In: PURI, H S (ed) Proc VII Forum Geol ind Mineral Geol Phos phate Dolomite, Limestone, Clay deposits, Tampa, Florida, 28 -30 April 1971, Florida Dept Nat Res, Spec Publ 17: 217 225
LOWENSTAM, H A (1972) Phosphatic hard tissues of marine invertebrates: their nature and mechanical function,-and some fossil implications. Chem Geol, 9: 153-166
LOWENSTAM, H A, WEINER, S (1989) On biomineralization. 324 pp
LUCAS, J, ABBAS, M (1989) Uranium in natural phosphosrites: The Syrian example. Sci Géol, Bull, 42: 223-236
LUCAS, J, EL FALEH, E M, PRÉVôT, L (1990) Experimental study of the substitution of Ca by Sr and Ba in synthetic apatites. Geol Soc London, Spec Publ, 52: 33-47
LUCAS, J, FLICOTEAUX, R, NATHAN, Y, PRÉVôT, L, SHAHAR, Y (1980) Different aspects of phosphorite weathering. Soc Econ Pal Min, Spec Publ, 29: 41-51
LUCAS, J, PRÉVôT, L (1981) Synthèse d'apatite à partir de matière organique phosphorée (ARN) et de calcite par voie bactérienne. C R Acad Sci Paris, 292: 1203-1208
LUCAS, J, PRÉVôT, L (1984a) Les synthèses de l'apatite, données nouvelles pour un modèle de genèse des phosphorites sédimentaires. Proc 27th Int Geol Cong, V 15: 173-186, Moscov
LUCAS, J, PRÉVôT, L (1984b) Synthèse d'apatite par voie bactérienne à partir de matière organique phosphatée et de divers carbonates de calcium dans des eaux douce et marine naturelles. Chem Geol, 42: 101-118

LUCAS, J, PRÉVôT, L (1985) The synthesis of apatite by bacterial activity: mechanism. Sci Géol, Mém, 77: 83-92

LUCAS, J, PRÉVôT, L (1993) Quelques réflexions sur la genèse de l'apatite sédimentaire et des séries phosphatées. Coll "Sédimenologie et Géochemie de la Surface" à la mémoire de Georges MILLOT, 243-257, Strasbourg

LUCAS, J, PRÉVôT, L, LAMBOY, M (1978) Les phosphorites de la marge nord de l'Espagne: Chimie, minéneralogie, genèse. Oceanol Acta, 1: 5-72

LUCAS, J, PRÉVôT, L, BENALIOULHAJ, N,, EL FALEH, E (1987) Relation between sodium and apatite in sedimentary phospho rites of Morrocco. Terra Cognita, 7: 410

MABIE, C P, HESS, H D (1964) Petrographic study and classification of western phosphate ores. Bureau of Mines, report of investigations, 6468: 95 pp

MACKENZIE, F T, VER, L M, SABINE, C, LANE, M, LERMAN, A (1993) C, N, P, S global biogeochemical cycles and modeling of global change. In: WOLLAST, R, MACKENZIE, F T, CHOU, L (eds) Interactions of C, N, P and S, biogeochemical cycles and global change, NATO ASI Series, I4: 1-61

MALLINSON, D J, COMPTON, J S, SNYDER, S W, HODELL, D A (1994) Strontium isotopes and Miocene sequence stratigraphy across the northeast Florida platform. J Sediment Res, B64: 392-407

MALONE, P H G, TOWE, K M (1970) Microbial carbonate and phosphate precipitates from seawater cultures. Mar Geol, 9: 301-309

MANHEIM, F T, POPENOE, P, SIAPNO, W, LANE, C (1982) Manganese-phosphorite deposits of the Blake Plateau. In: HALBACH, P, WINTER, P (eds) Marine mineral deposits: 9-44

MANHEIM, F T, PRATT, R M (1967) Geochemistry of manganese-phosphorite deposits on the Blake Plateau. Woods Hole Oceanogr Institut Report, 68-32: 48-49

MANHEIM, F T, PRATT, R M, McFARLIN, P F (1980) Composition and origin of phosphorite deposits of the Blake Plateau. Soc Econ Paleont Miner, Spec Publ, 29: 117-137

MARTENS, C S, HARRISS, R C (1970) Inhibition of apatite precipitation in the marine environment by magnesium ions. Geochim et Cosmochim Acta, 34: 621-625

MARTINDALE, S G (1986) Depositional environment and phosphatization of the Meade Peak Phosphatic Shale Tongue of the Phosphoria Formation, Leach Mountains, Nevada. Contributions to Geology, University of Wyoming, 24: 143-156

MARTíN-ALGARRA, A S, NCHEZ-NAVAS, A (1995) Phosphate stromatolites from condensed cephalopod limestones, Upper Jurassic, Southern Spain. Sedimentology, 42: 893-919

MARTíN-ALGARRA, A S, VERA, J A (1994) Mesozoic palagic phosphate stromatolites from the Penibetic (Betic Cordillera southern Spain). In: BERTAND-SARFATI, J, MONTY, C,: Phanerozoic stromatolites II, 345-391

MATTHEWS, A, NATHAN, Y (1977) The decarbonation of carbonate fluorapatite (francolite). Amer Mineral, 62: 565-573

MAUGHAN, E K (1965) The Goose Egg Formation in the Laramie Range and adjacent parts of southeastern Wyoming. US Geol Surv, Prof Pap, 501-B: B53-B60

MAUGHAN, E K (1976) Organic carbon and selected element distribution in the phosphatic shale members of the Permian Phosphoria Formation, eastern Idaho and parts of adjacent states. US Geol Surv, Open-File Report, 76-577: 92 pp

MAUGHAN, E K (1980) Relation of phosphorite, organic carbon, and hydrocarbon in the Permian Phosphoria Formation (western USA). BRGM, Doc, 24: 63-92

MAUGHAN, E K (1983) Geological setting and geochemistry of oil shales in the Permian Phosphoria Formation. In: MIKINS, F P, McKAY, J F (eds) Geochemistry and chemistry of oil shales: 199-224

MAUGHAN, E K (1984) Geological setting and some geochemistry of petroleum source rocks in the Permian Phosphoria Formation. In: WOODWARD, J, MEISSNER, F F, CLAYTON, J L (eds) Hydrocarbon source rocks of the greater Rocky Mountain region: 281-294

MAUGHAN, E K (1990) Summary of the ancestral Rocky Mountains epeirogeny in Wyoming and adjacent areas. US Geol Surv, Open-File Report, 90-447: 8 pp

MAUGHAN, E K (1994) Phosphoria Formation (Permian) and its resource significance in the Western Interrior, USA. In: EMBRY, A F, BEAUCHAMP, B, GLASS, D J (eds) Pangea: Global environments and resources, Canad Soc Petrol Geol, Mem 17: 479-495

MCARTHUR, J M (1978) Systematic variations in the contents of Na, Sr, CO_2, and SO4 in marine carbonate fluorapatite and their relation to weathering. Chem Geol 21: 41-52

MCARTHUR, J M (1980) Post-depositional alteration of the carbonate-fluorapatite phase of Moroccan phosphates. Soc Econ Pal Min, Spec Publ, 29: 53-60

MCARTHUR, J M (1985) Francolite geochemistry - compositional controls during formation, diagenesis, metamorphism and weathering. Geochim et Cosmochim Acta, 49: 21-35

MCARTHUR, J M (1990) Flourine-deficient apatite. Mineral Mag, 54: 508-510

MCARTHUR, J M, BENMORE, R A, COLEMAN, M L, SOLDI, C, YEH, H-W, O'BRIEN, G W (1986) Stable isotope characterization of francolite formation. Earth Planet Sci Lett, 77: 20-34

MCARTHUR, J M, SAHAMI, A R, THIRLWALL, M, HAMILTON, P J, OSBORN, A O (1990) Dating of phosphogenesis with strontium isotopes. Geochim et Cosmochim Acta, 54: 1343-1351

MCARTHUR, J M, THIRLWALL, M F, GALE, A S, KENNEDY, W J, BURNETT, J A, MATTEY, D, LORD, A R (1993) Strontium isotope stratigraphy for the Late Cretaceous: a new curve based on the English Chalk. In: HAILWOOD E A, KIDD, R B (eds) High Resolution Stratigraphy, Geol Soc, Spec Publ, 70: 195-209

MCARTHUR, J M, WALSH, J N (1984) Rare-earth geochemistry ofphosphorites. Chem Geol, 47: 191-220

MCCLELLAN, G H (1980) Mineralogy of carbonate fluorapatite. J geol Soc London, 137: 675-681

MCCLELLAN, G H, SAAVEDRA, F N (1986) Chemical and mineral characteristics of some Cambrian and Precambrian phosphorites. In: COOK, P J, SHERGOLD, J H (eds) Phosphorite deposits of the world, Vol 1: Precambrian and Cambrian phosphorites, 244-267

MCCLELLAN, G H, VAN KAUWENBERGH, S J (1990) Mineralogy of sedimentary apatites. Geol Soc London, Spec Publ 52: 23-31

MCCONNELL, D (1938) A structural investigation of the isomorphism of the apatite group. Amer Mineral, 23: 1-19

MCCONNELL, D (1971) The mineralogy of the apatites and their relation to biologic precipitation. Proc North American Paleontological Convention, K: 1525-1535

MCCONNELL, D (1973) Apatite Its crystal chemistry, mineralogy, Utilization, and geologic and biologic occurrences. 111 pp

MCCONNELL, D, FRAJOLA, W J, DEAMER, D W (1961) Relation between inorganic chemistry and biochemistry of bone mine ralization. Science, 133: 281-182

MCKEE, E D, ORIEL, S S (1967) Paleotectonic investigations of the Permian system in the United States. Geol Surv Prof Pap, 515: 271 pp

MCKEE, E D, ORIEL, S S, others (1967) Paleotectonic maps of the Permian system. US Geol Surv, Misc Geol Invest Maps, I-450: 164 pp, 20 pl

MCKELVEY, V E (1950) Rare earths in western phosphate rocks. US Geol Surv, Trace Element Mem Rept, pp 194 MCKELVEY, V E (1956) Uranium in phosphate rock. US Geol Surv, Prof Pap, 300: 477-481

MCKELVEY, V E (1967) Phosphate deposits. US Geol Surv, Bull, 1252-D: D121

MCKELVEY, V E (1973) Abundance and distribution of phosphorus in the lithosphere. In: Environmental Phosphorus Handbook: 13-31

MCKELVEY, V E, WILLIAMS, J S, SHELDON, R P, CRESSMAN, E R, CHENEY, T M, SWANSON, R W (1956) Summary description of Phosphoria, Park City, and Shedhorn formations in the Western Phosphate Field. Amer Assoc Petrol Geol, Bull, 40: 2826-2863

MCKELVEY, V, WILLIAMS, J S, SHELDON, R P, CRESSMAN, E R, CHENEY, J M (1959) The Phosphoria, Park City and Shedhorn Formations in the Western Phosphate Field. US Geol Surv, Prof Pap, 313-A: 1-47

MICHARD, A (1976) Eléments de géologie marocaine. Notes & Mém Serv géol Maroc, 252: 408 pp

MILLER, R G (1990) A paleoceanographic approach to the Kimmeridge Clay Formation. In: HUC, A Y (ed) Deposition of organic facies, Amer Assoc Petrol Geol, Studies in Geology, 30: 13-26

MILTON, C, BENNISON, A P (1969) Phosphate nodules from the Wills Point Formation, Hopkins County, Texas. US Geol Surv, Prof Pap, 600-B: B11-B15

MINSTER, T, FLEXER, A (1991) Cadmium-bearing sphalerite in Senonian section, northern Negev - remarks on its formation and technological implications. In: 1th Getner Symposium on Geoscience, Phosphorites and Black Shales 1991, Abs, 35

MIRTOV, Y V, KRASIL'NIKOVA, N A, CHIRKIN, A N (1976) Ultramicrocrystalline structure of basic types of phosphorites. Lith, Min Res, 1976/1: 85-94

MONCIARDINI, C (1989) The Senonian (Cretaceous) phosphatic chalks of the Paris Basin, France. In: NOTHOLT, A J G, SHELDON, R P, DAVIDSON, D F: Phosphate deposits of the world, Vol 2: 407-410

MOTT, L V, DREVER, J I (1983) Origin of uraniferous phosphatic beds in the Wilkins Peak Member of Green the River Formation, Wyoming. Amer Assoc Petrol Geol, Bull, 67: 70-82

MULLINS, H T, RASCH, R F (1985) Sea floor phosphorites along the central California continental margin. Econ Geol, 80: 695-715

MUNOZ CABEZON, B (1989) The Bu-Craa phosphate deposit Western Sahara, Morocco. In: NOTHOLT, A J G, SHELDON, R P, DAVIDSON, D F: Phosphate deposits of the world, Vol, 2: 176-183

MURRAY, J, RENARD, A F (1891) Deep sea deposits (rep of the scient results of the exploring voyage of HMS "Challenger" 1873-1876)

MÜCKENHAUSEN, E (1977) Entstehung, Eigenschaften und Systematik der Böden der Bundesrepublik Deutschland. 2 ed, 299 pp

NATHAN, Y (1990) Humic substances in phosphorites: occurrence, characterization and significance. Geol Soc London, Spec Publ, 52: 49-58

NATHAN, Y, LUCAS, J (1976) Expériences sur la précipitation directe de l'apatite dans l'eau de mer: implucations dans la genèse des phosphorites. Chem Geol, 18: 181-186

NATHAN, Y, NIELSEN, H (1980) Sulfur isotopes in phosphorites. Soc Econ Paleont Miner, Spec Publ, 29: 73-78

NATHAN, Y, SHILONI, Y RODED, R, GAL, I, DEUTSCH, Y (1979) The geochemistry of the northern and central Negev phosphorites, southern Isreal. Geol Surv Isreal, Bull, 73: 1-41

NATHAN, Y, SOUDRY, D, AVIGOUR, A (1990) Geological significance of carbonate substitution in apatites: Israeli phosphorite as an example. Geol Soc London, Spec Publ, 52: 179-191

NATHAN, Y, SOUDRY, D, DORFMAN, E, LEVY, Y, SHITRIT, D (1991) The geochemistry of cadmium in the Negev phosphorites. In: 1th Getner Symposium on Geoscience, Phosphoreites and Black Shales 1991, Abs, 35

NICHOLS, K M, SILBERLING, N J (1990) Delle Phosphatic Member: An anomalous phosphatic interval in the Mississippian (Osagean-Meramecian) shelf sequence of central Utah. Geology, 18: 46-49

NOTHOLT, A J G, JARVIS, I (eds)(1990) Phosphorite research and development. Geol Soc London, Spec Publ, 52: 326 pp

NOTHOLT, A J G, SHELDON, R P, DAVIDSON, D F (eds)(1989) Phosphate deposits of the world, Vol II: Phosphate rock recources. 566 pp

NRIAGU, J O, DELL, C I (1974) Diagenetic formation of iron phosphates in recent lake sediments. Am Mineral, 59: 934-946

NRIAGU, J O, MOORE, P H (eds)(1984) Phosphate minerals. 442 pp (Springer)

OBERLINDACHER, H P, ROBERTS-TOBEY, E D (1986) Stratigraphy, environment of deposition, and age of a phosphatic unit and adjacent rocks in the Wells Formation, southeastern Idaho, with evidence for a revised Pennsylvanian-Permian stratigraphic boundary. Contributions to Geology, University of Wyoming, 24: 237-241

O'BRIEN, G W, HARRIS, J R, MILNES, A R, VEEH, H H (1981) Bacterial origin of East Australian continental margin phosphorites. Nature, 294: 442-444

O'BRIEN, G W, MILNES, A R, VEEH, H H, HEGGIE, D T, RIGGS, S R, CULLEN, D J, MARSHALL, J F, COOK, P J (1990) Sedimentation dynamics and redox iron-cycling: controlling factors for the apatite-glauconite association on the East Australian continental margin. Geol Soc London, Spec Publ, 52: 61-86

O'BRIEN, G W, VEHH, H H, CULLEN, D J, MILNES,A R (1986) Uranium-series isotopic studies of marine phosphorites and associated sediments from the East Australian continental margin. Earth Planet Sci Let, 80: 19-35

ODIN, G S, LETOLLE, R (1980) Glauconitization and phosphatization environments: A tentative comparison. Soc Econ Paleont Miner, Spec Publ, 29: 227-237

ODIN, G S, MATTER, A (1981) De glauconarium origine: Sedimentology, 28: 611-641

ÖPIK and others (1957) The Cambrian geology of Queensland. Austr Bur Min Res, Bull, 49: 284 pp

OPPENHEIMER, C H (1958) Evidence for fossil bacteria in phosphate rock. Inst Mar Sci Publ, 5: 157-159

OSCHMANN, W (1988) Kimmeridge Clay, sedimentation -a new cyclic model. Palaeogeogr, Palaoeclimatol, Palaeoecol, 65: 217-251

OSCHMANN, W (1990) Environmental cycles in the late Jurassic northwest European epeiric Basin: interaction with atmospheric and hydrospheric circulations. Sediment Geol, 69: 313-332

PACEY, N R (1985) The mineralogy geochemistry and origin of pelletal phosphates in the English Chalk. Chem Geol, 48: 61-86

PANCZER, G, NATHAN, Y, SHILONI, Y (1989) Trace element characterization of phosphate nodules in Israel. Sci Géol, Bull, 42: 173-184

PAPROTH, E, ZIMMERLE, W (1980) Stratigraphic position, petrography, and depositional environment of phosphorites from the Federal Republic of Germany. Meded Rijks Geol Dienst, 32-11: 81-95

PARRISH, J T (1982) Upwelling and petroleum source beds, with reference to Paleozoic. Amer Assoc Petrol Geol Bull, 66: 750-774

PARRISH, J T, CURTIS, R L (1982) Atmospheric circulation, upwelling, and organic-rich rocks in the Mesozoic and Cenozoic eras. Palaeogeogr, Palaeoclimatol, Palaeoecol, 40: 31-66

PASCAL, M, TRAORE, H (1989) Eocene Tilemsi phosphorite deposits, eastern Mali. In: NOTHOLT, A J G, SHELDON, R P, DAVIDSON, D F: Phosphate deposits of the world, Vol 2: 226-232

PASCAL, M, SUSTRAC, G, BARTHELEMY, F, DIENG, M, FAYE, B, FAYA, C, KANDE, S, SECK, M (1989) Phosphorite deposits of Senegal. In: NOTHOLT, A J G, SHELDON, R P, DAVIDSON, D F: Phosphate deposits of the world, Vol 2: 233-246

PASSEGA, R (1957) Texture as a characteristic of clastic deposition. Amer Assoc Petrol Geol, Bull, 41: 1952-1984

PEDLEY, H M, BENNETT, S M (1985) Phosphorites, hardgrounds and syndepositional solution subsidence: a palaeoenvironmental from the Miocene of the Maltese Islands. Sediment Geol, 45: 1-34

PEPAULO, D J, INGRAM, B (1985) High resolution stratigraphy with strontium isotopes. Science, 227: 938-941

PETERSON, J A (1980a) Depositional history and petroleum geology of the Permian Phosphoria, Park City, and Shedhorn For mations, Wyoming and southeastern Idaho. US Geol Surv, Open-File Report, 80-667: 42 pp, 13 pl

PETERSON, J A (1980b) Permian paleogeography and sedimentary provinces, west-central United States. In: FOUCH, T D, MAGATHAM, E R: Paleozoic paleogeography of west-central United States: 271-292

PETERSON, J A (1984) Permian stratigraphy, sedimentary facies and petroleum geology, Wyoming and adjacent areas. Wyoming Geological Association, Guidebook of the 35th Annual Field Conference: 25-64

PEVEAR, D R (1967) The estuarin formation of United States Atlantic coastal plain phosphorite. Econ Geol, 61: 251-256

PICKARD, G L, EMERY, W J (1990) Descriptive physical oceanography, 5th edition. 320 pp

PIETZNER, H, RICHTER, G (1986) Rezente Phosphorit-Bildung im Auftriebsgebiet vor der NW-afrikanischen Küste. Senckenbergiana mart, 17: 333-377

PIPER, D Z (1991) Geochemistry of a Tertiary sedimentary phosphate deposit: Baja California Sur, Mexico. Chem Geol 92: 283-316

PIPER, D Z, BAEDECKER, P A, CROCK, J G, BURNETT, W C, LOEBNER, B J (1988) Rare earth elements in the phosphatic -enriched sediments of the Peru shelf. Marine Geology, 80: 269-286

PIPER, D Z, CODISPOTI, L A (1975) Marine phosphorite deposits and the nitrogene cycle. Science, 188: 15-18

PIPER, D Z, ISAACS, C M (1995) Geochemistry of minor elements in the Monterey Formation, California: seawater chemistry of deposition. US Geol Surv, Prof Pap, 1566: 41pp

PIPER, D Z, KOLODNY, Y (1987) The stable isotopic composition of a phosphate deposit: delta13C, delta14S and delta18 O. Deep Sea Res 34: 897-911

PIPER, D Z, LOEBNER, B, AHARON, P (1990) Physical and chemical properties of the phosphate deposit on Nauru, western equatorial Pacific Ocean. In: BURNETT, W C, RIGGS, S R: Phosphate deposits of the world, Vol 3, Neogene to modern phosphorites: 177-194

PIPER, D Z, MEDRANO, M D (1995) Geochemistry of the Phosphoria Formation at Monpellier Canyon, Idaho: Environment of deposition. US Geol Surv, Bull, 2023-B: B1-B28

POMONI-PAPAIOANNOU, F, SOLAKIUS, N (1991) Phosphatic hardgrounds and stromatolites from the limestone/shale boundary section, at Prossilion (Maastrichtian-Paleocene) in the Parnassus-Ghiona Zone, Central Greece. Palaeogeogr, Palaeoclimatol, Palaeoecol, 86: 243-254

POMONI-PAPAIOANNOU, F (1994) Palaeoenvironmental reconstruction of a condensed hardground-type depositional sequence at the Cretaceous-Tertiary contact in the Parnassus-Ghiona zone, central Greece. Sediment Geol, 93: 7-24

POND, S, PICKARD, G L (1991) Introductory dynamical oceanography, 2nd edition. 329 pp

POPENOE, P (1990) Paleoceanography and paleogeography of the Miocene of the southeastern United States. In: BURNETT, W C, RIGGS, S R (eds) Phosphate deposits of the world, Vol 3, Neogene to Modern phosphorites: 352-380

POPPE, L J, MANHEIM, F T, POENOE, P (1992) Late Cretaceous to Miocene phosphatic sediments in the Georges Basin, US North Atlantic outer continental shelf. Mar Geol, 107: 227-238

PORTER, K G, ROBBINS, E I (1981) Zooplankton fecal pellets link fossil fuel and phosphate deposits. Science, 212: 931-933

POTTS, P J, TINDLE, A G, WEBB, P C (1992) Geochemical reference material composition. CRC Press, Boca Raton, Fl

POWELL, T G, COOK, P J, McKIRDY, D M (1975) Organic Geochemistry of phosphorites: Relevance to petroleum genesis. Amer Assoc Petrol Geol, Bull, 59: 618-632

PRICE, N B, CALVERT, S E (1978) The geochemistry of phosphorites from the Namibian shelf. Chem Geol, 23: 151-170

PRÉVôT, L (1982) Proposal for normalized easy description of the so-called palaeophosphorites. IGCP Proj 156, Newsletters, 10: 24-31

PRÉVôT, L (1990) Geochemistry, petrography, genesis of Cretaceous-Eocene phosphorites. Soc Géol Fr, Mem, 158: 232 pp

PRÉVôT, L, EL FALEH, E, LUCAS, J (1989) Details on synthetic apatites formed through bacterial mediation Mineralogy and geochemistry of the products. Sci Géol, Bull, 42: 237-254

PRÉVôT, L, LUCAS, J (1979) Comportement de quelques éléments traces dans les phosphorites. Sci Géol, Bull, 32: 91-105

PRÉVôT, L, LUCAS, J (1980) Behavior of some trace elements in phosphatic sedimentary formations. Soc Econ Paleont Miner, Spec Publ, 29: 31-39

PRÉVôT, L, LUCAS, J (1985) Utilization of geochemistry to explain the setting of the phosphatic series of the Ganntour Basin (Morocco). Sci Géol, Mém, 77: 45-51

PRÉVôT, L, LUCAS, J (1986) Microstructure of apatite replacing carbonate in synthesized and natural samples. Jour Sed Petrol, 56: 153-159

PRÉVôT, L, LUCAS, J, DOUBINGER, J (1979) Une correspondance entre le contenu palynologique et la composition minéralogique et chimique d'une série phosphatée sédimentaire (Ganntour, Maroc). Sci Géol, Bull, 32: 69-90

PUURA, I, HOLMER, L E (1993) Lingulate brachiopods from the Cambrian-Ordovician boundary beds. Geol För Stockholm Förh, 115: 215-237

RAO, V P (1986) Phosphorites from the Error seamount, northern Arabian Sea. Mar Geol, 71: 177-186 RAO, V P, LAMBOY, M, NATARJAN, R (1992) Possible microbial origin of phosphorites on Error Seamount, northwestern Arabian Sea. Mar Geol, 106: 149-164

RAO, V P, NAIR, R R (1988) Microbial origin of the phosphorites of the western continental shelf of India. Mar Geol, 84: 105-110

RAUDSEP, R (1993) Development of the Estonian phosphate deposits. Intern Training Workshop, October 18-29, 1993, Intern Fertilizer Development Center (IFDC), Muscle Shoals, Alabama, 13 pp

RAUSCHER, R (1985) Les dinokystes, des outils stratigraphiques pour les Séries Phosphatées Application aux phosphorites du Maroc. Sci Géol, Mém, 77: 69-74

READ, J F (1982) Carbonate platforms of passive (extensional) continental margins: types, characteristics and evolution. Tectonophysics, 81: 195-212

READ, J F (1985) Carbonate platform facies models. Amer Assoc Petrol Geol, Bull, 69: 1-21

READ, J F (1995) Overview of carbonate platform sequences, cycle stratigraphy and reservoirs in greenhouse and icehouse worlds. In: READ, J F, KERANS, C, WEBER, L J (eds) Milancovitch sea-level changes, cycles, and reservoirs on carbonate platforms in greenhouse and icehouse worlds, Soc Econ Paleont Miner, Short Course, 35: 102pp

REHFELD, U, JANSSEN A W (1995) Development of phosphatized hardgrounds in the Miocene Globigerina Limestone of the Maltese Archipelago, including a description of Gamopleura melitensis sp nov (Gastropoda, Euthecosomata). Facies, 33: 91-106

REIF, W-E (1982) Muschelkalk/Keuper bone-beds (Middle Triassic, West-Germany) - Storm condensation in a regressive cycle. In: EINSELE, G, SEILACHER, A: Cyclic and event stratification: 299-325

REIMERS, C E, SUESS, E (1983) Spatial and temporal pattern of organic matter accumulation on the Peru continental margin. In: THIEDE, J, SUESS, E (eds), Coastal upwelling: Its sedimentary record, part B: Sedimentary records of ancient coastal upwelling, 311-346

REINECK, H-E, SINGH, I B (1980) Depositional sedimentary environments. 549 pp

REISS, Z (1962) Stratigraphy of phosphate deposits in Israel. Israel Geol Surv, Bull, 34: 1-23

RETLLACK, G J, VEEVERS, J J, MORANTE, R (1996) Global coal gap between Permian-Triassic extinction and Middle Triassic recovery of peat-forming plants. Geol Soc America, Bull 108: 195-207

RHOADS, D C, MORSE, J M (1971) Evolutionary and ecologic significance of oxygen-deficient marine basins. Lethaia, 4: 413-428

RIGGS, S R (1967) Phosphorite stratigraphy, sedimentation, and petrology of the Noralyn Mine, central Florida phosphate district. PhD thesis, University of Montana, XIII+197pp+36 pl, Missoula

RIGGS, S R (1979a) Petrology of the Tertiary phosphorite system of Florida. Econ Geol, 74: 195-220

RIGGS, S R (1979b) Phosphorite sedimentation in Florida - a model phosphogenetic system. Econ Geol, 74: 285-314

RIGGS, S R (1980) Intraclast and pellet phosphorite sedimentation in the Miocene of Florida. J Geol Soc London, 137: 741-748

RIGGS, S R (1984a) Paleoceanographic model of Neogene phosphorite deposition, US Atlantic continental margin. Science, 223: 123-131

RIGGS, S R (1984b) Patterns of Miocene phosphate sedimentation on the southeastern United States continental margin. Proc 27th Int Geol Cong, V 15: 201-222, Moscov

RIGGS, S R (1986) Proterozoic and Cambrian phosphorites - specialist studies: phosphogenesis and its relationship to exploration for Proterozoic and Cambrian phosphorites. In: COOK, J P, SHERGOLD, J H (eds) Phosphate deposits of the World, Vol 1: Proterozoic and Cambrian phosphories, 352-368

RIGGS, S R (1989) Phosphate deposits of the North Carolina Coastal plain, continental shelf and adjacent Blake Plateau, USA. In: NOTHOLT, A J G, SHELDON, R P, DAVIDSON, D F: Phosphate deposits of the world, Vol 2: 42-52

RIGGS, S R, MALLETTE, P M (1990) Patterns of phosphate deposition and lithofacies relationships within the Miocene Pungo River Formation, North Carolina continental margin. In: BURNETT, W C, RIGGS, S R (eds) Phosphate deposits of the world, Vol 3, Neogene to Modern phosphorites: 424-443

RIGGS, S R, MANHEIM, F T (1988) Mineral resources of the US Atlantic continental margin. In: SHERIDAN, R E, GROW, J A (eds) The geology of North America, Vol I-2, The Atlantic continental margin, US, 501-520

RIGGS, S R, SHELDON, R P (1990) Paleooceanographic and paleoclimatic controls of the temporal and geographic distribution of Upper Cenozoic continental margin phosphorites sites of the world, Vol 3, Neogene to Modern phosphorites: 207-222

RIGGS, S R, SNYDER, S W, O'BRIEN, G W, COOK, P J, HEGGIE D T (1989) Sedimentology of the Neogene to Modern glauconite-goethite-phosphate system: East Australian continental margin between 29 and 32 south latitude. Sci Géol, Bull, 42: 185-204

RIGGS, S R, SNYDER, St W, SNYDER, ScW, HINE, A C (1990) Stratigraphic framework for cyclical deposition of Miocene sediments in the Carolina Phosphogenic Province. In: BURNETT, W C, RIGGS, S R (eds) Phosphate deposits of the world, Vol 3, Neogene to Modern phosphorites: 381-395

ROE, K K, BURNETT, W C (1985) Uranium geochemistry and dating of pacific island apatite. Geochim Cosmochim Acta, 49: 1581-1592

ROGERS, J K, CRASE, N J (1980) The Phosphate Hill rock phosphate deposit, northwest Queensland, Australia - An outline of geological development. In: SHELDON, R P, BURNETT, W C: Fertilizer mineral potential in Asia and the Pacific. Proceedings of the Raw Materials Resources Workshop August 1979 Honolulu: 307-328

ROONEY, T P, KERR, P F (1967) Mineralogic nature and origin of phosphorite, Beaufort County, North Carolina. Geol Soc America, Bull, 78: 731-748

ROSS, C A, ROSS, J R P (1987) Biostratigraphic zonation of Late Paleozoic depositional sequences. Cushman Foundation, for Foraminiferal Research, Spec Publ, 24: 151-161

ROUGERIE, F, FAGERSTROM, J A (1994) Cretaceous history of Pacific Basin guyot reefs: a reappraisal based on geothermal endo-upwelling. Palaeogeogr, Palaeoclimatol, Palaeoecol, 112: 239-260

ROUGERIE, F, WAUTHY, B (1989) Une nouvelle hypothèse sur la genèse des phosphates d'atolls: les rôle du prossesus d'endo-upwelling. CR Acad Sci, II, 308: 1043-1047

ROUGERIE, F, JEHL, C, TRICHET, J (1997): Phosphorus pathways in atolls: interstitial nutrient pool, cyanobacterial accumulations and carbonate-fluor-apatite (CFA) precipitation. Mar Geol, 139: 201-217

RUNNELS, R T, SCHNEIDER, J A, VAN NORTWICK, H S (1953) Composition of some uranium-bearing phosphate nodules from Kansas shales. Geol Surv Kansas, Bull 102: 93-104

RUSSELL, R T, TRUEMAN, N A (1971) The geology of the Duchess phosphate deposits, northwest Queensland, Australia. Econ Geol, 66: 1186-1214

RUTTENBERG, K C, BERNER, R A (1993) Authigenic apatite formation and burial in sediments from non-upwelling, continental margin environments. Geochim Cosmochim Acta, 57: 991-1007

SALVAN, H M (1954) Les Invertébrés fossiles des phosphates marocains Tome II: Etude paléontologique. Notes & Mém Serv géol Maroc, 93: 258 pp

SALVAN, H M (1986) La genèse des phosphates de chaux sédimentaires et la formation de depots phosphates. In: Geologie des gites mineraux Marocains, Vol 2 (2nd ed), Notes & Memoires Serv Geol Maroc, 276: 312-344

SANDBERG, C A, GUTSCHICK, R C (1980) Sedimentation and biostratigraphy of Osagean and Meramecian starved basin and foreslope, western United States. In: FOUCH, T D, MAGATHAM, E R: 129-147

SANDBERG, C A, GUTSCHICK, R C (1989) Deep-water phosphorite in the early Carboniferous Deseret starved basin, Utah, USA. In: NOTHOLT, A J G, SHELDON, R P, DAVIDSON, D F: Phosphate deposits of the world, Vol II: 18-23

SANDSTORM, M W (1982) Diagenesis of organic phosphorus in marine sediments: Implications for global carbon and phosphorus cycles. In: GALBALLY, I W (ed) Workshop on interactions of the biogeochemical cycles of carbon, nitrogen and phosphorus. Australian Academy of Sciences, 133-142

SANDSTORM, M W (1986) Proterozoic and Cambrian phosphorites -specialist studies: Geochemistry of organic matter in Middle Cambrian phosphorites from the Georgina Basin, northeastern Australia. In: COOK, P J, SHERGOLD, J H: Phosphate deposits of the world, Vol 1: Proterozoic and Cambrian phosphorites: 268-279

SANDSTORM, M W (1990) Organic matter in modern marine phosphatic sediments from the Peruvian continental margin. In: BURNETT, W C, RIGGS, S R (eds) Phosphate Deposits of the World, Vol 3: Neogene to Modern Phosphorites, 33-45

SARG, J F (1988) Carbonate sequence stratigraphy. Soc Econ Paleont Miner Spec Publ, 42: 155-181

SARNTHEIN, M (1970) Sedimentologische Merkmale für die Untergrenze der Wellenwirkung im Persischen Golf. Geol Rdsch, 59: 649-666

SASSI, S (1980) Contexte paléogéographique des dépôts phosphatés de l'Eocène de Tunisie. BRGM, Doc, 14: 167-184

SAVENKO, V S (1978) Experimental studies on the conditions of chemical precipitation of calcium phosphates from seawater . Dokl Akad Nauk SSSR, 243: 1302-1305

SCHAFFER, G (1986) Phosphate pumps and shuttles in the Black Sea. Nature, 321: 515-517

SCHEERE, J, VAN TASSEL, R (1968) Phosphorites du passage Viséen-Namurien à Blaton, province de Hainaut et à Warnant, province de Namur. Bull Soc belge Géol, 77: 245-268

SCHIDLOWSKI, M (1985) Early life and mineral resources. Nature and Resources, 21: 11-17

SCHIDLOWSKI, M, WIGGERING, H (1988) Die Erdatmosphäre im Präkambrium. Geowissenschaften, 6: 212-217

SCHIEBER, J (1994) Evidence for high-energy events and shallow water deposition in the Chattanooga Shale, Devonian, central Tennessee, USA. Sediment Geol, 93: 193-208

SCHLAGER, W (1981) The paradox of drowned reefs and carbonate platforms. Geol Soc America, Bull, 92: 197-211

SCHLANGER, S O, JENKYNS, H C (1976) Cretaceous oceanic anoxic events: causes and consequences. Geol Mijnb, 55: 179-184

SCHMITT, M, SOUTHGATE, P N (1982) A phosphatic stromatolite (Ilicta cf composita Sidrov) from the Middle Cambrian, northern Australia. Archerigna, 6: 175-183

SCHRÖTER, T (1986) Die lithofazielle Entwicklung der oberkretazischen Phosphatgesteine Ägyptens - ein Beitrag zur Genese der Tethys-Phosphorite der Ostsahara. Berliner geowiss Abh, (A), 67: 105 pp

SCHRÖTER, T (1989) The Abu Tartur phosphorite deposit, Western Desert, Egypt. In: NOTHOLT, A J G, SHELDON, R P, DAVIDSON, D F: Phosphate deposits of the world, Vol 2: 194-199

SCHUFFERT, J D (1988) Multi-layered authigenic apatite formation off southern Baja California. EOS, 69: 1234

SCHWAB, R G, De OLIVEIRA, N P (1981) Zur Rolle des Phosphors bei lateritischer Verwitterung. Zbl Geol Paläont, Teil I, H3/4: 419-436

SCHWENNICKE, T (1992) Phosphoritführende Tief- und Flachwassersedimente aus dem Oberoligozän von Niederkalifornien, Mexiko - die San Juan-Einheit (El Cien-Formation). Diss, Univ Hannover, 163 pp

SCHWENNICKE, T (1994) Deep and shallow water phosphorite-bearing strata of the Upper Oligocene of Baja California, Mexico (San Juan Member, El Cien Formation). Zbl Geol Paläont Teil I, 1993 (1/2) 567-580

SCHWENNICKE, T (1995) Phosphatic grains of the upper Oligocene San Juan Member (El Cien Formation) of Baja California Sur, Mexico. Bol Depto Geol Geol Uni-Son, 12: 41-64

SCOTT, T M (1988) The lithostratigraphy of the Hawthorn Group (Miocene) of Florida. Fla Geol Survey, Bull, 59: 148 pp

SCOTT, T M (1990) The lithostratigraphy of the Hawthorn Group of peninsular Florida. In: BURNETT, W C, RIGGS, S R: Phosphate deposits of the world, Vol 3: Neogene to Modern phosphorites: 325-336

SEIBERTZ, E (1992) Modelle der Glaukonitgenese und ihre Aussagekraft für Radiometrie, Stratigraphie und Paläogeographie der Kreide Zentraleuropas. Habilitationsschrift FU Berlin, 172 pp, 12 pl

SHATSKII, N S (1955) Phosphorite bearing formations and classification of phosphorite deposits. USSR Academy of ep Science, Report on Conference on Sedimentary Rocks, 2, Moscow

SHELDON, R P (1957) Physical stratigraphy of the Phosphoria Formation in northwestern Wyoming. US Geol Surv, Bull, 1042-E: 105-185

SHELDON, R P (1963) Physical stratigraphy and mineral resources of the Permian rocks in western Wyoming. US Geol Surv, Prof Pap, 313-B: 49-273

SHELDON, R P (1964a) Exploration for phosphorites in Turkey - a case history. Econ Geol, 59: 1159-1175

SHELDON, R P (1964b) Paleolatitudinal and paleogeographic distribution of phosphorite. US Geol Surv, Prof Pap, 501-C: C106-C113

SHELDON, R P (1972) Phosphate deposition seaward of barrier islands at the edge of Phosphoria sea in northwestern Wyoming. Amer Assoc Petrol Geol, Bull, 56: 653

SHELDON, R P (1980) Episodicity of phosphate deposition and deep ocean circulation - a hypothesis. Soc Econ Paleont Miner Spec Pub, 29: 239-247

SHELDON, R P (1981) Ancient marine phosphorites. Ann Rev Earth Planet Sci, 9: 251-284

SHELDON, R P (1984a) Ice-ring origin of the earth's atmosphere and hydrosphere and late Proterozoic -Cambrian phosphogene sis. Geol Surv India, Spec Publ, 17: 17-21

SHELDON, R P (1984b) Polar control on sedimentation of Permian phosphorites of the Rocky Mountains, USA. Proc 27th Int Geol Cong, V 15: 223-243, Moscow

SHELDON, R P (1985) Equatorial upwelling origin of the early Tertiary equatorial-belt family of phosphorites Were they caused by the shadow of a tektite ring system orbiting the earth? Sci Géol, Mém, 77: 99-104

SHELDON, R P (1987a) Association of phosphatic and siliceous marine sedimentary deposits. In: HEIN, J R (ed) Siliceous sedimentary rock-hosted ores and petroleum, 58-80

SHELDON, R P (1987b) Association of phosphorites, organic-rich shales, chert and carbonate rocks. Carbonates and Evaporates, 2: 7-14

SHERGOLD J H, BRASIER M D (1986) Biochronology of Proterozoic and Cambrian phosphorites. In: COOK, J P, SHERGOLD J H (eds) Phosphate deposits of the world, Vol 1: Pro terozoic and Cambrian phosphorites, 295-326

SHERGOLD, J H, DRUCE, E C (1980) Upper Proterozoic and lower Paleozoic rocks of the Georgina basin. In: HENDERSON, R A, STEPHENSON, P J (eds):The geology and geo physics of northeastern Australia, Geol Soc Australia, Queensland Division: 149-174

SHERGOLD, J H, SOUTHGATE, P N (Eds)(1986) Middle Cambrian phosphatic and calcareous lithofacies along the eastern margin of the Georgina Basin, western Queensland. Australasian Sedimentologists Group, Field Guide Series 2, Geolo gical Society of Australia: 89 pp

SIEGMUND, H (1995) Fazies und Genese unterkambrischer Phosphorite und mariner Sedimente der Yangtze Plattform, Südchina Berliner geowiss Abh, (A), 173: 114pp

SILBERLING, N J, NICHOLS, K M (1991) Petrology and regional significance of the Mississippian Delle phosphatic member, Lakeside Mountains, northwestern Utah. In: COOPER, J D, STEVENS, C H: Paleozoic paleogeography of the western United States, II: 425-438

SILBERLING, N J, NICHOLS, K M (1995) A Mid-Mississippian episode of flexural loading in the Antler foreland, Eastern Great Basin. Amer Assoc Petrol Geol, Rocky Mountain section Meeting, Abs A11-A12

SILBERLING, N J, NICHOLS, K M, MACKE, D L, TRAPPE, J (1995) Upper Devonian-Mississippian stratigraphic sequences in the distal Antler foreland of western Utah and adjoining Nevada. US Geol Surv, Bull, 1988-H: H1-H33

SIMMONS, S P, SCHOLLE, P A (1990) Late Paleozoic uplift and sedimentation, northeast Bighorn Basin, Wyoming. Wyoming Geological Assiciation, Guidebook of the 41st Annual Field Conference, 39-55

SISODIA, M S, CHAUHAN, D S (1990) The influence of magnesium ions during the formation of stromatolitic phosphorites of Udaipur, Rajasthan, India. Geol Soc London, Spec Publ, 52: 313-320

SLANSKY, M (1980a) Géologie des phosphates sédimentaires. Mém BRGM, 114: 92pp

SLANSKY, M (1980b) Localisation des gisements de phosphates dans les bassins sédimentaires. BRGM, Doc, 24: 13-22

SLANSKY, M (1986) Geology of sedimentary phosphates. 210 pp

SLANSKY, M (1987) Apports d'études récentes de matière organique à la compréhension des mécanismes de formation des dépots phosphatés. Mem Soc géol France, N S, 151: 115-125

SLANSKY, M (1989a) The Eocene phosphate deposits of Togo. In: NOTHOLT, A J G, SHELDON, R P, DAVIDSON, D F: Phosphate deposits of the world, Vol 2: 258-261

SLANSKY, M (1989b) The Lower Carboniferous phosphate deposits of the Pyrenees and Montagne Noire (southern France). In: NOTHOLT, A J G, SHELDON, R P, DAVIDSON, D F: Phosphate deposits of the world, Vol 2: 403-406

SLOSS, L L (1963) Sequences in the cratonic interior of North America. Geol Soc America, Bull, 74: 93-114

SMITH, K G (1972) Stratigraphy of the Georgina Basin. Austr Bur Min Res, Bull, 111: 156 pp + 4 maps

SMITH, R L (1992) Coastal upwelling in the modern ocean. In: SUMMERHAYES, C P PRELL, W L, EMEIS, K C (eds) Upwelling systems: evolution since the Early Miocene. Geol Soc London, Spec Publ 64: 9-28

SMOOT, J P (1983) Depositional subenvironments in an arid closed basin; the Wilkins Peak Member of the Green River Formation (Eocene), Wyoming, USA. Sedimentology, 30: 801-828

SNYDER, S W, HALE, W R, RIGGS, S R, SPRUILL, R K, WATERS, V J (1984) Occurrence of clinoptilolite as moldic fillings in foraminifera tests in continental margin sediments. Geol Soc America, Abstracts with programs, 16: 662

SOUDRY, D (1987) Ultra-fine structures and genesis of the Campanian Negev high-grade phosphorites (southern Israel). Sedimentology, 34: 641-660

SOUDRY, D (1992) Primary bedded phosphorites in the Campanian Mishash Formation, Negev, southern Israel. Sediment Geol, 80: 77-88

SOUDRY, D (1993) Internal structure and growth of an intraformational concretional phosphorite from an early Tertiary starved-sediment sequence: Arava Valley, southern Israel. Kaupia, 2: 67-76

SOUDRY, D, CHAMPETIER, Y (1983) Microbial processes in the Negev phosphorites (southern Israel). Sedimentology, 30: 411-423

SOUDRY, D, LEWY, Z (1988) Microbially influenced formation of phosphate nodules and megafossil moulds (Negev, southern Israel). Palaeogeogr, Paleoclimatol, Palaeoecol, 64: 15-34

SOUDRY, D, LEWY, Z (1990) Omission-surface incipient phosphate crusts on early diagenetic calcareous concretions and their possible origin, Upper Campanian, southern Israel. Sediment Geol, 66: 151-163

SOUDRY, D, SOUTHGATE, P N (1989) Ultrastructure of a Middle Cambrian primary nonpelletal phosphorite and its early transformation into phosphate vadoids: Georgina Basin, Australia. J Sed Petrol, 59: 53-64

SOUTHAM, J R, PETERSON, W H, BRASS, G W (1982) Dynamics of anoxia. Palaeogeogr, Paleoclimatol, Palaeoecol, 40: 183-198

SOUTHGATE, P N (1980) Cambrian stromatolitic phosphorites from the Geogrina Basin, Australia. Nature, 285: 395-397

SOUTHGATE, P N (1983) Middle Cambrian phosphate and calcareous depositional environments, the Undilla region of the Georgina Basin. PhD Thesis, Australian National University, 363 pp, Canberra

SOUTHGATE, P N (1986a) Cambrian phoscrete profiles, coated grains, and microbial processes in phosphognesis: Georgina basin, Australia. J sedim Petrol, 56: 429-441

SOUTHGATE, P N (1986b) Proterozoic and Cambrian phosphorites - specialist studies: Middle Cambrian phosphatic hardgrounds, phoscrete profiles and stromatolites and their implication for phosphogenesis. In: COOK, P J, SHERGOLD, J H: Phosphate deposits of the world, Vol 1: Proterozoic and Cambrian phosphorites: 327-351

SOUTHGATE, P N (1986c) The Gowers Formation and Bronco Stromatolite Bed, two new stratigraphic units in the Undilla portion of the Georgina Basin. Queensland Government Mining Journal, 10/1986: 407-411

SOUTHGATE, P N (1988) A model for the development of phosphatic and calcareous lithofacies in the Middle Cambrian Thorntonia Limestone, northeast Georgina Basin, Australia. Austral Jour Earth Sci, 35: 111-130

SOUTHGATE, P N, SHERGOLD, J H (1991) Application of sequence stratigraphic concepts to Middle Cambrian phosphogenesis, Georgina basin, Australia. BMR Journal of Australian Geology, Geophysics, 12: 119-144

SPEARING, D R (1976) Upper Cretaceous Shannon Sandstone: an offshore shallow-marine sand body. Wyoming Geol Assoc, 28th Annual Guidebook: 65-72

STARINSKY, A, KATZ, A, KOLODNY, Y (1982) The incorporation of uranium into diagenetic phosphorite. Geochim Cosmochim Acta, 46: 1365-1374

STETS, J, WURSTER, P (1981) Zur Strukturgeschichte des Hohen Atlas in Marokko. Geol Rdsch, 70: 801-841

STILLE, P, CHAUDHURI, S, KHARAKA, Y K, CLAUER, N (1992) Neodymium, strontium, oxygen and hydrogen isotope compositions of waters in present and past oceans: A review. In: CLAUER, N, CHAUDHURI, S: Isotopic signatures and sedimen tary records. Leture Notes in Earth Sciences, 43: 389-410

STILLE, P, RIGGS, S R, CLAUER, N, AMES, D, CROWSON, R, SNYDER, S (1994) Sr and ND isotopic analysis of phosphorite sedimentation through one Miocene high-frequency depositional cycle on the North Carolina continental shelf. Mar Geol, 117: 253-273

STODDART, D R, SCOFFIN, T P (1983) Phosphate rock on coral reef islands. In: GOUDIE, A S, PYE, K (eds) Chemical sediments and geomorphology, 369-400

STRAKHOV, N M (1960) Fundamentals of the theory of lithogenesis. Acad Nauk SSR, Vol 1: 212pp

STRUCKMEIER, W (1982) Zur Geochemie, Mikroskopie und paläogeographischen Deutung der Liegenden Alaunschiefer (Dinantium II) und eingelagerter Phosphorite im Belecker und Warsteiner Sattel (nördliches Rheinisches Schiefergebirge). Fortschr Geol Rheinld u Westf, 30: 321-339

STURESSON, U (1995) Llanvirnian (Ord.) iron ooids in Baltoscandia: element mobility, REE distribution patterns, and origin of the REE. Chem Geol, 125: 45-60

SUHR, P (1991) Allochthone phosphoritisierte Ichnofossilien aus den Böhlner Schichten der Weißelstersenke. Mauritania, 13: 225-232

SUMMERHAYES, CP, NUTTER, A H, TOOMS, J S (1972) The distribution and origin of phosphate in sediments off northwest Africa. Sediment Geol, 8: 3-28

SWANSON, R W (1970) Mineral Resources in Permian Rocks of Southwest Montana. US Geol Surv, Prof Pap, 313-E: 661-777

SWANSON, R W (1973) Geology and Phosphate Deposits of the Permian Rocks in Central Montana. U S Geol Surv, Prof Pap, 313-F: 779-833

SWETT, K, CROWDER, B K (1982) Primary phosphatic oolites from the Lower Cambrian of Spitsbergen. J Sediment Petrol, 52: 587-593

SWIFT, D J P (1974) Continental shelf sedimentation. In: BURK, C A, DRAKE, C L (eds) The geology of continental marigns, 117-136

SWIRYDCZUK, K, WILKINSON, B H, SMITH, G R (1981) Synsedimentary lacustrine phosphorites from the Pliocene Glenns Ferry Formation of southwestern Idaho. Jour Sed Petrology, 51: 1205-1214

TEWARI, V C (1981) The systematic study of Precambrian stromatolites from Gangolihat Dolomite, Kumaon Himalaya. Himalayan Geol, 11: 119-146

TEWARI, V C (1984) Discovery of Lower Cambrian stromatolites from Mussourie Tal phosphorite. India Curr Sci, 53: 319-321

TEWARI, V C (1989) Upper Proterozoic - Lower Cambrian stromatolites and Indian stratigraphy. Himalayan Geol, 13: 143-180

TEWARI, V C (1993) Precambrian and Lower Cambrian stromatolites of the Lesser Himalaya. Geophytol, 23: 19-39

THORNBURG, J M (1990a) Discovery of relict evaporites in Permian (Guadalupian) cherts, sandstones, and dolostones of the uppermost Phosphoria rock complex, southwest Montana (abs). Geol Soc America, Abstracts with Programs, 22/7: A314-A315

THORNBURG, J M (1990b) Petrography and sedimentology of a phosphatic shelf deposit: The Permian Shedhorn Sandstone and associated rocks in southwest Montana and northwest Yellowstone National Park. PhD Thesis University of Colorado, 620 pp

TIMMERMANN, A (1974) Phosphoritische Mineralisationen im Vorflysch des Thüringisch-Fränkischen Schiefergebirge. Z angew Geol, 20: 401-410

TIMMERMANN, A (1976) Vergleichende Betrachtungen der Phosphoritmineralisation des ostthüringischen Dinants mit Mineralisation und Lagerstätten gleichen Typs. Z geol Wiss, 4: 1023-1031

TOOMS, J S, SUMMERHAYES, C P, CRONAN, D D S (1969) Geochemistry of marine phosphate and manganese deposits. Oceanogr Marine Biol, Ann Rev, 1: 49-100

TOURTELOT, H A, COBBAN, W A (1968) Stratigraphic significance and petrology of phosphate nodules at the base of Niobrara Formation, east flank of Black Hills, South Dakota. US Geol Surv, Prof Pap, 594-L: L1-L22

TRAPPE, J (1989) Das marine Alttertiär im Westlichen Hohen Atlas - Mikrofazies, Paläogeographie, Phosphoritgenese. Thesis, 219 pp, Bonn

TRAPPE, J (1991) Stratigraphy, facies distribution and Paleogeography of the marine Paleogene from the Western High Atlas, Morocco. N Jb Geol Paläont Abh, 180: 279-321

TRAPPE, J (1992a) Facies zonation and spatial evolution of carbonate ramp: Marginal Moroccan phosphate sea during the Paleogene. Geol Rdsch, 81: 105-126

TRAPPE, J (1992b) Mechanisms of phosphate formation on the northwestern margin of Pangea. IAS, 13th Regional Meeting on Sedimentology, Abstracts (Supplement) 2 pp

TRAPPE, J (1992c) Paläogeographie und Akkumulationsprozesse klastischer Phosphatlagerstätten am Beispiel der Phosphoria Formation (PRC, Perm, USA) - Die Erweiterung eines Lagerstättenkonzepts. unpubl DFG-Report, VIII+126 pp, Bonn

TRAPPE, J (1992e) Sea level controlled formation of clastic phosphorites during the Late Paleozoic along the western margin of North America. (Abs) In: Sea Level Changes - Processes and Products, Profil, 1: 48-49

TRAPPE, J (1993) Phosphogenesis and depositional environment of phosphorites from the Phosphoria Rock Complex - A new perspective. Carboniferous to Jurassic Pangea, Ann Meeting, Canad Soc Petrol Geol, Program with Abstracts: 312

TRAPPE, J (1994) Pangean phosphorites - ordinary phosphorite formation in an extraordinary world? In: BEAUCHAMP, B, EMBRY, A, GLASS, D: Carboniferous to Jurassic Pangea: Canad Soc Geol, Mem, 17: 469-478

TRAPPE, J, ELLENBERG, J (1994) The phosphorites from the Ordovician iron ore deposits in Thuringia, Germany: Transgressive amd regressive systems-shift facies. Zbl Geol Paläont, Teil 1, 1991,11/12: 1387-1402

TRAPPE, J, GREWE, C, LEMPIO, A, ZIMMERLE, W (1995) Lower Carboniferous concretionary phosphorites in Germany - a model for phosphogenic pathways? 1st Soc Econ Paleont Miner Congress on Sedimentary Geology "Linked Earth Systems", St. Pete Beach, Florida, August 13-16, 1995

TRICHET, J, RACHADI, M, BELAYOUNI, H (1990) Organic geochemistry of phosphorites: relative behaviors of phosphorus and nitrogen during the formation of humic compounds in phosphate bearing sequences. In: BURNETT, W C, RIGGS, S R (eds) Phosphate Deposits of the World, Vol 3: Neogene to modern phosphorites, 87-96

TRUEMAN, N A (1965) The phosphate, volcanic and carbonate rocks of Christmas Island (Indian Ocean). J geol Soc Aust, 12: 261-283

TUCKER, D H, WYATT, B W, DRUCE, E C, MATHUR, S P, HARRISON, P L (1979) The upper crustal geology of the Georgina Basin region. BMR Jour Austr Geol Geophy, 4: 209-226

TUCKER, M E (1992) The Precambrian-Cambrian boundary: seawaterchemistry, ocean circulation and nutrient supply in metazoanevolution, extinction and biomineralisation. J Geol SocLondon, 149: 655-668

TUREKIAN, K K, WEDEPOHL, K H (1961) Distribution of some major elements of the Earth's crust. Geol Soc America, Bull, 72: 172-195

TYSON, R V, PEARSON, T H (1991):Modern and ancient continental shelf anoxia: an overview. In: TYSON, R V, PEARSON, T H: Modern and ancient contineta1 shelf anoxia . Geol Soc London, Spec Publ 58: 1-24

TYSON, R V, WILSON, R C L, DOWNIE, C (1979) A stratified water colum model for the type Kimmeridge Clay. Nature, 277: 377-380

VAIL, P R, MITCHUM jr, R M, THOMPSON III, S (1977) Seismic stratigraphy and global changes of sea level, part 4: Global cycles of relative changes of sea level. Amer Assoc Petrol Geol, Mem, 26: 83-97

VAILLANT, B (1987) Étude experimentale du role de flores bactériennes dans la formation de l'apatite. PhD Thesis, Université de Nancy, Nancy

VALDIYA, K S (1972) Origin of phosphorite of the Late Precambrian ?angolihat dolomite of Pithograrh, Kumaun Himalaya, India. Sedimentology, 19: 115-128

VAN CAPPELLEN, P, BERNER, R A (1988) A mathematical model for the early diagenesis of phosphorus and fluorine in marine sediments: apatite precipitation. Amer J Sci, 28 288: 289-333

VAN CAPPELLEN, P, BERNER, R A (1991) Fluorapatite crystal growth from modified sea water. Geochim et Cosmochim, 55: 1219-1234

VAN CAPPELLEN, P (1991) The formation of marine apatite: A kinetic study. PhD Tesis, Yale University, 240pp

VAN WAGONER, J C, MITCHUM, R M, CAMPION, K M, RAHMANIAN, V D (1990) Siliciclastic Sequence stratigraphy in well logs, cores, and outcrops: Concepts for high-resolution correlation of time and facies. Amer Soc Petrol Geol, Methods in Exploration Series, 7: 55 pp

VAN WAGONER, J C, POSAMENTIER, H W, MITCHUM, R M, VAIL, P R, SARG, J F, LOUTIT, T S, HARDENBOL, J (1988) An overview of the fundamentals of sequence stratigraphy and key definitions. Soc Econ Paleont Miner, Spec Publ, 42: 39-45

VEEH, H H, BURNETT, W C, SOUTAR, A (1973) Contemporary phosphorites on the continental margin of Peru. Science, 181: 844-845

VEEH, H H, CALVERT, S E, PRICE, N B (1974) Accumulation of uranium in sediments and phosphorites on the south-west African shelf. Mar Chem, 2 188-202

VEEVERS, J J (1990) Tectonic-climatic supercycle in the billion-year plate-tectonic eon: Permian Pangean icehouse alternates with Cretaceous dispersed-continent greenhouse. Sediment Geol, 68: 1-16

VEIZER, J (1989) Strontium isotopes in seawater through time. Ann Rev Earth Planet Sci, 17: 141-167

VERMA, K K (1984) Biogenic concept for the origin of stromatolitic Precambrian phosphorites of western and central India. Geol Surv India, Spec Publ 17: 41-46

VINOGRADOV, A P (1953) The elementary chemical composition of marine organisms. Sear Foundation of Marine Research, Mem 2, Yale University

WARDLAW, B R (1979) Transgressions of the Retort Phosphatic Shale Member of the Phosphoria Formation (Permian) in Idaho, Montana, Utah, and Wyoming. US Geol Surv, Prof Pap, 1163-A: 1-4

WARDLAW, B R (1980) Middle-Late Permian paleogeography of Idaho, Montana, Nevada, Utah, and Wyoming. In: FOUCH, T D, MAGATHAM, E R: Paleozoic paleogeography of west-central United States: 353-361

WARDLAW, B R (1995) Permian conodonts. In: SCHOLLE, P A, PERYT, T M, ULMER-SCHOLLE, D S (eds) The Permian of northern Pangea, Vol 1:Paleogeography, Paleoclimates, Stratigraphy: 187-195

WARDLAW, B R, COLLINSON, J W (1978) Stratigraphic relations of Park City Group (Permian) in eastern Nevada and western Utah. Amer Assoc Petrol Geol, Bull, 62/7: 1171-1184

WARDLAW, B R, COLLINSON, J W (1979a) Biostratigraphic zonation of the Park City Group. US Geol Surv, Prof Pap, 1163-D: 17-22

WARDLAW, B R, COLLINSON, J W (1979b) Youngest Permian conodont faunas from the Great Basin and Rocky Mountain regions. Brigham Young University Geology Studies, 26/3: 151-159

WARDLAW, B R, COLLINSON, J W (1986) Paleontology and deposition of the Phosphoria Formation. Contributions to Geology, University of Wyoming, 24: 107-142

WARDLAW, B R, COLLINSON, J W, KETNER, K B (1979) Regional relations of Middle Permian rocks in Idaho, Nevada, and Utah. In: NEWMAN, G W, GOODE, H D: Basin and Range Symposium. Rocky Mountain Association of Geologists, 277-283

WARDLAW, B R, SNYDER, W S, SPINOZA, C, GALLEGOS, D M (1995) Permian of the Western United States. In: SCHOLLE, P A, PERYT, T M, ULMER-SCHOLLE, D S (eds) The Permian of northern Pangea, Vol 2: Sedimentary basins and economic resources: 23-40

WIGNALL, P B (1991) Model for transgressive black shales? Geology, 19: 167-170

WIGNALL, P B (1994) Black shales. Oxford Monographs on Geology and Geophysics, 30: 127pp

WIGNALL, P B, HALLAM, A (1992) Anoxia as a cause of the Permian/Triassic extinction: facies evidence from northern Italy and the western United states. Palaeogeogr, Palaeoclimat, Palaeoecol, 93,: 21-46

WILCOX, N R (1953) The origin of beds of phosphatic chalk with special reference to those at Taplow, England. 19th Sess Intl Geol Congress, Algier, sec 11, 119-134

WILSON, J L (1975) Carbonate Facies in Geologic History. 471 pp

WINNOCK, E (1980) Les dépôts de l'Eocène au Nord de l'Afrique: Aperçu paléogeographique de l'ensemble. Doc du BRGM, 24: 219-243

WRIGHT, J, SCHRADER, H, HOLSER, W T (1987) Paleoredox variations in ancient oceans recorded by rare earth elements in fossil apatite. Geochim Cosmochim Acta, 51: 631-644

WRIGHT, J, SEYMOUR, R S, SHAW, H F (1984) REE and Nd isotopes in conodont apatite: Variations with geological age and depositional environment. Geol Soc America, Spec Pap, 196: 325-340

WÜRZBURGER, U S (1953) A survey of phosphate deposits of Israel. In: Proceedings of a seminar on sources of mineral raw materials for the fertilizer industry in Asia and Far East, United Nations Resource Development Series, 32: 152-165

YANG, H, YANG, S (1994) The Shanwang fossil biota in eastern China: a Miocene Konservat Lagerstätte in lacustrine deposits. Lethaia, 27: 345-354

YANSHIN, A L (1986) Proterozoic and Cambrian phosphorites - regional overview: Asian part of the USSR and the Mongolian People's Republic. In: COOK, P J, SHERGOLD, J H (eds) Phosphate deposits of the world, Vol1: Proterozioc and Cambrian phosphorites, 63-69

YOCHELSON, E L (1968) Biostratigraphy of the Phosphoria, Park City, and Shedhorn Formation. US Geol Surv, Prof Pap 313-D: 571-660

YOUNG, R A (1975) Some aspects of crystal structural modeling of biological apatites. In: Physico-chime et cristallographie des apatites d'intérêt biologique, CNRS, 21-40

ZANIN, Y (1987) Ultramicrostructures of phosphorites (atlas of pictures). NAUKA, 223 pp (in Russian)

ZANIN, Y N, GILINSKAYA, L G, KRASIL'NIKOVA, N A, KRIVOPUTSKAYA, L M, MIRTOV, Y V, STOLPOVSKAYA, V N (1985) Calcium phosphates in phosphorites of different types. Int Geol Rev, 27: 1212-1229

ZANIN, Y N, LETOV, S V, KRASIL'NIKOVA, N A, MIRTOV, Y V (1985) Phosphatized bacteria from Cretaceous phosphorites of east-European platform and Paleocene phosphorites of Morocco. Sci Géol, Mém, 77: 79-81

ZANIN, Y N, ZVEREV, K V, SOLOTCHINA, E P (1990) Clay minerals and phosphorite genesis in the Upper Cretaceous of the northern Siberian Platform. Geol Soc London, Spec Publ, 52: 223-235

ZIEGLER, P (1989) Evolution of Laurussia. 102pp+14pl

ZIEGLER, P (1990) Geological atlas of western and central Europe. 2nd ed 239pp+56pl.

ZIMMERLE, W (1982) Die Phosphorite des nordwestdeutschen Apt und Alb. Geol Jb, A65: 159-244

ZIMMERLE, W (1985) New aspects on the formation of hydrocarbon source rocks. Geol Rundsch, 74: 385-416

ZIMMERLE, W (1992) Lithologie und Genese der Phosphorit-Konkretionen. In: THOMAS, E (ed) Oberdevon und Unterkarbon von Aprath im Bergischen Land: 240-267

ZIMMERLE, W (1994) Aptian and Albian phosphorites of north-western Germany (with emphasis on the biogenic aspect of phosphorite formation). Kaupia, 4: 79-102

ZIMMERLE, W, EMEIS, K (1983) Nachweis von fossilen Krebsbauten und Mikroben anus dem Unter-Alb von Vöhrum bie Peine. Ber Naturhist Ges Hannover, 126: 67-89

ZIMMERLE, W, GAIDA, K-H, DEDENK, R, KOCH, R, PAPROTH, E (1980) Sedimentological, mineralogical and organic-geochemical analyses of Upper Devonian and Lower Carboniferous strata of Riescheid, Federal Republic of Germany. Meded Rijks Geol Dienst, 32-5: 34-43

Subject index

(italized number refer to figures or tables)

Appendix

Color plates to figures:
4.1a., 4.1b., 4.7a., 4.7b., 4.7c., 4.8., 4.9a., 4.9b.,
6.16, 6.17., 6.18.,
7.8., 7.9., 7.22., 7.23., 7.26..

Fig. 4.1a.

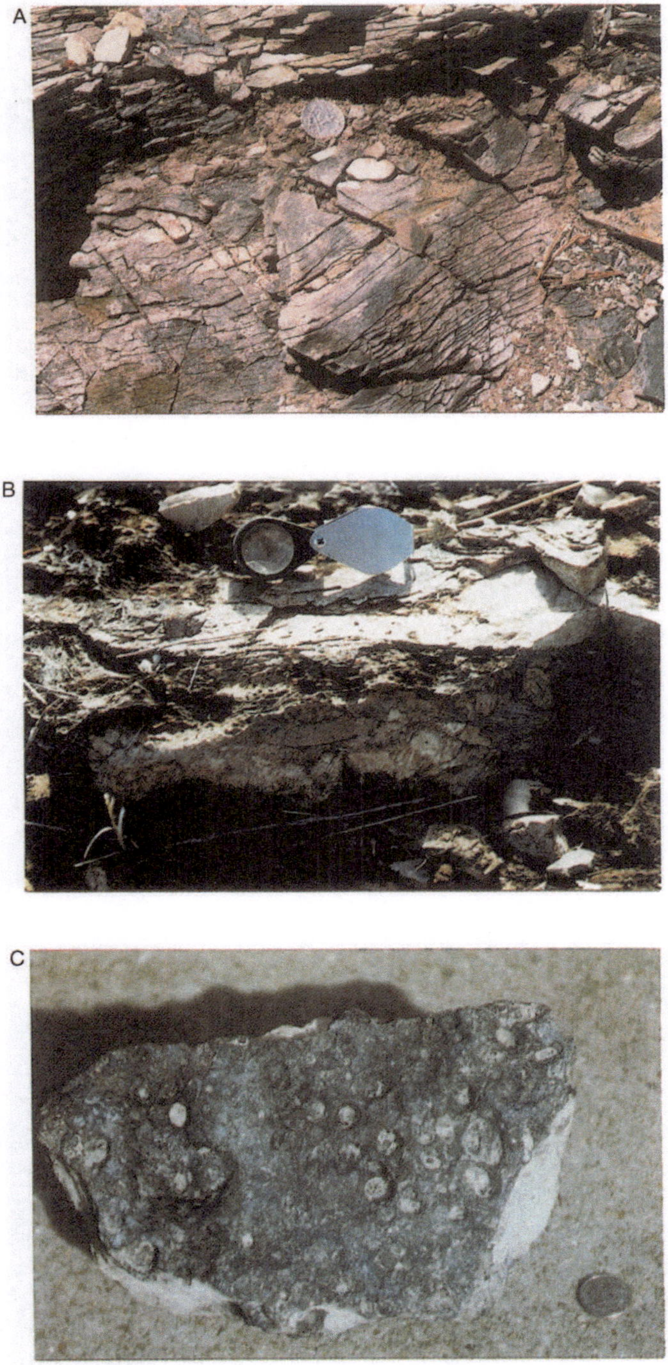

Fig. 4.1b.

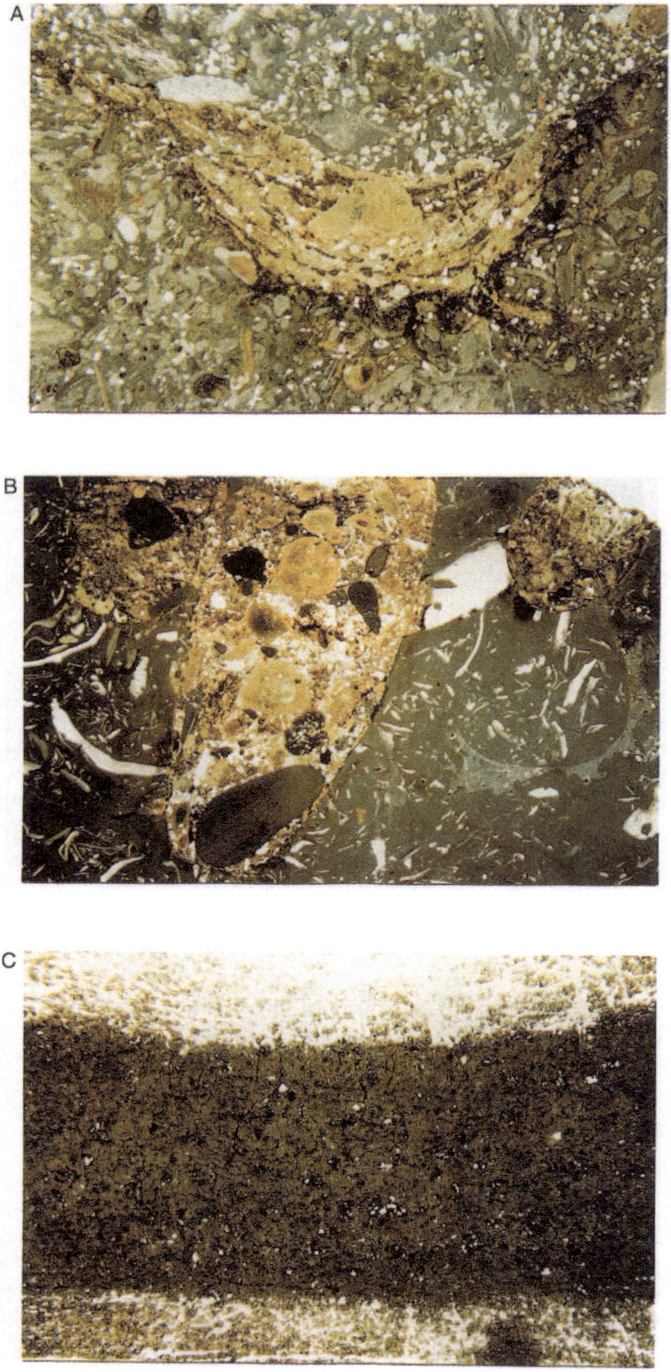

Fig. 4.7a.

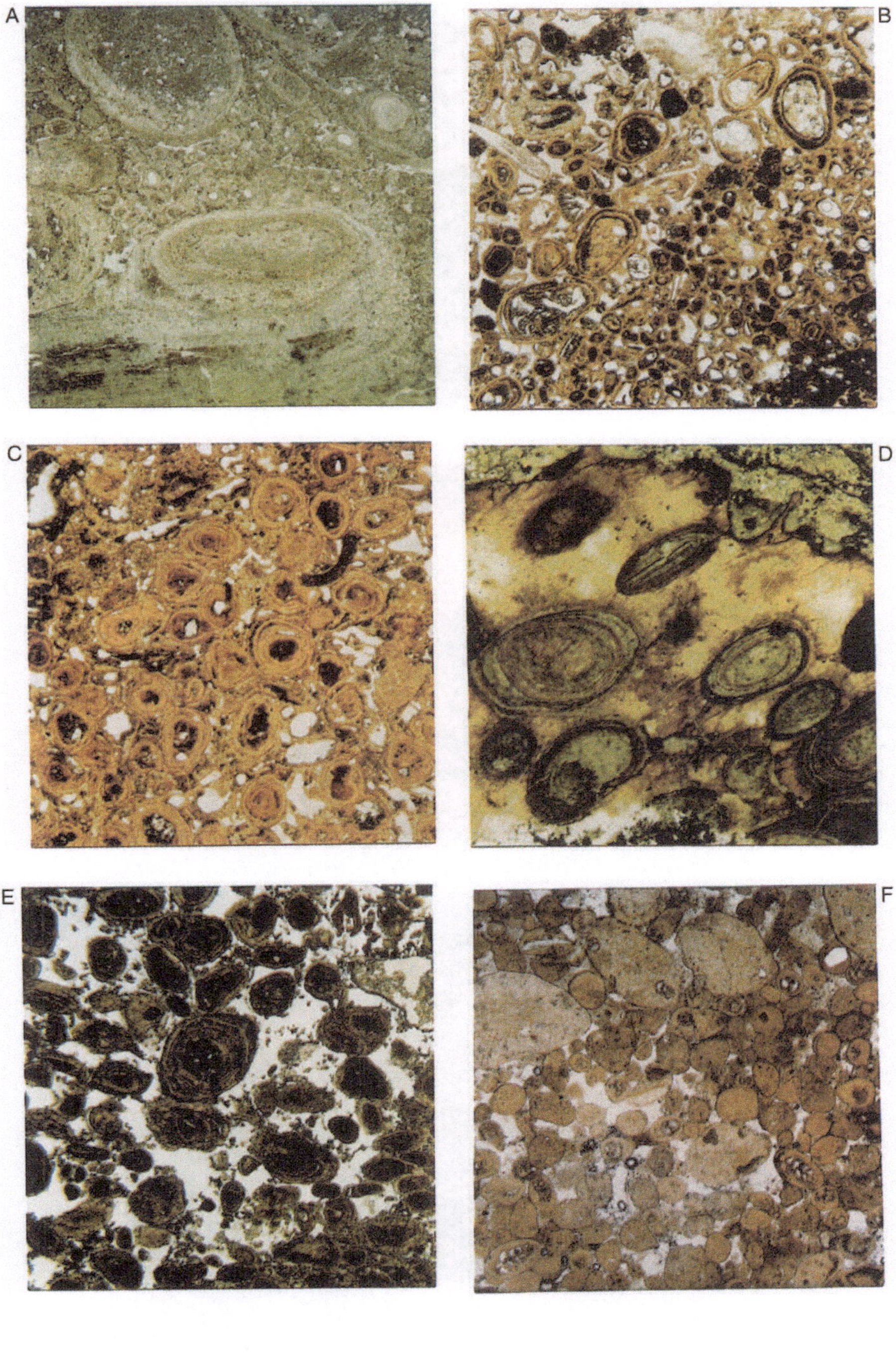

Fig. 4.7b.

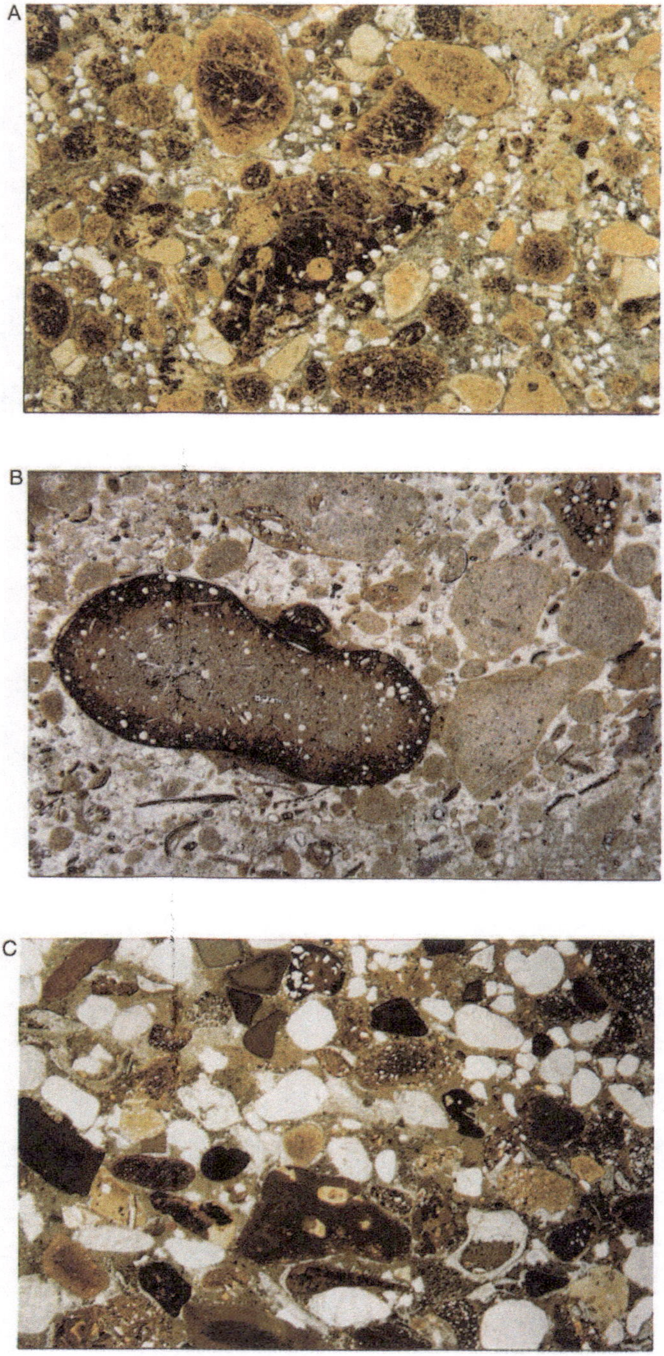

Fig. 4.7c.

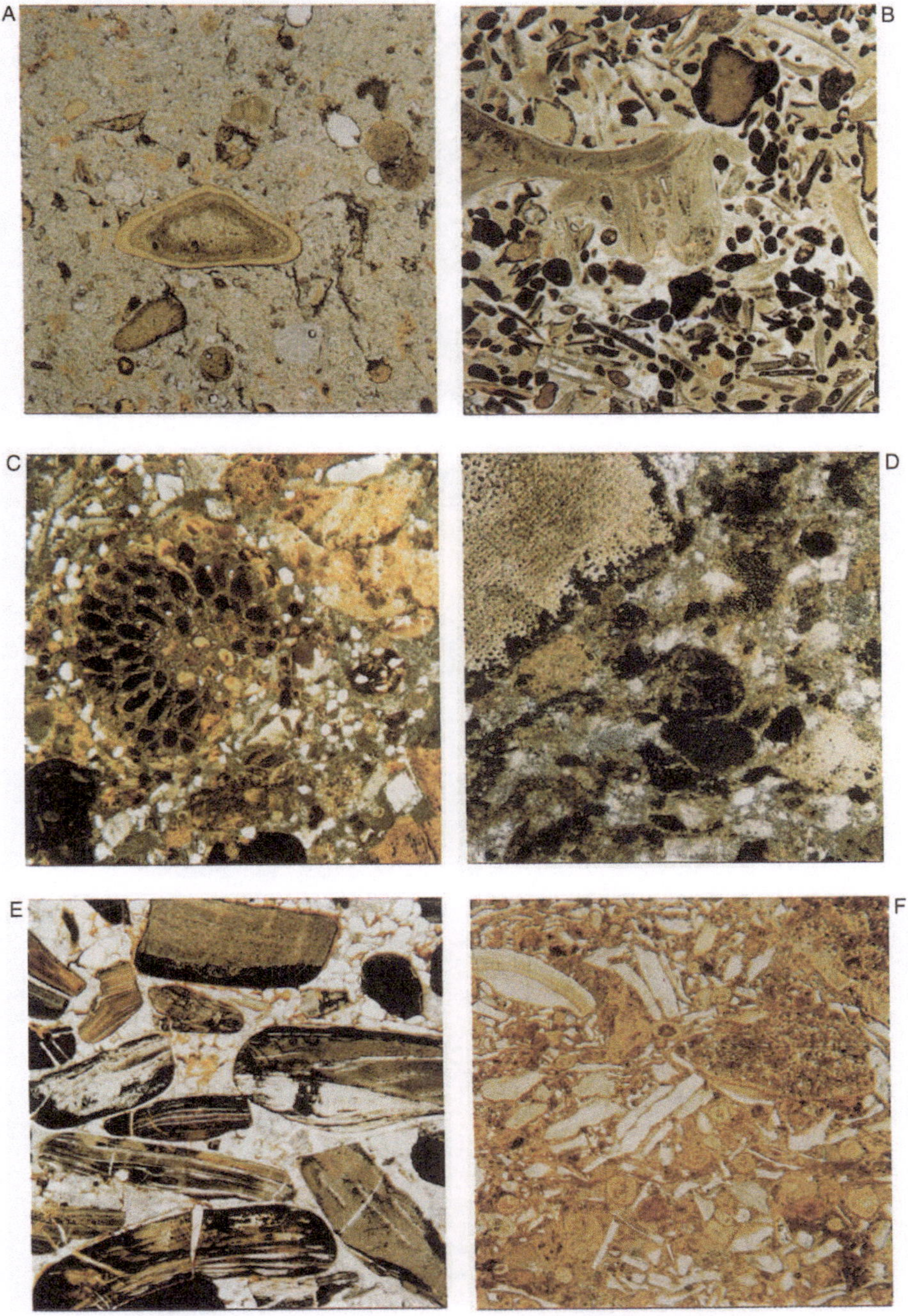

Fig. 4.8.

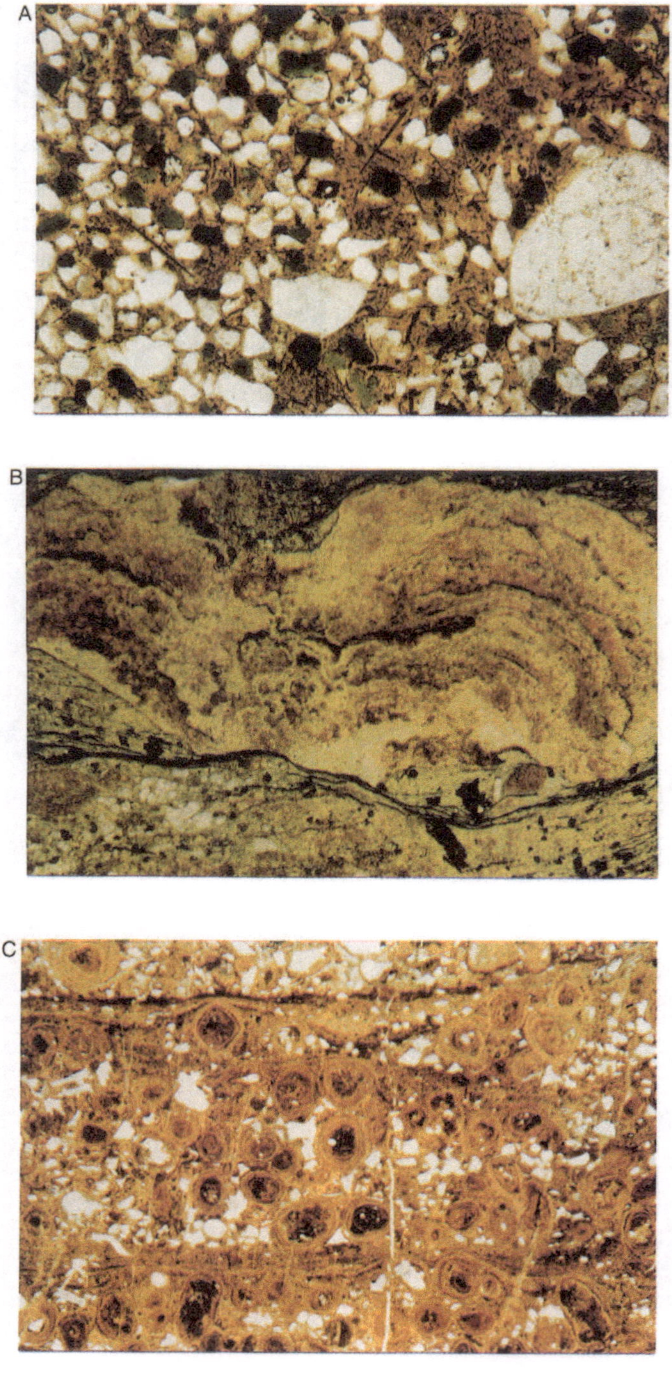

Fig. 4.9a.

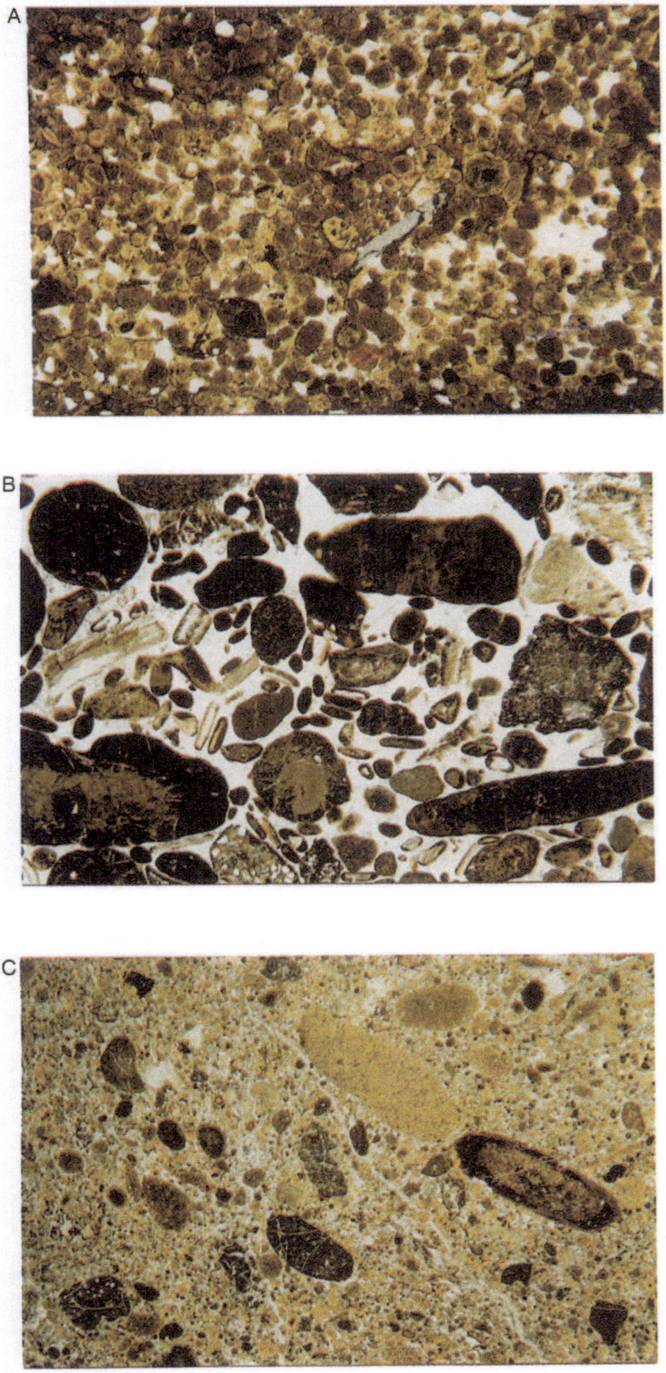

Fig. 4.9b.

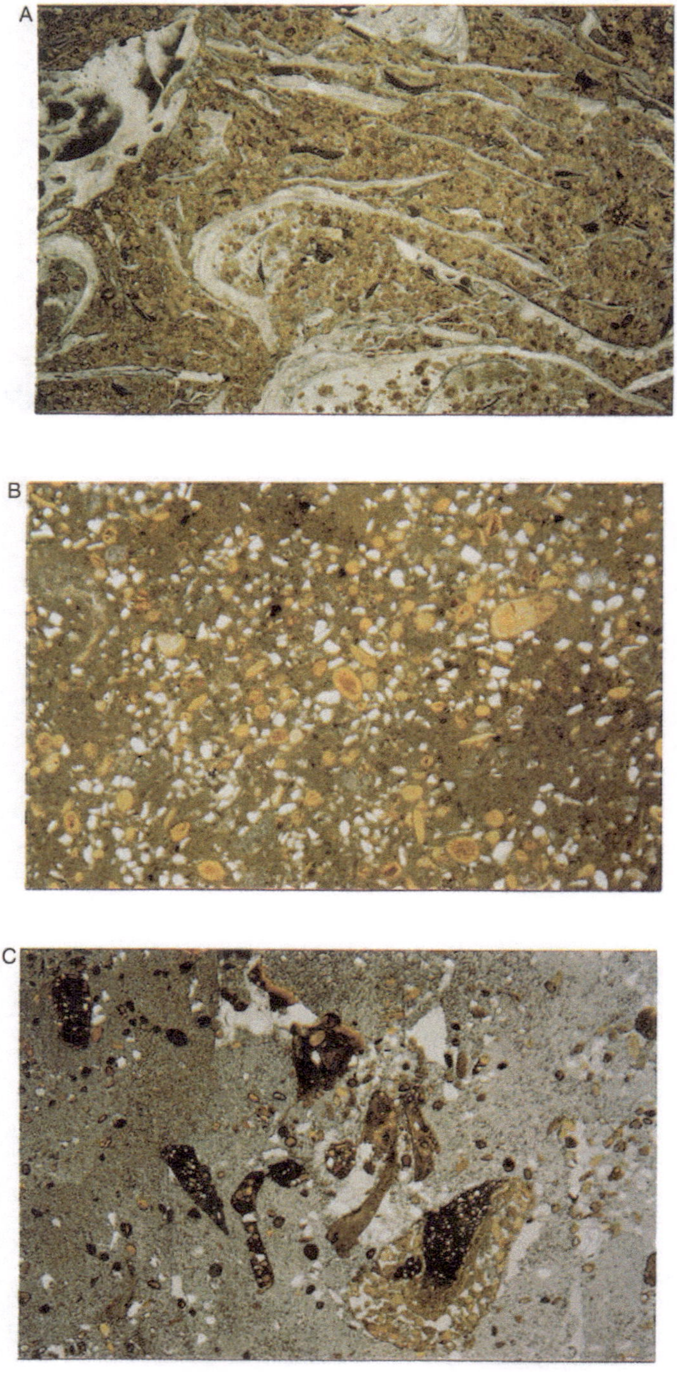

Fig. 6.16.

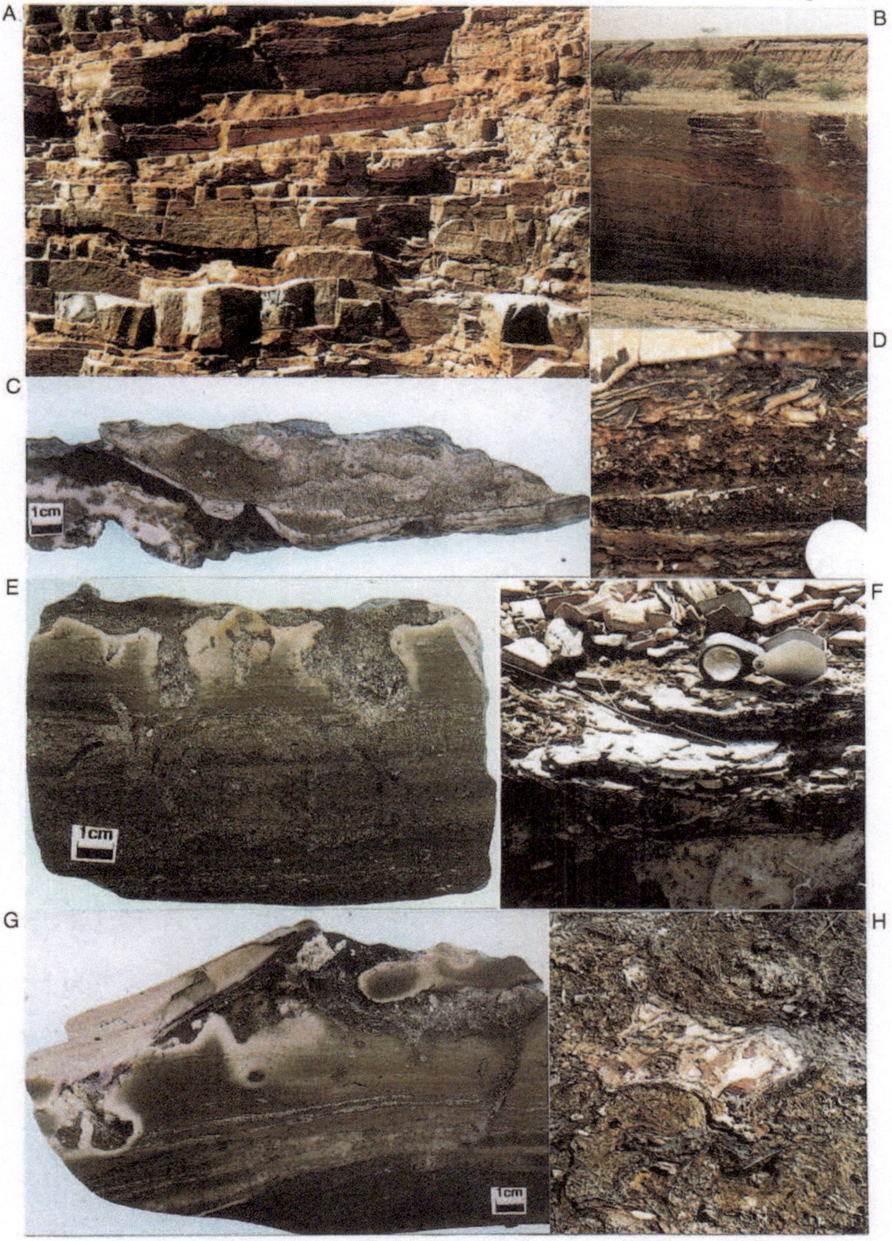

Fig. 6.17.

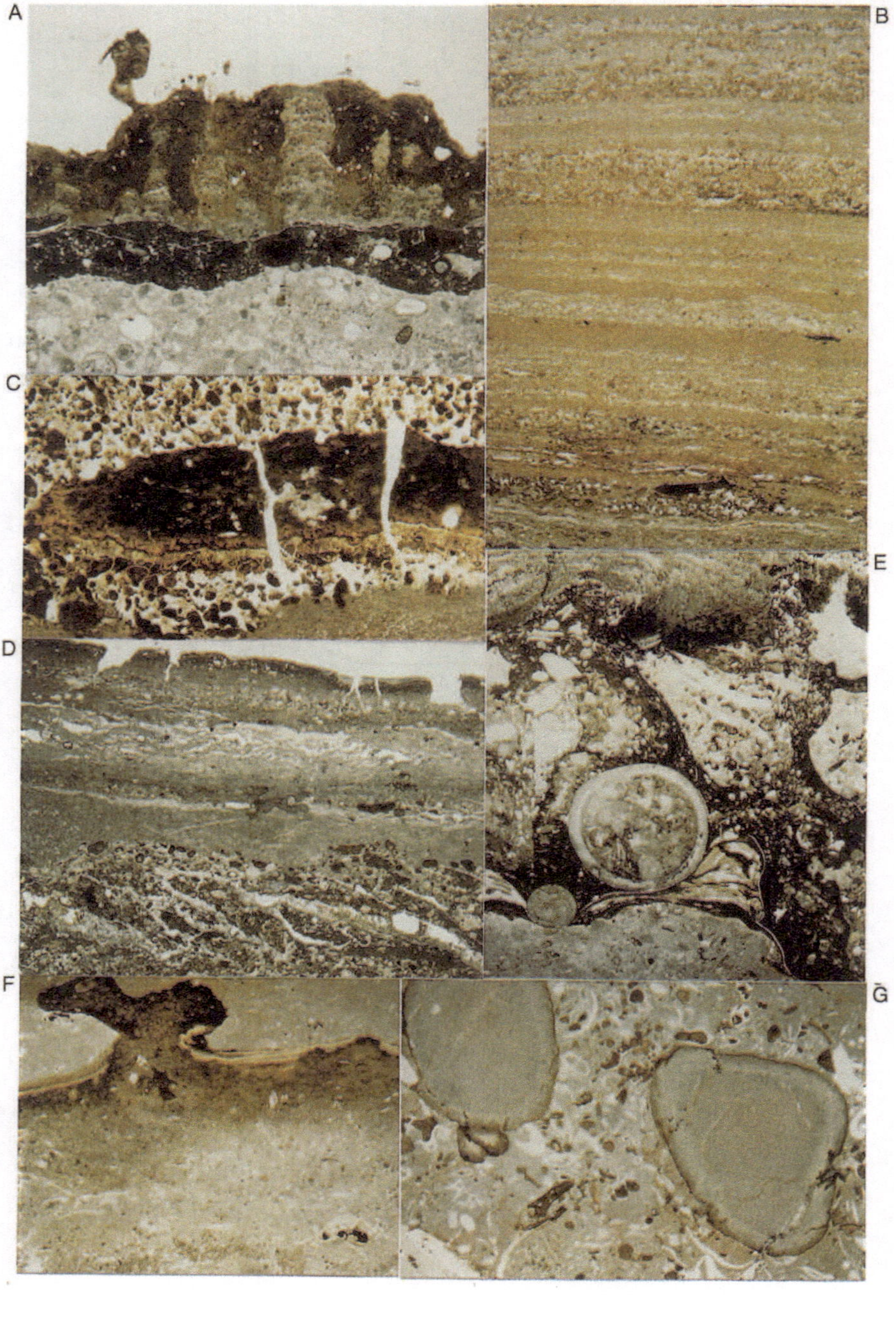

Fig. 6.18.

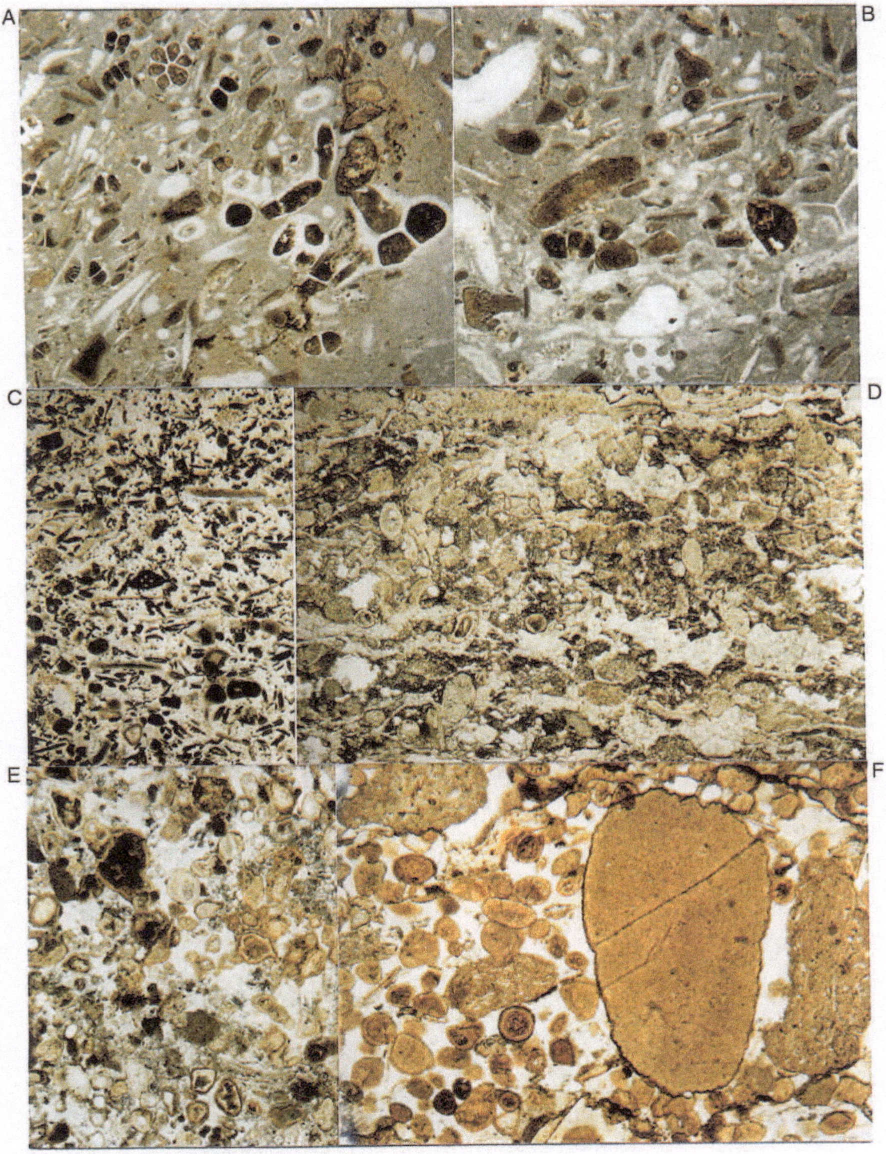

Fig. 7.8.

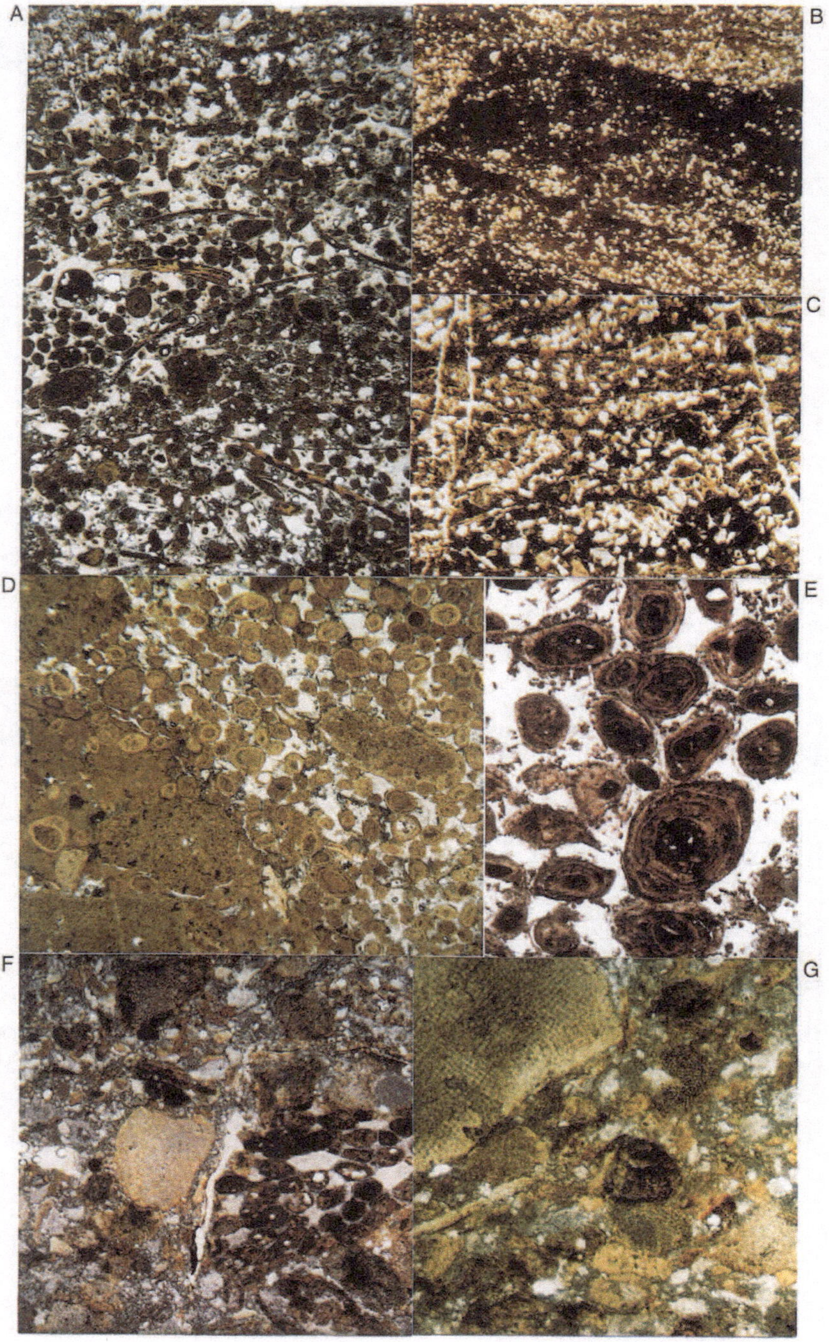

Fig. 7.9.

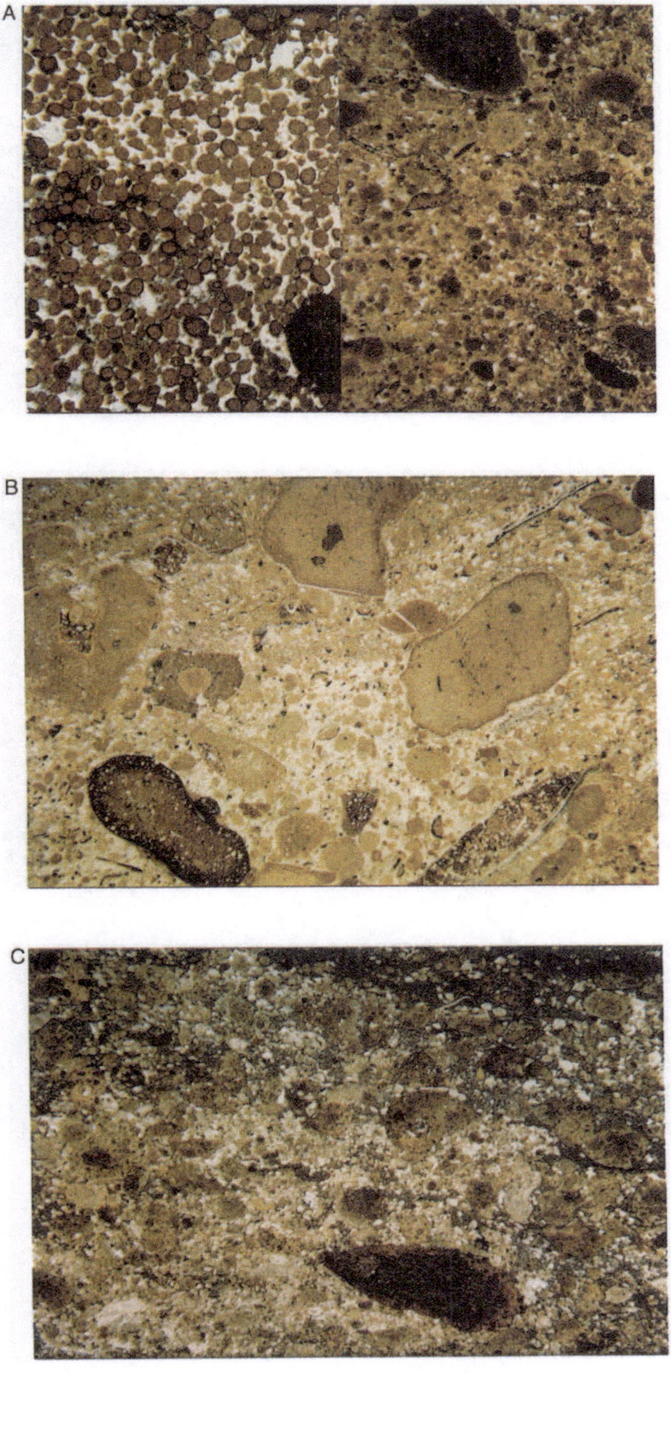

Fig. 7.22.

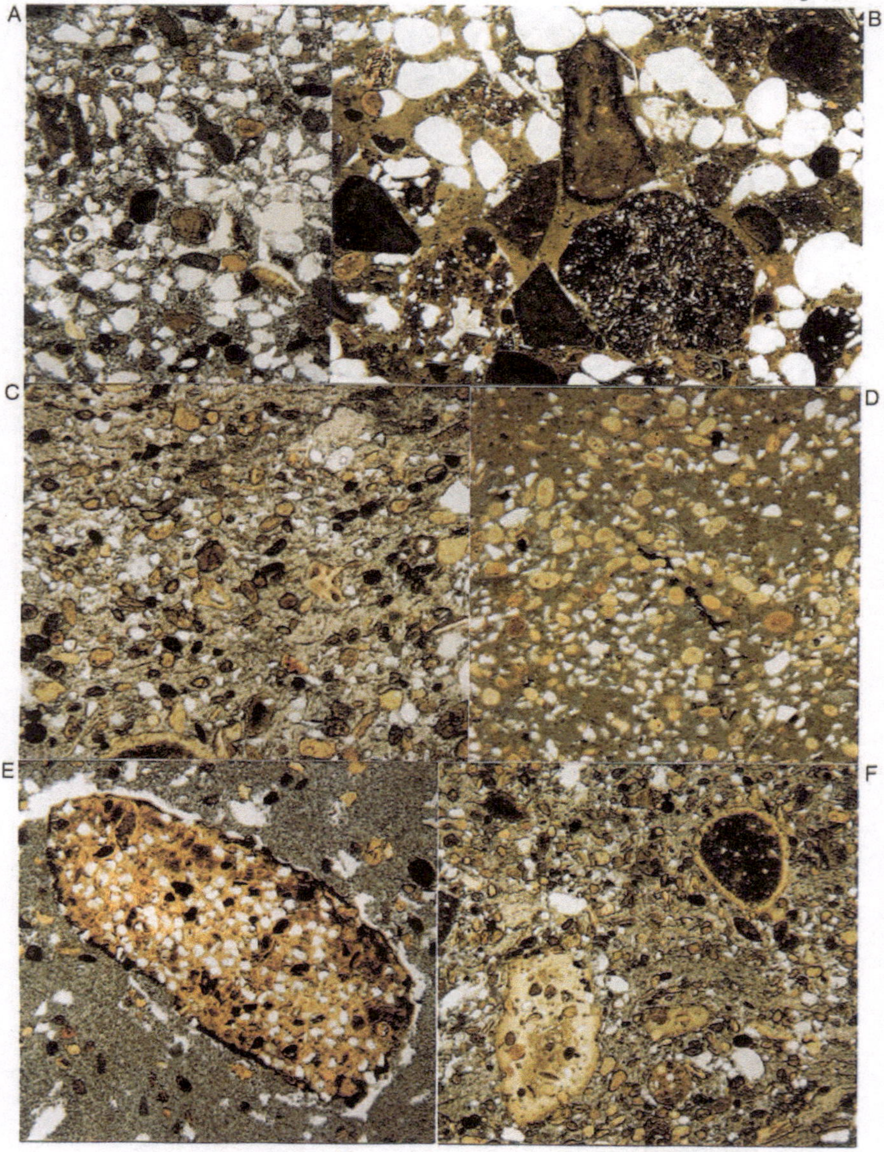

Fig. 7.23.

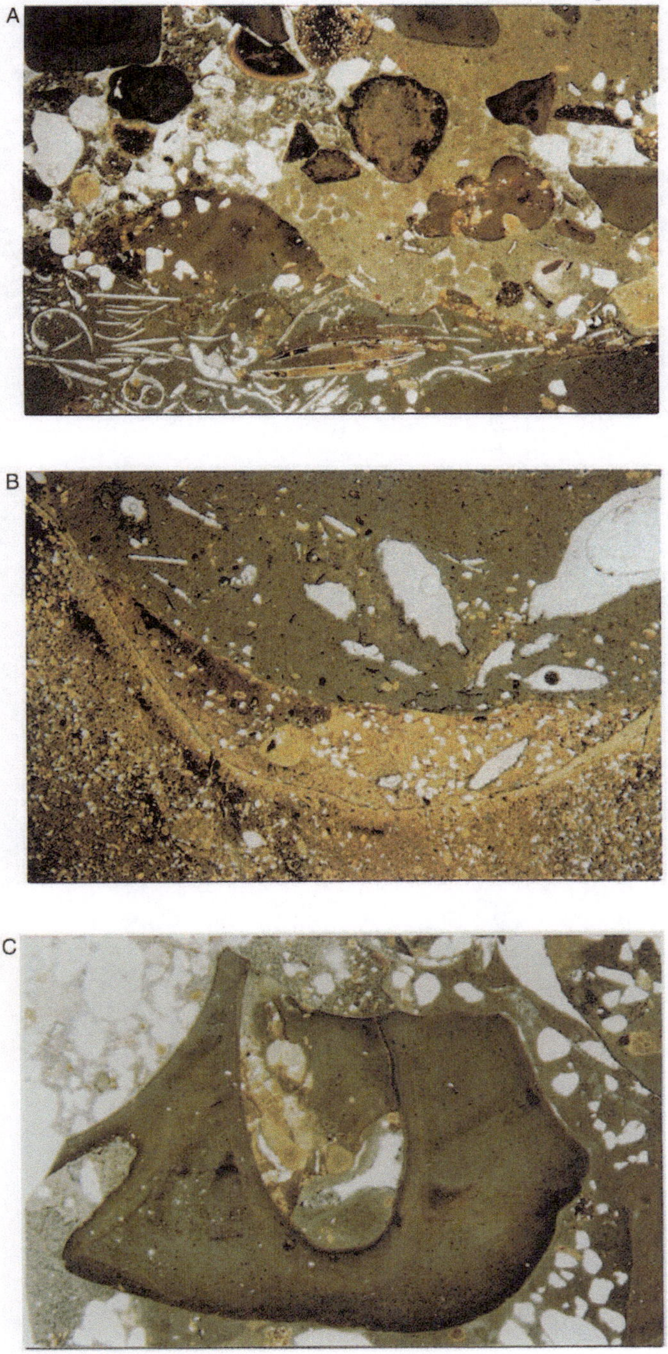

Fig. 7.26.

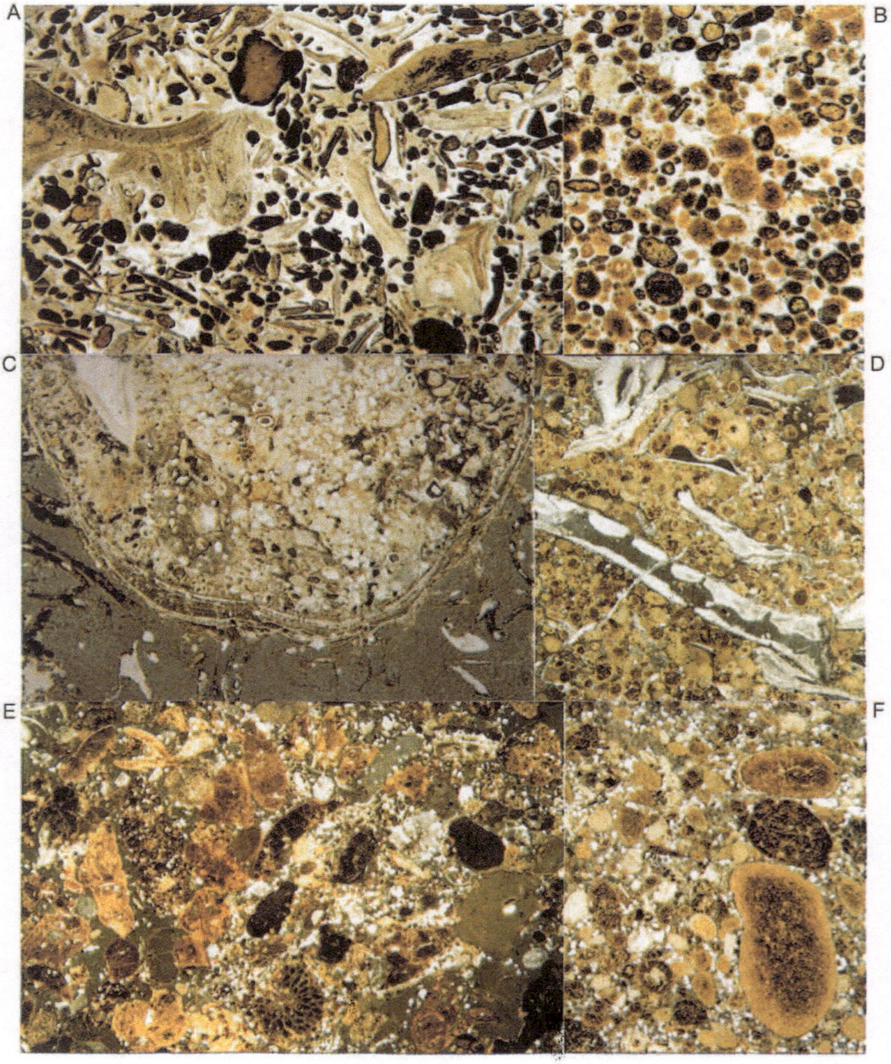

Lecture Notes in Earth Sciences

Vol. 1: Sedimentary and Evolutionary Cycles. Edited by U. Bayer and A. Seilacher. VI, 465 pages. 1985. (out of print).

Vol. 2: U. Bayer, Pattern Recognition Problems in Geology and Paleontology. VII, 229 pages. 1985. (out of print).

Vol. 3: Th. Aigner, Storm Depositional Systems. VIII, 174 pages. 1985.

Vol. 4: Aspects of Fluvial Sedimentation in the Lower Triassic Buntsandstein of Europe. Edited by D. Mader. VIII, 626 pages. 1985. (out of print).

Vol. 5: Paleogeothermics. Edited by G. Buntebarth and L. Stegena. II, 234 pages. 1986.

Vol. 6: W. Ricken, Diagenetic Bedding. X, 210 pages. 1986.

Vol. 7: Mathematical and Numerical Techniques in Physical Geodesy. Edited by H. Sünkel. IX, 548 pages. 1986.

Vol. 8: Global Bio-Events. Edited by O. H. Walliser. IX, 442 pages. 1986.

Vol. 9: G. Gerdes, W. E. Krumbein, Biolaminated Deposits. IX, 183 pages. 1987.

Vol. 10: T.M. Peryt (Ed.), The Zechstein Facies in Europe. V, 272 pages. 1987.

Vol. 11: L. Landner (Ed.), Contamination of the Environment. Proceedings, 1986. VII, 190 pages.1987.

Vol. 12: S. Turner (Ed.), Applied Geodesy. VIII, 393 pages. 1987.

Vol. 13: T. M. Peryt (Ed.), Evaporite Basins. V, 188 pages. 1987.

Vol. 14: N. Cristescu, H. I. Ene (Eds.), Rock and Soil Rheology. VIII, 289 pages. 1988.

Vol. 15: V. H. Jacobshagen (Ed.), The Atlas System of Morocco. VI, 499 pages. 1988.

Vol. 16: H. Wanner, U. Siegenthaler (Eds.), Long and Short Term Variability of Climate. VII, 175 pages. 1988.

Vol. 17: H. Bahlburg, Ch. Breitkreuz, P. Giese (Eds.), The Southern Central Andes. VIII, 261 pages. 1988.

Vol. 18: N.M.S. Rock, Numerical Geology. XI, 427 pages. 1988.

Vol. 19: E. Groten, R. Strauß (Eds.), GPS-Techniques Applied to Geodesy and Surveying. XVII, 532 pages. 1988.

Vol. 20: P. Baccini (Ed.), The Landfill. IX, 439 pages. 1989.

Vol. 21: U. Förstner, Contaminated Sediments. V, 157 pages. 1989.

Vol. 22: I. I. Mueller, S. Zerbini (Eds.), The Interdisciplinary Role of Space Geodesy. XV, 300 pages. 1989.

Vol. 23: K. B. Föllmi, Evolution of the Mid-Cretaceous Triad. VII, 153 pages. 1989.

Vol. 24: B. Knipping, Basalt Intrusions in Evaporites. VI, 132 pages. 1989.

Vol. 25: F. Sansò, R. Rummel (Eds.), Theory of Satellite Geodesy and Gravity Field Theory. XII, 491 pages. 1989.

Vol. 26: R. D. Stoll, Sediment Acoustics. V, 155 pages. 1989.

Vol. 27: G.-P. Merkler, H. Militzer, H. Hötzl, H. Armbruster, J. Brauns (Eds.), Detection of Subsurface Flow Phenomena. IX, 514 pages. 1989.

Vol. 28: V. Mosbrugger, The Tree Habit in Land Plants. V, 161 pages. 1990.

Vol. 29: F. K. Brunner, C. Rizos (Eds.), Developments in Four-Dimensional Geodesy. X, 264 pages. 1990.

Vol. 30: E. G. Kauffman, O.H. Walliser (Eds.), Extinction Events in Earth History. VI, 432 pages. 1990.

Vol. 31: K.-R. Koch, Bayesian Inference with Geodetic Applications. IX, 198 pages. 1990.

Vol. 32: B. Lehmann, Metallogeny of Tin. VIII, 211 pages. 1990.

Vol. 33: B. Allard, H. Borén, A. Grimvall (Eds.), Humic Substances in the Aquatic and Terrestrial Environment. VIII, 514 pages. 1991.

Vol. 34: R. Stein, Accumulation of Organic Carbon in Marine Sediments. XIII, 217 pages. 1991.

Vol. 35: L. Håkanson, Ecometric and Dynamic Modelling. VI, 158 pages. 1991.

Vol. 36: D. Shangguan, Cellular Growth of Crystals. XV, 209 pages. 1991.

Vol. 37: A. Armanini, G. Di Silvio (Eds.), Fluvial Hydraulics of Mountain Regions. X, 468 pages. 1991.

Vol. 38: W. Smykatz-Kloss, S. St. J. Warne, Thermal Analysis in the Geosciences. XII, 379 pages. 1991.

Vol. 39: S.-E. Hjelt, Pragmatic Inversion of Geophysical Data. IX, 262 pages. 1992.

Vol. 40: S. W. Petters, Regional Geology of Africa. XXIII, 722 pages. 1991.

Vol. 41: R. Pflug, J. W. Harbaugh (Eds.), Computer Graphics in Geology. XVII, 298 pages. 1992.

Vol. 42: A. Cendrero, G. Lüttig, F. Chr. Wolff (Eds.), Planning the Use of the Earth's Surface. IX, 556 pages. 1992.

Vol. 43: N. Clauer, S. Chaudhuri (Eds.), Isotopic Signatures and Sedimentary Records. VIII, 529 pages. 1992.

Vol. 44: D. A. Edwards, Turbidity Currents: Dynamics, Deposits and Reversals. XIII, 175 pages. 1993.

Vol. 45: A. G. Herrmann, B. Knipping, Waste Disposal and Evaporites. XII, 193 pages. 1993.

Vol. 46: G. Galli, Temporal and Spatial Patterns in Carbonate Platforms. IX, 325 pages. 1993.

Vol. 47: R. L. Littke, Deposition, Diagenesis and Weathering of Organic Matter-Rich Sediments. IX, 216 pages. 1993.

Vol. 48: B. R. Roberts, Water Management in Desert Environments. XVII, 337 pages. 1993.

Vol. 49: J. F. W. Negendank, B. Zolitschka (Eds.), Paleolimnology of European Maar Lakes. IX, 513 pages. 1993.

Vol. 50: R. Rummel, F. Sansò (Eds.), Satellite Altimetry in Geodesy and Oceanography. XII, 479 pages. 1993.

Vol. 51: W. Ricken, Sedimentation as a Three-Component System. XII, 211 pages. 1993.

Vol. 52: P. Ergenzinger, K.-H. Schmidt (Eds.), Dynamics and Geomorphology of Mountain Rivers. VIII, 326 pages. 1994.

Vol. 53: F. Scherbaum, Basic Concepts in Digital Signal Processing for Seismologists. X, 158 pages. 1994.

Vol. 54: J. J. P. Zijlstra, The Sedimentology of Chalk. IX, 194 pages. 1995.

Vol. 55: J. A. Scales, Theory of Seismic Imaging. XV, 291 pages. 1995.

Vol. 56: D. Müller, D. I. Groves, Potassic Igneous Rocks and Associated Gold-Copper Mineralization. 2nd updated and enlarged Edition. XIII, 238 pages. 1997.

Vol. 57: E. Lallier-Vergès, N.-P. Tribovillard, P. Bertrand (Eds.), Organic Matter Accumulation. VIII, 187 pages. 1995.

Vol. 58: G. Sarwar, G. M. Friedman, Post-Devonian Sediment Cover over New York State. VIII, 113 pages. 1995.

Vol. 59: A. C. Kibblewhite, C. Y. Wu,Wave Interactions As a Seismo-acoustic Source. XIX, 313 pages. 1996.

Vol. 60: A. Kleusberg, P. J. G. Teunissen (Eds.), GPS for Geodesy. VII, 407 pages. 1996.

Vol. 61: M. Breunig, Integration of Spatial Information for Geo-Information Systems. XI, 171 pages. 1996.

Vol. 62: H. V. Lyatsky, Continental-Crust Structures on the Continental Margin of Western North America. XIX, 352 pages. 1996.

Vol. 63: B. H. Jacobsen, K. Mosegaard, P. Sibani (Eds.), Inverse Methods. XVI, 341 pages, 1996.

Vol. 64: A. Armanini, M. Michiue (Eds.), Recent Developments on Debris Flows. X, 226 pages. 1997.

Vol. 65: F. Sansò, R. Rummel (Eds.), Geodetic Boundary Value Problems in View of the One Centimeter Geoid. XIX, 592 pages. 1997.

Vol. 66: H. Wilhelm, W. Zürn, H.-G. Wenzel (Eds.), Tidal Phenomena. VII, 398 pages. 1997.

Vol. 67: S. L. Webb, Silicate Melts. VIII. 74 pages. 1997.

Vol. 68: P. Stille, G. Shields, Radiogenetic Isotope Geochemistry of Sedimentary and Aquatic Systems. XI, 217 pages. 1997.

Vol. 69: S. P. Singal (Ed.), Acoustic Remote Sensing Applications. XIII, 585 pages. 1997.

Vol. 70: R. H. Charlier, C. P. De Meyer, Coastal Erosion – Response and Management. XVI, 343 pages. 1998.

Vol. 71: T. M. Will, Phase Equilibria in Metamorphic Rocks. XIV, 315 pages. 1998.

Vol. 72: J. C. Wasserman, E. V. Silva-Filho, R. Villas-Boas (Eds.), Environmental Geochemistry in the Tropics. XIV, 305 pages. 1998.

Vol. 73: Z. Martinec, Boundary-Value Problems for Gravimetric Determination of a Precise Geoid. XII, 223 pages. 1998.

Vol. 74: M. Beniston, J. L. Innes (Eds.), The Impacts of Climate Variability on Forests. XIV, 329 pages. 1998.

Vol. 75: H. Westphal, Carbonate Platform Slopes – A Record of Changing Conditions. XI, 197 pages. 1998.

Vol. 76: J. Trappe, Phanerozoic Phosphorite Depositional Systems. XII, 316 pages. 1998.